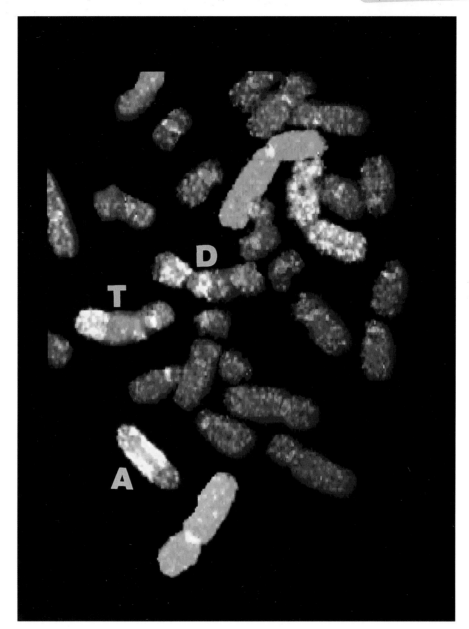

Partial metaphase spread of a human fibroblast that had been exposed to 4 Gy of X-rays. Using the fluorescent *in situ* hybridization (FISH) technique, chromosome #1 has been stained orange-yellow and chromosome #2 green. All other chromosomes are blue. Centromeres are stained red, except in chromosomes #2 where they appear yellow. Both #2 chromosomes are seen intact, as is one #1. The other #1 chromosome has, in response to radiation damage, undergone a complex interchange (termed 3O according to the nomenclature of Savage and Simpson) involving at least four breaks in three chromosomes. A break either side of the chromosome #1 centromere has resulted in an inserted dicentric chromosome (D), a translocated chromosome (T) and an acentric fragment (A). At a subsequent mitosis the dicentric chromosome and acentric fragment would fail to partition correctly and inevitably lead to cell death. From P J Simpson and J R K Savage, *Int J Radiat Biol* 69: 429–36 (1996), with permission.

Basic Clinical Radiobiology

Basic Clinical Radiobiology

SECOND EDITION

Edited by
G. GORDON STEEL

A member of the Hodder Headline Group
LONDON • NEW YORK • SYDNEY • AUCKLAND
Co-published in the USA by Oxford University Press, Inc., New York

First published in Great Britain in 1997 by
Arnold, a member of the Hodder Headline Group,
338 Euston Road, London NW1 3BH

Co-published in the United States of America by
Oxford University Press, Inc.,
198 Madison Avenue, New York, NY100016
Oxford is a registered trademark of Oxford University Press

© 1997 Arnold

All rights reserved. No part of this publication may be reproduced or
transmitted in any form or by any means, electronically or mechanically,
including photocopying, recording or any information storage or retrieval
system, without either prior permission in writing from the publisher or a
licence permitting restricted copying. In the United Kingdom such licences
are issued by the Copyright Licensing Agency: 90 Tottenham Court Road,
London W1P 9HE.

Whilst the advice and information in this book is believed to be true and
accurate at the date of going to press, neither the authors nor the publisher
can accept any legal responsibility or liability for any errors or omissions
that may be made. In particular (but without limiting the generality of the
preceding disclaimer) every effort has been made to check drug dosages;
however it is still possible that errors have been missed. Furthermore,
dosage schedules are constantly being revised and new side-effects
recognized. For these reasons the reader is strongly urged to consult the
drug companies' printed instructions before administering any of the drugs
recommended in this book.

British Library Cataloguing in Publication Data
A catalogue record for this book is available from the British Library

Library of Congress Cataloguing-in-Publication Data
A catalog record for this book is available from the Library of Congress

ISBN 0 340 70020 3

Typeset in 10/11pt Times by Saxon Graphics, Derby.
Printed and bound in Great Britain by The Bath Press, Bath.

Contents

List of contributors — vii
Preface — viii

1. Introduction: The significance of radiobiology for radiotherapy *GG Steel* — 1
2. The growth rate of tumours *GG Steel* — 8
3. Cell proliferation in tumours *AC Begg* — 14
4. Proliferative and cellular organization of normal tissues *FA Stewart and AJ van der Kogel* — 24
5. Radiation response and tolerance of normal tissues *AJ van der Kogel* — 30
6. Clonogenic cells and the concept of cell survival *GG Steel* — 40
7. Models of radiation cell killing *MC Joiner* — 52
8. DNA damage and cell killing *TJ McMillan and GG Steel* — 58
9. Genetic control of the cellular response to ionizing radiation *TJ McMillan* — 70
10. Dose–response relationships in radiotherapy *SM Bentzen* — 78
11. Clinical manifestations of normal-tissue damage *SM Bentzen and J Overgaard* — 87
12. Time–dose relationships in radiotherapy *SM Bentzen and J Overgaard* — 98
13. The linear-quadratic approach to fractionation and calculation of isoeffect relationships *MC Joiner and AJ van der Kogel* — 106
14. Hyperfractionation and accelerated radiotherapy *MC Joiner* — 123
15. The oxygen effect *MR Horsman and J Overgaard* — 132
16. Overcoming hypoxic cell radioresistance *J Overgaard and MR Horsman* — 141
17. The radiobiology of tumours *GG Steel* — 152
18. The dose rate effect: brachytherapy *GG Steel* — 163
19. Particle beams in radiotherapy *MC Joiner* — 173
20. Combination of radiotherapy and chemotherapy: principles *GG Steel* — 184

21.	**Combined radiotherapy and chemotherapy: clinical application and evaluation** *FA Stewart and MI Saunders*	195
22.	**Re-treatment tolerance of normal tissues** *FA Stewart*	203
23.	**Hyperthermia** *J Overgaard and MR Horsman*	212
24.	**Targeted radiotherapy** *TE Wheldon*	224
25.	**Individualization of radiotherapy** *AC Begg*	234
Glossary		246
Index		250

Contributors

AC Begg PhD
Head, Division of Experimental Therapy, The Netherlands Cancer Institute, Amsterdam, The Netherlands

S M Bentzen PhD DSc
Associate Professor, Department of Experimental Clinical Oncology, University Hospital of Aarhus, Aarhus, Denmark

MR Horsman PhD DMSc
Associate Professor, Danish Cancer Society, Department of Experimental Clinical Oncology, Aarhus University Hospital, Aarhus, Denmark

MC Joiner PhD
Professor and Head, Experimental Oncology Group, Gray Laboratory Cancer Research Trust, Mount Vernon Hospital, Northwood, Middlesex, UK

AJ van der Kogel PhD
Professor of Clinical Radiobiology, Institute of Radiotherapy, University Hospital Nijmegen, Nijmegen, The Netherlands

TJ McMillan PhD
Professor of Cancer Biology, Division of Biological Sciences, Institute of Environmental and Biological Sciences, University of Lancaster, UK

J Overgaard MD DMSc FRCR
Professor and Head of Department, Danish Cancer Society, Department of Experimental Clinical Oncology, Aarhus University Hospital, Aarhus, Denmark

MI Saunders MD FRCR
Reader in Oncology, Marie Curie Research Wing, Mount Vernon Hospital, Northwood, Middlesex, UK

GG Steel DSc
Professor of Radiobiology Applied to Radiotherapy, Department of Radiotherapy Research, The Institute of Cancer Research, Sutton, Surrey, UK

FA Stewart PhD
Division of Experimental Therapy, The Netherlands Cancer Institute, Amsterdam, The Netherlands

TE Wheldon PhD
Department of Radiation Oncology, University of Glasgow, CRC Beatson Laboratories, Glasgow, Scotland, UK

Preface

Developments in cancer therapy are increasingly arising out of studies in basic science and it is best for their implementation to be in the hands of clinicians who are familiar with the relevant scientific areas. This book deals with the biological aspects of radiotherapy. It seeks to present in a concise and interesting way the main ideas and significant scientific developments that underlie current attempts to improve the radiotherapeutic management of cancer.

The first edition of the book was published in 1993 with the intention to revise it at frequent intervals in order to keep it up-to-date. This second edition has been carefully and extensively revised. New chapters have been added on three clinical topics: dose–response relationships, re-treatment tolerance and combined radiotherapy and chemotherapy. The chapter on molecular and genetic aspects of radiobiology is now enlarged and covered in two chapters. The text of the other chapters has been updated and many illustrations have been redrawn to a more uniform style.

The book is directed at an international audience. It has arisen out of teaching courses organized by the European Society for Therapeutic Radiation Oncology (ESTRO) and the material will also be of use to readers in North America and the rest of the world.

For the benefit of readers who have a visual memory, and particularly for those whose first language is not English, we have chosen to use a high ratio of charts to text. We have also added a substantial glossary of scientific terms. At the end of each chapter is a list of prominent published material, mainly books and reviews, that are recommended for further reading. Citations in the text sometimes refer to these works.

G. Gordon Steel

1

Introduction: The significance of radiobiology for radiotherapy

G. Gordon Steel

1.1 The role of radiotherapy in the management of cancer

Radiotherapy is one of the two most effective treatments for cancer. Surgery, which of course has the longer history, is in the majority of cases the primary form of treatment and it leads to good therapeutic results in a range of early non-metastatic tumours. Radiotherapy has replaced surgery for the long-term control of many tumours of the head and neck, cervix, bladder, prostate and skin, in which it often achieves a reasonable probability of tumour control with good cosmetic result. Wide-field irradiation is also successful in Hodgkin's disease and other lymphomas. In addition to these examples of the curative role of radiation therapy, a large proportion of patients with a wide variety of types of malignant disease receive valuable palliation by radiation.

Chemotherapy is the third most important treatment modality at the present time. Following the early use of nitrogen mustard during the 1920s it has emerged to the point where upwards of 30 drugs are available for the management of cancer, although no more than 10–15 are in common use. A large proportion of patients receive chemotherapy at some point in their management and useful symptom relief and disease arrest are often obtained.

Reliable information about the relative roles of surgery, radiotherapy and chemotherapy is difficult to obtain, but is necessary whenever priorities for research funding in this area are under review. Often quoted is the analysis by DeVita *et al* (1979) of the probable outcome of treatment of the 700 000 newly diagnosed cases in the USA. DeVita estimated that local treatment, which includes surgery and/or radiotherapy, could be expected to be successful in approximately 280 000 (~40%) of these cases. In perhaps 15% of the total, radiotherapy would be the principal form of treatment. In contrast, of the cases where chemotherapy was the main-line treatment, DeVita estimated that only around 2% (of the total 700 000 cases) were likely to achieve long-term survival. Similar figures from a review by Souhami and Tobias (1986) are shown in Figure 1.1. They estimated that apart from skin cancers and *in situ* cervix carcinomas probably 30% of cancers are cured by surgery, radiotherapy or both. Many patients (perhaps 40%) do receive chemotherapy but their contribution to the overall cure rate may be only around 2%, with some prolongation of life in perhaps another 10%.

If these figures are correct, it may be that around 7 times as many patients currently are cured by radiotherapy as by chemotherapy. This is not to undervalue the important benefits of chemotherapy in a number of chemosensitive diseases, but to stress the greater role of radiotherapy as a curative agent (Tubiana, 1992).

Considerable efforts are being devoted at the present time to the improvement of radiotherapy and chemotherapy. Wide publicity is given to the newer areas of drug development such as lymphokines, growth factors, anti-oncogenes and gene therapy. But if we were to imagine aiming to increase the results shown in Figure 1.1 to the point where, say, 2% more patients are cured, it would seem on a realistic estimation that this would more likely be achieved by increasing the results of radiotherapy from 15% to 17% than by *doubling* the results achieved by chemotherapy.

There are three main ways in which such an improvement in radiotherapy might be obtained:

2 Introduction: The significance of radiobiology for radiotherapy

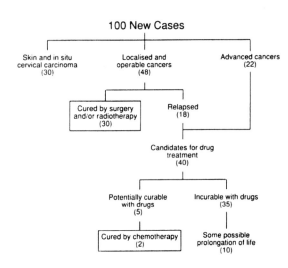

Figure 1.1 The relative roles of local and systemic treatment in the management of cancer. From Souhami and Tobias (1986), with permission.

(i) by raising the standards of radiation dose prescription and delivery to those currently in use in the best radiotherapy centres;
(ii) by improving radiation dose distributions beyond those that are conventionally achieved, either using techniques of conformal radiotherapy with photons, or ultimately by the use of proton beams;
(iii) by exploiting radiobiological initiatives.

The proportion of radiotherapists worldwide who work in academic centres is probably less than 5%. They are the clinicians who may have access to large new treatment machines, for instance for proton therapy, or to new radiosensitizers or to new agents for targeted therapy. Chapters of this book allude to these exciting developments which may well have impact on treatment success in the future. But it should not be thought that the improvement of radiation therapy lies exclusively with clinical research in the specialist academic centres. It has widely been recognized that by far the most effective way of improving cure rates nationwide or Europe-wide is by quality assurance in the prescription and delivery of radiation treatment. Chapters 10–14 of this book deal with the principles on which fractionation schedules should be optimized, including how to respond to unavoidable gaps in treatment. For many radiotherapists this will be the most important part of this book, for in even the smallest department it is possible without access to greatly increased funding to move closer to optimum fractionation practices.

1.2 The role of radiation biology

Experimental and theoretical studies in radiation biology contribute to the development of radiotherapy at three different levels, moving in turn from the most general to the more specific.

Ideas – providing a conceptual basis for radiotherapy, identifying mechanisms and processes that underlie the response of tumours and normal tissues to irradiation and which help to explain observed phenomena. Examples are knowledge about hypoxia, reoxygenation, tumour cell repopulation or mechanisms of repair of DNA damage.

Treatment Strategy – development of specific new approaches in radiotherapy. Examples are hypoxic cell sensitizers, high-LET radiotherapy, accelerated radiotherapy, hyperfractionation.

Protocols – advice on the choice of schedules for clinical radiotherapy. For instance, conversion formulae for changes in fractionation or dose rate, or advice on whether to use a chemical radiosensitizer at the start or at the end of a course of radiotherapy. We may also include under this heading methods for predicting the best treatment for the individual patient (*individualized radiotherapy*).

There is no doubt that radiobiology has been very fruitful in the generation of new ideas and in the identification of potentially exploitable mechanisms. A variety of new treatment strategies have been produced, but unfortunately few of these have so far led to demonstrable clinical gains. In regard to the third of the levels listed above, the newer conversion formulae based on the linear-quadratic equation seem to be successful and there are encouraging signs that some approaches to the individualization of radiotherapy may be working. But beyond this, the ability of laboratory science to guide the radiotherapist in the choice of specific protocols is limited by the inadequacy of the theoretical and experimental models: it will always be necessary to rely on clinical trials for the final choice of a protocol.

1.3 The time-scale of effects in radiation biology

Irradiation of any biological system generates a succession of processes that differ enormously in time-scale. This is illustrated in Figure 1.2 where these processes are divided into three phases (Boag, 1975).

The Physical Phase consists of interactions between charged particles and the atoms of which the tissue is composed. A high-speed electron takes about 10^{-18} seconds to traverse the DNA molecule and about 10^{-14} seconds to pass across a mammalian cell. As it does so it interacts mainly with orbital electrons, ejecting some of them from atoms (*ionization*) and raising others to higher energy levels within an atom or molecule (*excitation*). If sufficiently energetic, these secondary electrons may excite or ionize other atoms near which they pass, giving rise to a cascade of ionization events. For 1 Gy of absorbed radiation dose, there are in excess of 10^5 ionizations within the volume of every cell of diameter 10 μm.

The Chemical Phase describes the period in which these damaged atoms and molecules react with other cellular components in rapid chemical reactions. Ionization and excitation lead to the breakage of chemical bonds and the formation of broken molecules (*free radicals*). These are highly reactive and they engage in a succession of reactions that lead eventually to the restoration of electronic charge equilibrium. Free radical reactions are complete within approximately 1 ms of radiation exposure. An important characteristic of the Chemical Phase is the competition between *scavenging* reactions, for instance with sulphydryl compounds that inactivate the free radicals, and *fixation* reactions that lead to stable chemical changes in biologically important molecules.

The Biological Phase includes all subsequent processes. These begin with enzymatic reactions that act on the residual chemical damage. The vast majority of lesions, for instance in DNA, are successfully repaired. Some rare lesions fail to repair and it is these that lead eventually to cell death. Cells take time to die; indeed after small doses of radiation they may undergo a number of mitotic divisions before dying. It is the killing of stem cells and the subsequent loss of the cells that they would have given rise to that causes the early manifestations of normal-tissue damage during the first weeks and months after radiation exposure. Examples are breakdown of the skin or mucosa, denudation of the intestine and haemopoietic damage (Section 5.2). A secondary effect of cell killing is compensatory cell proliferation, which occurs both in normal tissues and in

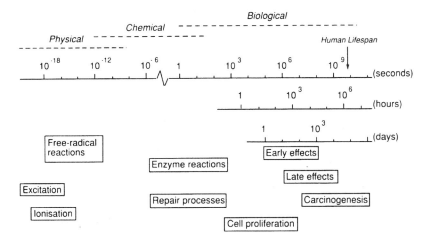

Figure 1.2 Time-scale of the effects of radiation exposure on biological systems.

tumours. At later times after the irradiation of normal tissues the so-called *late reactions* appear. These include fibrosis and telangiectasia of the skin, spinal cord damage and blood-vessel damage. An even later manifestation of radiation damage is the appearance of second tumours (*radiation carcinogenesis*): the time-scale of the observable effects of ionizing radiation may extend up to many years after exposure.

1.4 Response of normal and malignant tissues to radiation exposure

Much of the text of this book will focus on effects of radiation exposure that become apparent to the clinician or the patient during the weeks, months and years after radiotherapy. These effects are seen both in tumour tissues and in the normal tissues that surround a tumour and which are unavoidably exposed to radiation. The primary tasks of radiation biology as applied to radiotherapy are to *explain* observed phenomena, and to suggest improvements to existing therapies (as set out in Section 1.2).

The response of a tumour is seen by *regression*, often followed by *regrowth* (or *recurrence*), but perhaps with failure to regrow during the normal lifespan of the patient (which we term *cure* or *local control*). These italicized terms describe the tumour responses that we seek to understand. The relationship between regression and regrowth is illustrated graphically in Figure 2.7. The cellular basis of tumour response, including tumour control, is dealt with in Section 6.6.

The responses of normal tissues to therapeutic radiation exposure range from those that cause mild discomfort to others that are life-threatening. The speed at which the response develops varies widely from one tissue to another and often depends on the dose of radiation which the tissue receives. Generally speaking, the haemopoietic and epithelial tissues manifest radiation damage within weeks of radiation exposure, while damage to connective tissues becomes important at later times. A major development in the radiobiology of normal tissues during the 1980s was the realization that *early* and *late* normal-tissue responses are differently modified by a change in dose fractionation and this has given rise to the current interest in hyperfractionation (Section 14.2).

The first task of a radiobiologist is to *measure* a tissue response accurately and reliably. The term *assay* is used to describe such a system of measurement. Assays for tumour response are described in Section 17.3. For normal tissues, the following three general types of assay are available:

Scoring of Gross Tissue Effects. It is possible to grade the severity of damage to a tissue using an arbitrary scale as is done in Figure 5.2. In superficial tissues this approach has been remarkably successful in allowing isoeffect relationships to be determined.

Assays of Tissue Function. For certain tissues, functional assays are available that allow radiation effects to be documented. Examples are the use of breathing rate as a measure of lung function in mice (Figure 5.5), EDTA clearance as a measure of kidney damage (Figure 13.3), or blood counts as an indicator of bone marrow function.

Clonogenic Assays. In some tumours and some normal tissues it has been possible to develop methods by which the colony of cells that derive from a single irradiated cell can be observed. In tumours this is particularly important because of the belief that regrowth of a tumour after subcurative treatment is caused by the proliferation of a small number of tumour cells that retain colony-forming ability. This important area of radiation biology is introduced in Chapter 6.

1.5 Response curves, dose–response curves and isoeffect relationships

The damage that is observed in an irradiated tissue increases over a certain time period and it may subsequently decline (Figure 1.3A). How should we quantify the magnitude of this response? We could use the measured response at some chosen time after irradiation, such as the time of maximum response, but the timing of the

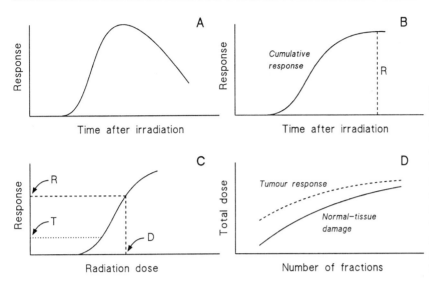

Figure 1.3 Four types of chart leading to the construction of an isoeffect plot. **A** time-course of development of radiation damage in a normal tissue. **B** the cumulative response. **C** a dose–response relationship, constructed by measuring the response (R) for various radiation doses (D). **D** isoeffect plot for a fixed level of normal-tissue damage (also a similar plot for tumour response).

peak may change with radiation dose and this would lead to some uncertainty in the interpretation of the results. A common device is to calculate the *cumulative* response by integrating this curve from left to right (Figure 1.3B). Some normal-tissue responses give a cumulative curve that rises to a plateau, and the height of the plateau is a good measure of the total effect of that dose of radiation on the tissue. Other normal-tissue responses, in particular the late responses seen in connective and nervous tissues, are progressive and the cumulative response curve will continue to rise (Figure 11.2). In such cases there is no satisfactory alternative to choosing a fixed time at which to evaluate a response.

The next stage in a study of the radiation response of a tissue will be to vary the radiation dose and thus investigate the *dose–response relationship* (Figure 1.3C). Many examples of such curves are given in this book (for instance, Figures 11.3, 11.6, 17.1); cell survival curves (Section 6.3) are further examples of dose–response curves that are widely used in radiobiology. The position of the curve on the dose scale indicates the sensitivity of the tissue to radiation; its *steepness* also gives a direct indication of the change in response that will accompany an increase or decrease in radiation dose (Sections 10.4, 10.5).

The foregoing paragraphs have for simplicity referred to 'dose' as though we are concerned only with single radiation exposures. It is a well-established fact that multiple radiation doses given over a period of a few weeks give a better curative response than can be achieved with a single dose (Chapter 12). Diagrams similar to Figures 1.3A, B, and C can also be constructed for fractionated radiation treatment, although the results are easiest to interpret when the fractions are given over a time that is short compared with the time-scale of development of the response. If we change the schedule of dose fractionation, for instance by giving a different number of fractions, changing the fraction size or radiation dose rate, we can then investigate the therapeutic effect in terms of an *isoeffect plot* (Figure 1.3D). Experimentally this is done by performing multiple studies at different doses for each chosen schedule and calculating a dose–response curve. We then select some particular level of effect (T in Figure 1.3C) and read off the total radiation dose that gives this effect. For effects on normal tissues the isoeffect will often be some upper limit of *tolerance* of the tissue, perhaps expressed as a probability of tissue failure (Sections 5.1, 11.2). The isoeffect plot shows how the total radiation dose for the chosen level of effect varies with dose schedule. Examples are Figures 12.3, 13.1, and recommendations for tolerance calculations are set out in Chapter 13. The dashed line in Figure 1.3D illustrates how therapeutic conclusions may be drawn from isoeffect curves. If the curve for tumour response is flatter than for normal-tissue tolerance, then there is a therapeutic advantage in using large fraction numbers: a tolerance dose given using small fraction numbers will be far short of the tumour-effective dose, whereas for large fraction numbers it may be closer to an effective dose.

1.6 The concept of therapeutic index

Discussion of the possible benefit of a change in treatment strategy must always consider simultaneously the effects on tumour response and on normal-tissue damage. A wide range of factors enter into this assessment. In the clinic, in addition to quantifiable aspects of tumour response and toxicity, there may be a range of poorly quantifiable factors such as new forms of toxicity or risks to the patient, or practicability and convenience to hospital staff, and also cost implications. These must be balanced in the clinical setting. The function of radiation biology is to address the *quantifiable biological aspects* of a change in treatment.

In the laboratory this can be done by considering dose–response curves. As radiation dose is increased, there is always a tendency for tumour response to increase, and the same is true of normal-tissue damage. If, for instance, we measure tumour response by determining the proportion of tumours that are controlled, then we may expect a sigmoid relationship to dose (for fractionated radiation treatment we could consider the total dose or any other measure of treatment intensity). This is illustrated in Figure 1.4A. If we quantify normal-tissue damage in some way for the same treatment schedule, there will also be a rising curve of toxicity (Fgiure 1.4B). The shape of this curve is unlikely to be the same as that for tumour response and we probably will not wish to determine more than the initial part of this curve (since a high frequency of damage is unacceptable). By analogy with what must be done in the clinic, we can then fix a notional upper limit of *tolerance* (Section 5.1). This fixes, for that treatment schedule, the upper limit of radiation dose that can be tolerated, for which the tumour response is indicated by the point in Figure 1.4A labelled A.

Consider now the effect of adding to radiation therapy a cytotoxic drug. We expect that this will increase the tumour response for any radiation dose and this will be seen as a movement to the left of the curve for tumour control (Figure 1.4). There will probably also be an increase in damage to normal tissues which again will consist of a leftward movement of the toxicity curve. The relative displacement of the curves for the tumour and normal tissues will usually be different. But in this simple case there is a straightforward way of asking whether the combined treatment is better than

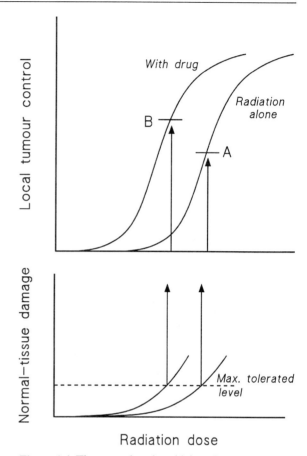

Figure 1.4 The procedure by which an improvement in *therapeutic index* might be identified, as a result of adding chemotherapy to radiotherapy. See also Figures 11.3, 20.1.

radiation alone: for the same tolerance level of normal-tissue damage the maximum radiation dose will be lower and the corresponding level of tumour control is indicated by point B in Figure 1.4. If B is higher than A then the combination is better than radiation alone. Improvement in therapeutic index has often been achieved in studies on experimental tumours in animals but the confirmation of this in the clinic is unfortunately rare (Chapter 21).

This example indicates the radiobiological concept of *therapeutic index*: it is the tumour response for a fixed level of normal-tissue damage (Section 10.6). The concept can in principle be applied to any therapeutic situation or to any appropriate measures of tumour response or toxicity; the application of this in the clinic is not a straightforward matter, as indicated in Section 20.1. Therapeutic index carries the notion of 'cost–benefit' analysis. It is impossible to reliably discuss the potential benefit of a new treatment without reference to its effect on therapeutic index.

1.7 The importance of radiation biology for the future development of radiotherapy

Many developments in radiotherapy have resulted from new technologies or have been made empirically by clinicians; there are few examples of developments that have begun in the radiobiological laboratory and been carried through to the point where patient survival has significantly improved. The role of oxygen is one positive example that has led to some benefits (Chapter 16), also the promising results with accelerated fractionation and hyperfractionation (Chapter 14).

Compared with chemotherapeutic agents, radiation is now a well understood cytotoxic agent. Its access to tumour cells is just a matter of dosimetry, independent of the transport mechanisms that largely determine the effectiveness of chemical agents. The sequence of processes listed in Section 1.3 above are well described for radiation; some of them are equally relevant to the response of tissues to cytotoxic drug treatment, and thus research into radiation biology has brought benefits to other areas of therapeutic cancer research.

The future is likely to require greater and greater dependence on basic science. The simple empirical things have mostly been fully exploited and increasing knowledge about the cellular and molecular nature of radiation effects will undoubtedly lead to developments for which the radiotherapist will require a grounding in fundamental mechanisms. That is the purpose of this book.

Key points

1. Radiotherapy is an important curative and palliative modality in the treatment of cancer. Significant gains are still to be made by the optimization of biological and physical factors.
2. Therapeutic index is 'the name of the game' in curative cancer therapy.
3. The effects of radiation on mammalian tissues should be viewed as a succession of processes extending from microseconds after exposure to months and years thereafter. In choosing one end-point of effect, it is important not to overlook the rest of this whole process.

BIBLIOGRAPHY

Boag JW (1975). The time scale in radiobiology. 12th Failla memorial lecture. In: *Radiation Research*. (Eds) Nygaard OF, Adler HI and Sinclair WK. Academic Press; San Diego

DeVita VT, Goldin A, Oliverio VT et al (1979). The drug development and clinical trials programs of the Division of Cancer Treatment, National Cancer Institute. *Cancer Clinical Trials* 2:195–216.

Souhami R and Tobias J (1986). *Cancer and its Management*. Blackwell; Oxford.

Tubiana M (1992). The role of local treatment in the cure of cancer. *Eur J Cancer* 28A:2061–69.

2

The growth rate of tumours

G. Gordon Steel

2.1 Introduction

The speed of development of the disease process in patients with cancer depends to a large extent on the growth rate of primary and metastatic tumours. In patients in whom treatment is unsuccessful the speed of recurrence and the survival of the patient also depend on tumour growth rate. Growth rate is not the *only* determinant of the time-course of cancer: the location, the invasiveness and the biochemical properties of the lesions may also play a part.

2.2 Measurement of tumour size

The precision and frequency with which tumour size can accurately be measured varies widely from one anatomical site to another. At the time of surgery a single measurement can often be made. The size of superficial lesions can be measured with engineer's callipers, within the constraints of skin thickness and depth of the lesion. Chest X-rays, CT and NMR scanning and other imaging techniques also allow repeated measurements of tumours in some sites. The greatest amount of published data on tumour growth has been on primary and metastatic tumours in the lung. However, even in this site where the difference of tissue density between tumour and surrounding tissues is considerable, the size of lesion that can be discriminated is limited. Spratt *et al* (1963) tested the acuity of radiologists by placing plastic balls of different sizes over various parts of the chest of patients before taking X-rays and summarized the results of their studies as follows:

The radiologists could locate the opacities of lucite balls 10 – 12 mm in diameter regularly regardless of their location. They could locate the radiopacities of balls 6 mm in diameter only when they were located in intercostal spaces contrasted against aerated lung. They could distinguish the radiopacities of 3 mm diameter balls from normal pulmonary shadows in these same areas of favourable contrast only when they were shown precisely where to look.

Careful studies of the growth and regression of lung metastases were made by Breur (1966) using a series of circles engraved on Perspex, increasing in diameter by 0.5 or 1.0 mm. Thomlinson (1982) made very precise measurements of primary breast tumours using callipers (Figure 2.6). Various formulae have been used to transform linear dimensions into volumes, for instance:

$V = \pi/6 \times$ (mean diameter)^3

$V = 0.5 \times$ length $\times$ (width)^2 [taking $\pi = 3$]

and provided the same formula is used consistently there is probably little to choose between them. For the measurement of tumours in laboratory animals, the calibration curve method (Steel, 1977) has a lot to commend it. Animals are taken with tumours of widely varying size. The external dimensions of each tumour are measured, after which it is excised and weighed. Tumour weight is then plotted against an external measurement (for instance the product of length and width) and a smooth curve through these data comprises a calibration curve that can be used to interpret any further external measurements on non-excised tumours. This approach does not depend on any assumptions about the geometric form of the tumours, nor about the effect of skin thickness, and the scatter of points around the calibration curve also gives a direct indication of precision.

The size and growth rate of tumours can be measured indirectly by quantifying tumour products in blood: CEA for choriocarcinomas, immunoglobulins for plasmacytomas, AFP or HCG in testicular tumours (Price *et al*, 1990a and b).

2.3 Exponential and non-exponential growth

Exponential growth is where tumour volume increases by a constant *fraction* in equal intervals of time. Thus, the time for tumour volume to double (the *volume doubling time*, T_d) is the same for lesions of size 1–2 g or 10–20 g or 100–200 g etc. The equation of exponential growth is:

$$V = \exp(0.693 \cdot \text{time} / T_d)$$

where 0.693 is $\log_e 2$. The logarithm of tumour volume increases linearly with time. It is conventional to plot tumour growth curves on a logarithmic scale of volume so that departures from exponential growth can easily be seen.

Why is the idea of exponential growth so important? This is, in fact, the *simplest* mode of growth. If cells are allowed to proliferate under constant conditions, with no loss or infertility, their number will increase exponentially. It is *departure* from exponential growth that we have to explain! As indicated in Section 3.2, there are two principal processes that cause tumours to grow with a doubling time that is longer than the cell cycle time: cell loss and decycling (*i.e.* proliferating cells moving into a non-proliferating state). Non-exponential growth can thus arise by any combination of three factors: increasing cell cycle time, decreasing growth fraction and increasing rate of cell loss (Steel, 1977). The nomogram in Figure 2.1 indicates the relation between tumour volume, cell number and the number of doublings, starting from a single cell. Cells are assumed to have a mass of 10^{-9} g. The nomogram is correct for any mode of growth; exponential is the special case where the doubling time is constant.

Figure 2.2 illustrates an important feature of exponential growth. The same exponential line is drawn on a *linear* scale of volume (Figure 2.2A) or on a *logarithmic* scale (Figure 2.2B). On the linear scale (which is what a clinical observer will tend to see) there appears to be a long 'silent interval' or

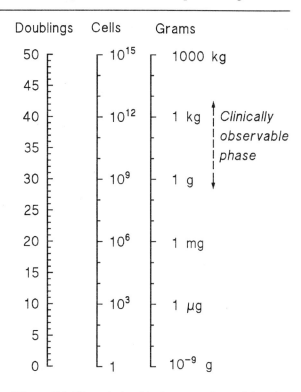

Figure 2.1 The relationship between the weight of a tumour, the number of cells it contains (assuming 10^9 per gram) and the number of doublings from a single cell.

latent period where no growth is seen. But during this time the tumour is growing regularly and with a constant doubling time. Once the tumour becomes detectable (at a size of perhaps 1 g) its size on a linear scale appears to sweep upwards, steeper and steeper. This is only a subjective and misleading impression, for growth is in fact perfectly regular and exponential.

Exponential growth of tumours in laboratory animals is uncommon. It is more usual to find that the doubling time increases progressively as the tumour gets bigger. This is illustrated in Figure 2.3, which shows a number of well-measured growth curves for tumours in rats and mice. The points are averages for a group of similar tumours. In each case the curves (on the logarithmic scale of volume) bend downwards, showing that the tumours grew progressively more slowly. The volume doubling time can be judged at any point by drawing a tangent to the curve and reading off its doubling time.

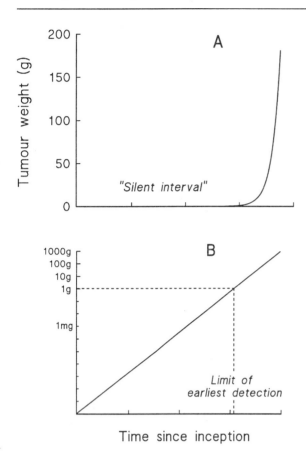

Figure 2.2 An exponential growth curve plotted on a linear scale (**A**) or on a logarithmic scale (**B**). The clinical phase of growth is a minor part of the whole life history of the tumour.

Growth curves of the type shown in Figure 2.3 have often been described by the Gompertz equation:

$$V = V_0 \exp\left[A/B (1 - e^{-Bt}) \right]$$

Here V_0 is the volume at time zero and A and B are parameters that determine the growth rate. At very early time intervals (t small) the equation becomes exponential: $V = V_0 \exp(A.t)$. At long time intervals $\exp(-B.t)$ becomes small compared to 1.0 and the volume tends to a maximum value of $V_0 \exp(A/B)$. It can be seen in Figure 2.3 that the curves for the two rat tumours (fibroadenomas and Walker tumours) tend towards a larger maximum than for the mouse tumours, suggesting that this maximum size bears some relation to the host body weight (rats weigh up to ~300 g, mice ~30 g). The Gompertz equation is not a unique description of such growth curves. For a fuller discussion see Steel (1977).

2.4 The growth rate of human tumours

Some examples of carefully measured human lung tumours are shown in Figure 2.4; the lines are straight or nearly so and these represent good examples of exponential tumour growth. Some human lung tumours show a Gompertzian pattern of growth, and irregular growth (sudden increase or decrease in growth rate) is not uncommon.

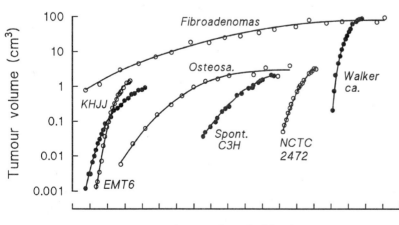

Figure 2.3 Growth curves for tumours in rats and mice. The primary breast fibroadenomas and the Walker tumours were in rats, the others in mice. The fitted curves are Gompertz equations. From Steel (1977), with permission.

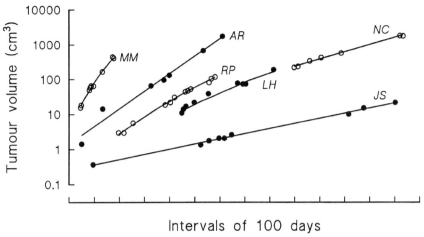

Figure 2.4 Growth curves for primary human lung tumours. Data of Schwartz, redrawn by Steel (1977), with permission.

Published data on the growth rate of tumours were reviewed by Steel (1977). Within any one tumour type there is a wide range of volume doubling times. For instance, the range of values for lung metastases of adenocarcinoma is shown in Figure 2.5. Some double their volume in a week, some in a year or more, and the median is around 90 days. This median value is typical of other classes of human tumour. Table 2.1 gives mean values of doubling time for tumours of various types. Lymphomas, teratomas and superficial breast metastases grow faster than the average; primary lung adenocarcinomas and colon tumours grow more slowly.

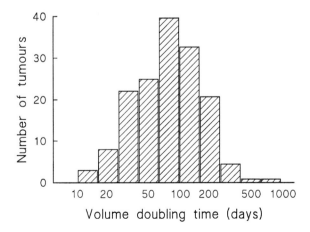

Figure 2.5 The distribution of volume doubling times for 159 lung metastases of adenocarcinoma from various primary sites. From Steel (1977) with permission.

2.5 The speed of tumour regression

After treatment, some tumours show a rapid volume response and others respond much more slowly. It is important to distinguish between *speed of shrinkage* and the *probability of local tumour control*, for some rapidly shrinking tumours recur early. Some clinical studies have shown a correlation between shrinkage rate and local control, but this is not the case in all clinical situations. Careful studies by Thomlinson (1982) showed that among primary breast tumours there was a 50-fold range of regression halving times (Figure 2.6). The rapidly shrinking tumours tended to be highly cellular, while those that shrank slowly had a large amount of connective tissue. He also found that in tumours that were treated by radiotherapy as well as by chemotherapy the rate of regression was independent of the treatment: it was characteristic of the biology of the tumour.

The overall volume response of a tumour to treatment is illustrated in Figure 2.7. There are two components of response, regression and regrowth, and the shape of the regrowth curve is a matter of considerable current debate (Sections 14.3, 25.4). The nadir (*i.e.* minimum tumour volume) will depend upon both components. For judging the effectiveness of tumour treatment it is therefore wise to choose a measure that reflects the regrowth component rather than the speed of regression, for it is the regrowth component that depends upon the degree of tumour cell kill. *Partial remission* is an unsatisfactory criterion. *Duration of disease-free*

12 The growth rate of tumours

Table 2.1 Volume doubling times for human tumours

Site and type	No. of tumours measured	Mean volume doubling time*	Confidence limits on mean
Metastases in lung			
Colon-rectum, adenocarcinoma	56	95	84–107
Breast, adenocarcinoma	44	74	56–98
Kidney, adenocarcinoma	14	60	37–98
Thyroid, adenocarcinoma	16	67	44–103
Uterus, adenocarcinoma	15	78	55–111
Head & neck, sq. cell carcinoma	27	57	43–75
Fibrosarcoma	28	65	46–93
Osteosarcoma	34	30	24–38
Teratoma	80	30	25–36
Lymphoma	11	27	19–39
Superficial metastases			
Breast carcinoma	66	19	16–24
Primary tumours			
Lung, adenocarcinoma	64	148	121–181
Lung, sq. cell carcinoma	85	85	75–95
Lung, undifferentiated	55	79	67–93
Colon-rectum	19	632	426–938
Breast	17	96	68–134

* Geometric mean volume doubling time in days.
From a review of early data on the growth rate of human tumours (Steel, 1977).

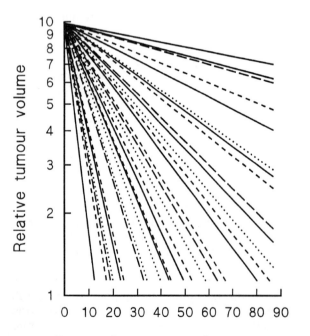

Figure 2.6 The range of regression rates of some of the 78 primary human breast tumours studied by Thomlinson (1982).

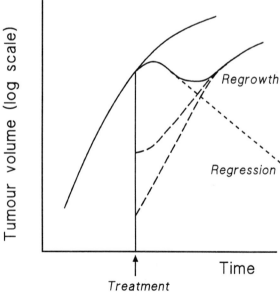

Figure 2.7 The volume response of an uncontrolled tumour is the resultant of two processes: regression and regrowth. Repopulation during the period of regression may take place at a rate that may differ from the growth rate of the untreated tumour.

interval is to be preferred. For measurable lesions in animals the preferred parameter is the *tumour growth delay*: this indicates the difference between the times at which treated or untreated tumours reach a fixed multiple of the pre-treatment tumour size (*e.g.* twice the size at the start of treatment, Figure 17.2A).

Key points

1. The accurate measurement of tumour size greatly helps clinical judgement on the rate of progression or response to treatment of tumours.
2. Tumour growth curves should always be plotted on a logarithmic scale in order to compare the data with exponential growth.
3. The rate of regression following treatment often varies widely among tumours of the same histopathological type. Slow regression does not necessarily mean little response.
4. The growth rate of tumours can vary widely among tumours of the same type. Volume doubling times in the region of 3 months are common.

BIBLIOGRAPHY

Breur K (1966). Growth rate and radiosensitivity of human tumours. *Eur J Cancer* 2:157–71.

Price P, Hogan SJ and Horwich A (1990a). The growth rate of metastatic non-seminomatous germ cell testicular tumours measured by marker production doubling time. I: Theoretical basis and practical application. *Europ J Cancer* 26:450–53.

Price P, Hogan SJ, Bliss JM and Horwich A (1990b). The growth rate of metastatic non-seminomatous germ cell testicular tumours measured by marker production doubling time. II: Prognostic significance in patients treated with chemotherapy. *Europ J Cancer* 26:453–56.

Spratt JS, Ter-Pogossian M and Long RTL (1963). The detection and growth of intrathoracic neoplasms. *Archs Surg* 86:283–88.

Thomlinson RH (1982). Measurement and management of carcinoma of the breast. *Clin Radiol* 33:481–93.

FURTHER READING

Steel GG (1977). *The Growth Kinetics of Tumours*. Oxford University Press; Oxford.

3

Cell proliferation in tumours

Adrian C. Begg

3.1 Cell kinetic compartments of a tumour

The neoplastic cells within a tumour can be divided into four compartments based on their kinetic properties. The most important compartment is that of actively dividing cells. All cells in this compartment are going through the cell cycle and can be distinguished using cell-labelling techniques. All new tumour cells are produced from this compartment which is therefore the major contributor to growth of the tumour volume. This compartment is called the *growth fraction*, and the cells within it are sometimes called 'P' (for proliferating) cells.

In addition to actively dividing cells, there are two other compartments in the majority of tumours which contain non-proliferating cells. These are the *resting* (or G0) compartment and the *sterile* (or differentiated) compartment. The major difference between them is that G0 cells are capable of re-entering the cell cycle, while sterile or differentiated cells are no longer capable of division. G0 cells are often called 'Q' (for quiescent) cells. Some G0 cells may be clonogenic (and thus capable of repopulating a tumour) and are therefore dangerous and need to be killed by the applied therapy. It is often not easy to distinguish G0 cells from sterile cells on the basis of kinetic techniques, although terminally differentiated cells can sometimes be distinguished morphologically in well-differentiated tumours. Differentiated cells are no longer a serious danger to the patient, although if present in large numbers their bulk can cause problems. The other main contributor to tumour bulk is the stroma: normal-tissue cells such as blood cells and fibroblasts that in some cases can exceed the number of neoplastic cells.

In addition, there exists in most tumours a compartment consisting of dead and dying cells. Necrosis is a characteristic of tumours and is the result of an inadequate blood supply. The volume occupied by necrosis varies widely from one tumour to another but it can be extensive (Figures 15.4, 15.5).

Transfer of cells from one compartment to another occurs continuously. Movement of cells from the Q to the P compartment, which can occur after or during some treatments, is called *recruitment*. Transfer from P to Q must also occur, for otherwise the proportion of Q-cells would decline towards zero in a growing tumour, due to the multiplication of P cells. Some cells may have an inadequate supply of nutrients such as oxygen and may fail to divide further; cells can also enter the differentiated compartment due to natural differentiation processes. Finally, cells can leave the volume of a primary tumour mass, either as viable cells (this can lead to metastases) or by death followed by the resorption of their constituents. These processes lead to the general phenomenon of *cell loss* from tumours.

3.2 Factors affecting tumour growth rate

The volume doubling time of a tumour (T_d, Section 2.3) is determined by three main factors: the cell cycle time (T_c), the growth fraction (GF) and the rate of cell loss. Tumours grow faster if: the cycle time is short, the growth fraction is high, and cell loss is low (Table 3.1). The potential doubling time (T_{pot}) is defined as the time within which the cell population of the tumour would double if there were no cell loss (Steel, 1977). It depends on the cell cycle time and the growth fraction. The potential doubling time can be obtained from the thymidine labelling index (LI) or S-phase fraction by the relation:

$$\text{Potential doubling time } (T_{\text{pot}}) = \lambda \frac{T_s}{LI} \quad \text{(Eqn 3.1)}$$

where T_s is the duration of the S-phase. The parameter λ corrects for the non-rectangular age distribution of growing cell populations and usually lies between 0.7 and 1.0. The S-phase duration can be measured by thymidine-analogue labelling techniques (Section 3.4). The volume doubling time (T_d) is often found using callipers for superficial tumours, or from radiographs or from CT scans. Cell loss from a tumour can be estimated from the cell loss factor:

$$\text{Cell loss factor} = 1 - \frac{T_{\text{pot}}}{T_d} \quad \text{(Eqn 3.2)}$$

(Steel, 1977). This factor is simply the cell loss rate as a fraction of the cell birth rate. The slow growth of many human tumours is largely the result of a high rate of cell loss. The principal mechanisms of cell loss are necrosis and differentiation: well-differentiated carcinomas (by definition) maintain some features of the hierarchical tissue structure of the tissue of origin (Section 4.1) and cell turnover by this normal pathway probably continues in the malignant state. In many tumours it is possible to recognize individual pyknotic or apoptotic cells scattered throughout the tissue.

Evidence from studies on tumours in mice and rats suggests that the slowing of growth rate as tumours increase in size (Section 2.3) is associated with a progressive increase in the rate of cell loss, a decrease in the growth fraction, and a lengthening of the mean cell cycle duration (Steel, 1977).

3.3 Values for kinetic parameters in human tumours

Studies have been carried out in which ^{3}H-thymidine was given to patients and multiple biopsies

Table 3.1 Kinetic parameters of a typical human tumour

Cell cycle time (~2 d)	Potential doubling time (~5 d)	Volume doubling time (~60 d)
Growth fraction (~40%)		
Cell loss (~90%)		

taken in order to measure cell cycle parameters. Radioactive thymidine is specifically incorporated into DNA, although some of it is catabolized and lost from the tissue. Many early studies of cell proliferation were based on thymidine labelling of S-phase cells. More recently, much information on the labelling index and T_s has been obtained from *in vivo* labelling with thymidine analogues, together with flow cytometry (see below). Cell cycle times for human tumours from these studies have ranged from 15 hours to more than 100 hours with an average of ~2.3 days (Steel, 1977). These values are similar to cell population doubling times found for human tumour cells grown in culture. As described in Section 2.4, the volume doubling times for human tumours are much longer, ranging from 4 days to over a year, around a median of roughly 3 months. There are large differences between individual tumours, even within a particular histological type. Growth fractions for the relatively few human tumours studied range between 6% and 90%, with most solid tumours having values well below 50%. Cell loss factors are almost always high in carcinomas, usually ranging from 70% to over 90% (Table 3.2). This is apparent from average T_{pot} values of a few days compared with T_d values of a few months: for instance, if T_{pot} = 5 days and T_d = 70 days, the cell loss factor = 1 − (5/70) = 0.93, or 93%.

3.4 Cell kinetic methods

PERCENT LABELLED MITOSES

With this technique, cells are pulse labelled with ^{3}H-thymidine (^{3}H-TdR), a radioactively labelled nucleic acid precursor that is specifically incorporated into DNA, not RNA. The labelling is done either by injection into an animal or patient (which as a result of the rapid metabolism of thymidine gives almost instantaneous labelling of cells in the S-phase), or for cells in tissue culture by washing unincorporated tracer out after a few minutes. Samples of the cell population are taken over a range of times thereafter and labelling of mitotic figures is detected by autoradiography. The fraction of mitoses which are labelled will rise and fall as the cohort of ^{3}H-labelled cells, initially in S, move into and through mitosis. This pattern will be repeated one cell cycle later. From the frequency

Table 3.2 Cell loss calculations for human tumours

	Thymidine labelling index (%) (median and range)	Volume doubling time (days) (median and range)	T_{pot} (days)	Cell loss factor (%)
Colorectal carcinoma	15 (10 – 22)	90 (60 – 170)	3.1	96
Squamous cell carcinoma of head and neck	6.9 (5 – 17)	45 (33 – 150)	6.8	85
Undifferentiated bronchial carcinoma	19 (8 – 23)	90 (40 – 160)	2.5	97
Melanoma	3.3	52 (20 – 150)	14	73
Sarcoma	2.0 (0.3 – 6)	39 (16 – 78)	23	40
Lymphoma	3.0 (0.4 – 13)	22 (15 – 70)	16	29
Childhood tumours	13 (10 – 25)	20	3.6	82

From Steel (1977).

and width of the waves of labelled mitoses, the lengths of all phases of the cycle, together with their variations, can be determined (Steel, 1977).

Other methods for measuring cell kinetics using radiolabelled thymidine include continuous labelling and double labelling. In continuous labelling, labelled thymidine is given continuously *in vitro* or *in vivo*, whereby the percentage of labelled cells rises progressively as G1 cells enter S and as labelled cells divide. From the rate of increase, T_s and T_{pot} can be estimated. Double labelling employs both ³H- and ¹⁴C-labelled thymidine, which can be distinguished from each other in an autoradiograph. A pulse label of the first, followed a few hours later by a pulse label of the second, results in cells containing no label, one label or both labels. The proportion of singly labelled cells increases with time as the first labelled cohort progresses out of S into G2, allowing T_s to be estimated from the ratio of single- to double-labelled cells (Steel, 1977).

The disadvantages of these methods for studies on cancer patients are the need to administer radioactivity, the requirement for repeated biopsies, the long exposure times of the autoradiographs and the labour and skill required for the scoring. Alternative more rapid methods using thymidine analogues and flow cytometry are now used routinely instead (see below).

The Stathmokinetic Technique

This technique uses compounds which block cells in mitosis, such as colchicine, colcemid, vincristine or vinblastine. Under this block, the fraction of cells in metaphase rises with time. If the dose of the blocking agent is correctly chosen, the rate of increase of the metaphase index gives a measure of the mitotic rate, *i.e.* the rate at which cells are entering mitosis. Satisfactory accumulation only continues for a limited time, after which some arrested cells begin to degenerate. If every mitosis produces two daughter cells, the mitotic rate equals the cell birth rate. It is thus possible to calculate the potential doubling time from the relation:

$$T_{pot} = \frac{\log_e 2}{mitotic\ rate} \quad \text{(Eqn 3.3)}$$

In addition, the duration of mitosis (T_m) can also in principle be obtained. The advantage of this technique is that no radioactivity or special staining are required, although it is not suitable for routine clinical use due to the toxicity of the arrest agents.

Thymidine Analogues

Replacement of the methyl group of thymidine (Figure 3.1) with halogen atoms of similar size such as iodine or bromine creates the analogues iodo- or bromo-deoxyuridine (IUdR, BrUdR). Enzymes responsible for DNA synthesis cannot easily distinguish between thymidine and its analogues and so these are incorporated into DNA via the same pathway as thymidine. Once in the DNA, they can be detected with specific antibodies which recognize the small distortions in the DNA molecule caused by their incorporation. The development of these antibodies has led to their widespread use for cell kinetic studies. The antibodies can in turn be labelled with an enzyme (usually a peroxidase)

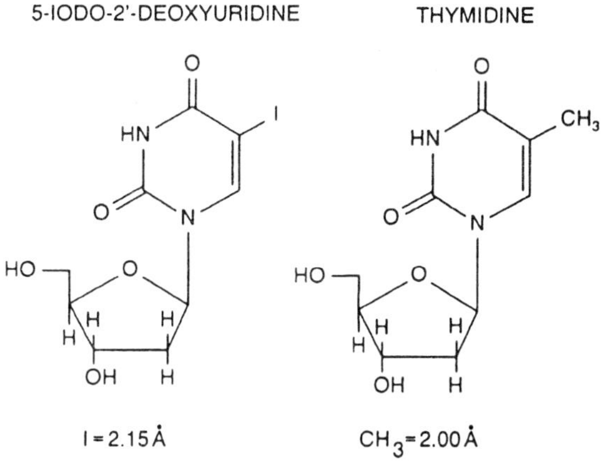

Figure 3.1 The structure of thymidine and its analogue, IUdR. The comparative diameters of the iodine atom and the methyl group are indicated.

which allows immunocytochemical staining, or a fluorescent label (usually the green fluorescent molecule FITC) for flow cytometry. In this way, all cells that have incorporated the analogue, and were therefore in the S-phase at the time of its administration, can be detected by a brown colour on immunoperoxidase-stained sections or by green fluorescence with flow cytometry.

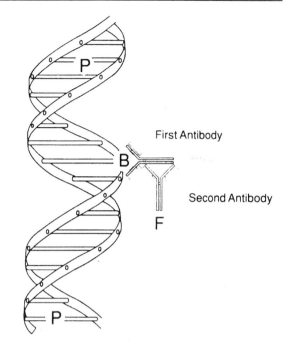

Figure 3.2 The principle of BrUdR staining of DNA. B = an incorporated BrUdR molecule; P = an intercalated propidium iodide molecule; F = FITC conjugated to the second antibody.

FLOW CYTOMETRY

For cell kinetic studies using flow cytometry, BrUdR-labelled cells are stained with a mouse antibody specific for the DNA-incorporated analogue, followed by an anti-mouse antibody conjugated with FITC (Figure 3.2). Cells are counter-stained for total DNA content using propidium iodide (PI, a red fluorescent stain). In the flow cytometer, the stained cells flow through a laser beam which excites the fluorochromes, giving green fluorescent emissions from FITC and red fluorescence from the PI (Hall, 1988). These colours can be separated using optical filters and their intensities measured for each cell using photomultiplier tubes (Figure 3.3). The data are displayed as two-parameter cytograms, for instance green versus red fluorescence intensities. Three subpopulations, representing G1 cells (low green and low red), G2/M cells (low green and high red) and S-phase cells (high green), can readily be seen. Up to 10 000 cells per second can be analysed in this way, making this a rapid and quantitative method which also avoids the use of radioactivity.

An example of the flow cytometer output is shown in Figure 3.4. This shows green (IUdR) versus red (DNA) fluorescence cytograms for Chinese hamster cells at different times after pulse labelling *in vitro* with IUdR. Movement of labelled and unlabelled cells through the cell cycle can be seen from the changing patterns. At time zero, *i.e.* immediately after pulse labelling, the labelled cells should be distributed equally between the G1 and G2 positions. In Figure 3.4, movement of cells from G2 into G1 is already visible by 2 hours, and the 'leaning' to the right of the dot pattern indicates movement through the S-phase. This becomes more pronounced with increasing time. By 6 hours, all the unlabelled cells that were in G1 have moved into the S-phase, and all the labelled cells have moved out of S and are now in G2 or have divided and have returned to G1. This gives an almost inverted picture compared with that at 1 hour. All phases of the cell cycle can be obtained

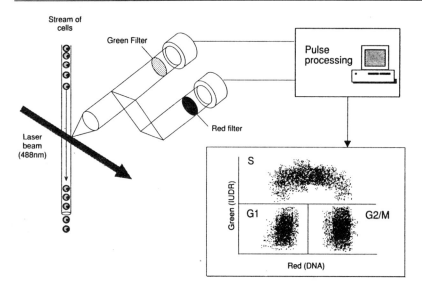

Figure 3.3 The principle of flow cytometry.

from these pictures, in a manner similar to the analysis of per cent labelled mitoses curves, except that the fraction of labelled cells in mid-S is measured instead of the fraction of labelled cells in mitosis.

Obtaining Kinetic Information from One Sample

The relative movement method allows T_s, LI and T_{pot} to be estimated from one tissue sample (Begg et al, 1985). This is useful for clinical application where it is often difficult to take more than one biopsy. Several hours after i.v. injection of BrUdR or IUdR a tumour biopsy is taken, fixed in ethanol, and a suspension of nuclei is subsequently made, stained and analysed by flow cytometry. The average position (red fluorescence) of the labelled cells which have not yet divided, relative to the positions of G1 and G2, is measured using computer-drawn windows around the appropriate subpopulations (Figure 3.5A). The relative movement parameter (RM) is a DNA-content parameter defined to be zero for cells in G1 and 1.0 for cells in G2. Since on average the labelled cells immediately after staining will be homogeneously distributed within the S-phase, it is assumed that RM at that time is 0.5. Subsequently, the cells progress towards G2 and RM increases. The procedure is to measure RM in a sample taken a few hours (t) after labelling, then plot a line between RM = 0.5 at t = 0 and the mea-

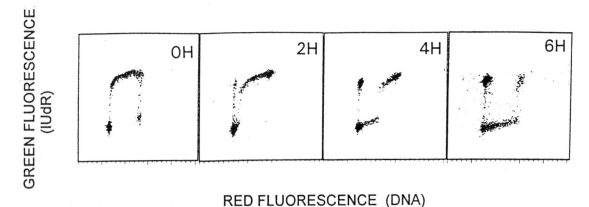

Figure 3.4 Flow cytometer traces from samples taken at various intervals from an *in vitro* cell population that was pulse labelled with IUdR at time zero. Green fluorescence (IUdR content) is plotted vertically; red fluorescence (DNA content) horizontally.

sured RM at time t. This is extrapolated to the time required for RM to reach 1.0, which gives the estimate of T_s (Figure 3.5B). Panel C in Figure 3.5 shows an example of data obtained in this way for cells growing at different rates due to different culture temperatures. It can be concluded that the duration of the S-phase approximately doubled as a result of lowering the temperature. The calculation assumes that the plot of RM versus time is linear, although theoretically it is curved. There are better, more rigorous mathematical ways to calculate T_s from RM, taking into account this curvature (Terry et al, 1991). For most accurate T_s measurements, the time interval between injection and biopsy should be longer than G2+M; 6–8 hours is recommended for most human tumours.

Proliferation Markers

The use of thymidine analogues can provide dynamic, or rate, information about the cell cycle. A disadvantage for clinical studies is the need to administer a potentially toxic compound *in vivo*, although no acute or late toxicity has yet been observed for IUdR or BrUdR when used at low doses for cell kinetic studies. Additionally, extra work is involved for doctors and nurses. Alternative assays for proliferation have therefore been sought which do not require drug administration. There are several proteins which appear to be expressed in proliferating but not quiescent cells. Examples of these are DNA polymerase, histones H2–4, Ki67, proliferating cell nuclear antigen (PCNA), and some of the cyclins. Antibodies are available to these proteins which can therefore be used to assess proliferation by immunochemical or flow cytometry methods.

The most commonly used marker at present is Ki67, a nuclear protein associated with proliferating cells and which can be detected using the originally discovered antibody (Ki67), or by MIB1, an antibody against the same protein but which works better on paraffin embedded material. Flow cytometry can also be used for detection. The function of the molecule is presently unknown, although progress is being made since the gene has now been cloned. It may not give accurate growth fraction estimates under all circumstances, particular after cytotoxic treatments, although it is clearly closely proliferation-related and is widely used to rank proliferative fractions of tumours and normal tissues under non-perturbed conditions.

PCNA is a protein involved in both DNA repair and DNA synthesis, being an auxiliary protein for DNA polymerase delta. When PCNA is tightly bound in the nucleus it is associated with DNA replication. The fraction of PCNA-positive cells is therefore a measure of the S-phase fraction. PCNA is expressed in other phases of the cycle, however, where it is loosely bound. Measurement of total (tight + loose) PCNA is then in principle a measure of the growth fraction (*i.e.* cells in all cycle phases). Different tissue preparation techniques resulting in the removal or not of loosely bound protein will therefore result in different PCNA indices, necessitating care in interpretation. As with Ki67, immunochemical methods or flow cytometry can be used for detection. Figure 3.6

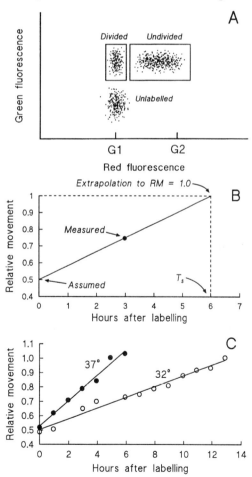

Figure 3.5 Principle of the relative movement method. **A**: the two-dimensional flow-cytometer display; **B**: the principle of the method; **C**: an example of a study on cells at two different temperatures; the difference in slopes indicates different S-phase durations.

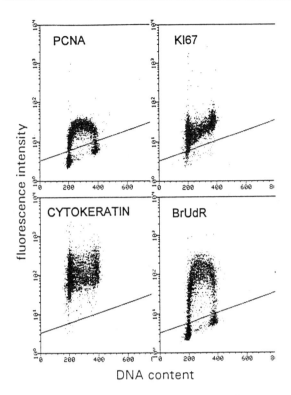

Figure 3.6 Flow cytometry detection of proliferation markers in the human non-small cell lung cancer line MR65. Each marker was detected with a fluorescent antibody (Y-axis) and plotted against total DNA content (X-axis). PCNA and BrUdR show S-phase-specific labelling, in contrast to Ki67 which is found in all phases. Cytokeratin, a potential tumour marker, is also expressed in all phases. (From Schutte et al, 1995, with permission).

shows flow cytometry data for dual staining using a marker (fluorescent antibody) and total DNA content (propidium iodide). PCNA shows predominant staining in S, similar to BrUdR, while Ki67 is expressed in all phases of the cycle. These lung tumour cells are also positive for cytokeratin, which could be used as a tumour marker.

Several genes are only expressed in particular phases of the cycle, allowing their use as phase markers. Histones 3 and 4 are expressed in S-phase, since they are necessary for nucleosome assembly on newly synthesized DNA. The expression of histone H3 has therefore been used as a marker for the S-phase and has been shown to give similar information (*i.e.* fraction of cells in S) as a thymidine or BrUdR labelling index. In this method, gene expression is monitored by measuring mRNA levels using *in situ* hybridization (ISH), whereby an oligonucleotide or ribonucleotide probe is used which specifically hybridizes to the mRNA of histone H3. Use of a biotin- or digoxygenin-labelled probe allows detection with standard immunoperoxidase or immunofluorescence methods. The method can also be combined with BrUdR labelling to give dynamic as well as static information on proliferation.

Cyclins are a class of molecule that are usually expressed in a cyclical fashion during the cycle, hence their name. They are instrumental in controlling the transitions between cycle phases, *e.g.* G1 to S or G2 to M. Cyclins activate particular kinases (enzymes which phosphorylate proteins) by forming complexes with them. Cyclins D and E are G1 cyclins; cyclin A is mostly found in the S-phase; cyclin B is found mostly in G2. There are several subtypes of the B and D cyclins. The availability of antibodies to almost all known cyclins allows detection by immunochemical or flow methods. In an asynchronous population, the expression of particular cyclins can indicate the approximate position of individual cells in the cycle. Cyclin B, for example, has been used as a marker of G2 cells, and is also useful in distinguishing doublets of G1 cells from true G2 cells in flow cytometry. Figure 3.7 shows flow cytometry data for two cyclins in leukaemic cells, showing the G2 specificity of cyclin B and the relatively greater S-phase specificity of cyclin A. The fraction of cells containing a particular cyclin can be used as a proliferative index, *e.g.* to rank different tumours, assuming it is proportional to the growth fraction. The phase specificity of the cyclins, however, renders this inaccurate and a marker such as Ki67 would be preferred for this purpose.

It should be noted that these proliferation markers on their own yield static parameters, for instance estimates of the S-phase fraction, G2-fraction or growth fraction, and do not give information on the rate of progression of the cells through the cycle. This is in contrast to the situation with labelled thymidine or thymidine analogues, which can yield both phase fractions and phase times. The most reliable, flexible and powerful methods are probably those using thymidine analogues, although the use of the markers described above can often provide useful information in clinical material, where there are considerable constraints concerning number of samples, drug administration and workload.

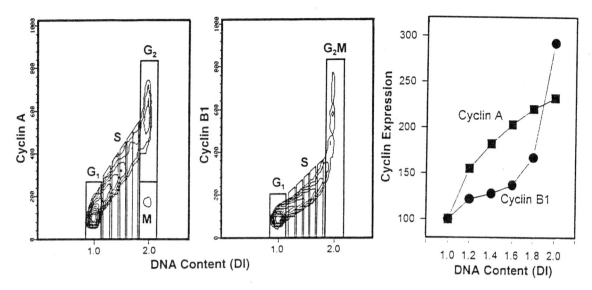

Figure 3.7 Cyclin A and B expression in MOLT4 leukaemic cells detected using fluorescent antibodies and flow cytometry. Cyclin fluorescence in different phases of the cycle (within the vertical narrow windows drawn in the figures) was plotted in the right-hand panel, showing the S-phase specificity of cyclin A and the G2-phase specificity of cyclin B. (From Gong et al, 1995, with permission).

HUMAN TUMOUR KINETICS *IN VIVO* USING THYMIDINE ANALOGUE LABELLING

Some kinetic information can be obtained by labelling human tumour biopsy material *in vitro* with ^{3}H-TdR or a thymidine analogue, thus obtaining the labelling index (*i.e.* fraction of cells synthesizing DNA). Better information can be obtained by *in vivo* labelling, and this avoids some potential artefacts of the *in vitro* procedure. Human tumours can be labelled by *in vivo* injection of IUdR or BrUdR because these analogues produce little or no toxicity at the low doses required for pulse-labelling kinetic studies. An example of the flow cytometer outputs following *in vivo* labelling of a human tumour with IUdR and biopsy a few hours later is shown in Figure 3.8. Panel B shows the distribution of DNA content: the peak on the left is a diploid peak that indicates the presence of non-malignant stromal cells. The large peak is due to G1 cells and the smallest peak to G2 tumour cells. Panel A shows IUdR uptake plotted vertically, against DNA content. The normal cells are again visible as the left-hand cluster of dots. These are not included in the analysis. Two subpopulations of labelled cells can be seen several hours after labelling, one small one at the tumour G1 position and a larger one covering most of the S-phase. The kinetic parameters labelling index (LI), DNA synthesis time (T_s), and the potential doubling time can be obtained from these data. This method is now being used as a predictive test for rapidly and slowly repopulating tumours (Section 25.4).

The presence of non-malignant cells in a tumour biopsy can disturb the accuracy of the estimates for the tumour cells. Attempts to overcome this have been made by using flow cytometry to measure T_s (described above) and combining it with the BrUdR labelling index of tumour cells counted in tissue sections using immunoperoxidase, where tumour areas can be distinguished from stromal areas morphologically (Bennett et al, 1992). Alternatively, one can use a second antibody, if available, as a tumour marker, and measure the BrUdR parameters only in cells that are positive for the tumour marker. One such marker is cytokeratin for carcinomas (Schutte et al, 1995; Begg et al, 1991).

No method to directly measure T_c from one biopsy has been published, although if the growth fraction were known it could be calculated from T_{pot}. A purported growth-fraction marker such as Ki67 (or MIB1) could be used for this purpose, although it should be realized that these may not give accurate growth fraction estimates in human tumours.

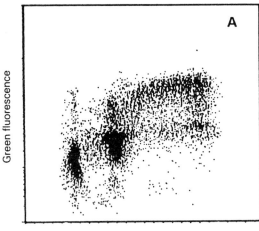

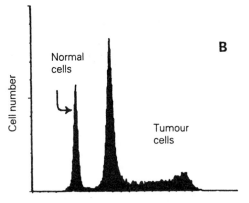

Figure 3.8 Flow cytometry output from a human lung tumour biopsy a few hours after labelling with IUdR. **A**: the two-dimensional display of red-green fluorescence; **B**: the derived distribution of DNA contents.

Bibliography

Begg AC and Hofland I (1991). Cell kinetic analysis of mixed populations using three-color fluorescence flow cytometry. *Cytometry* 12:445–54.

Begg AC, McNally NJ, Shrieve DC and Karcher H (1985). A method to measure the duration of DNA synthesis and the potential doubling time from a single sample. *Cytometry* 6:620–26.

Bennett MH, Wilson GD, Dische S *et al* (1992). Tumour proliferation assessed by combined histological and flow cytometric analysis: implications for therapy in squamous cell carcinoma in the head and neck. *Brit J Cancer* 65:870–78.

Gong J, Traganos F and Darzyntriewcz Z (1995). Discrimination of GZ and mitotic cells by flow cytometry based on different expression of cyclins A and B1. *Exp Cell Res* 220(1):226–31.

Hall EJ (1988). *Radiobiology for the Radiologist*. Third edition. Chapter 11. Lippincott; Philadelphia.

Schutte B, Tinnemans MMFJ, Pijpers GFP, Lenders MJH and Ramaekers FCS (1995). Three parameter flow cytometric analysis for simultaneous detection of cytokeratin, proliferation associated antigens and DNA content. *Cytometry* 21:177–86.

Terry NHA, White RA, Meistrich ML and Calkins DP (1991). Evaluation of flow cytometric methods for determining population potential doubling times using cultured cells. *Cytometry* 12:234–41.

Waldman FM, Dolbeare F and Gray J (1988). Clinical application of the bromodeoxyuridine assay. *Cytometry* 3:65–72.

Wilson GD, McNally NJ, Dische S *et al* (1988). Measurement of cell kinetics in human tumours in vivo using bromodeoxyuridine incorporation and flow cytometry. *Brit J Cancer* 58:423–31.

> **Key points**
>
> 1. Tumour growth rate is determined by the cell cycle time, the growth fraction, and the rate of cell loss. Potential doubling time indicates the cell proliferation rate and is the predicted doubling time of the cell population in the absence of cell loss.
> 2. Cell cycle times average around 2 days in human tumours, compared with volume doubling times of over 2 months. Cell loss is high in many human tumours, particularly in carcinomas.
> 3. There is a wide variation of kinetic parameters among human tumours, even between tumours of the same histological type.
> 4. The movement of cells through the cell cycle can be measured using radiolabelled thymidine or the thymidine analogues bromo-deoxyuridine or iodo-deoxyuridine. Detection methods include autoradiography, immunocytochemistry and flow cytometry.
> 5. Antibodies against a variety of proteins are available which can be used as proliferation markers. These give estimates of phase fractions or growth fraction, but not rates of progression through the cell cycle. Their advantage is that no drug needs to be administered.
> 6. Human tumour cell kinetics can be measured by *in vitro* labelling of biopsies, by proliferation marker measurements on biopsies, or by injecting the patient with low doses of thymidine analogues. Flow cytometric analysis using analogues allows calculation of the potential doubling time from one biopsy.

Further Reading

Steel GG (1977). *The Growth Kinetics of Tumours*. Oxford University Press; Oxford.

Tannock IF and Hill RP (1992). *The Basic Science of Oncology*. Second edition. Chapter 10. McGraw-Hill; New York.

4

Proliferative and cellular organization of normal tissues

Fiona A. Stewart and Albert J. van der Kogel

4.1 Proliferative organization of tissues

Cell proliferation in normal tissues is, in contrast to tumours, highly organized with cell production homeostatically controlled. In adult tissues under non-pathological conditions the production of cells is exactly balanced by the loss of differentiated mature cells; the cell loss factor (Section 3.2) is therefore 1.0. The degree of organization of cells within proliferative and functional compartments has important consequences for the response of tissues to radiation. Broadly speaking, tissues can be divided into two main categories: (1) tissues with a clearly recognizable separation between the stem cell compartment, the amplification compartment (usually proliferating rapidly) and the post-mitotic compartment of mature functional cells (Figure 4.1A); and (2) tissues without a recognizable separation between these compartments, in which at least some of the functional cells also have the capacity for cell renewal (Figure 4.1B).

Tissues with rapid cell turnover, such as skin, mucosae, intestinal epithelia and the haemopoietic system, are examples of a *hierarchical* organization, with separate stem cell and functional cell compartments. The self-renewing stem cells in these tissues may comprise only a small fraction (less than a few per cent) of the proliferating cells, while the bulk of the proliferating cells that make up the amplification compartment are involved in production of the cells that are needed for final maturation into functional cells. Following exposure to

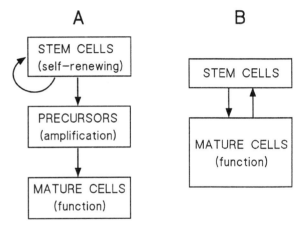

Figure 4.1 Schematic outline of the proliferative organization of **A**: hierarchical and **B**: flexible normal-tissue systems.

radiation or a cytotoxic drug, the time at which tissue failure occurs is determined by the lifespan of the mature cells, which may vary from a few days (in the case of granulocytes or intestinal villus cells) to more than 100 days (for example erythrocytes).

Tissues with slow cell turnover, such as liver, kidney, lung and central nervous system, are examples of a *flexible* organization. Some degree of proliferative hierarchy may exist but this is not as well defined as in the rapidly renewing tissues. In these slow turnover tissues there is no separate stem cell compartment and (some of) the functional cells are also capable of proliferation, although this normally takes place only slowly.

4.2 Radiation response in relation to proliferative organization

The most important mode of cell death following irradiation is mitotic cell death resulting from damage to the proliferative process. Irradiation therefore tends not to damage the mature (non-proliferating) functional cells of hierarchical tissues. Lymphocytes are an exception in that they undergo interphase death after irradiation, the phenomenon known as *apoptosis* (Sections 8.6, 9.3). In tissues with a hierarchical organization (*e.g.* skin), the time between irradiation and tissue response is mostly determined by the lifespan of the mature cells and is independent of the radiation dose. The rate of recovery is, however, inversely dependent on dose. After high doses, the rapidly proliferating precursor pool is severely depleted and first needs to be replenished by the more slowly proliferating stem cells. This is clearly demonstrated in irradiated mouse skin where loss of the differentiating (keratinized) layers is reflected by the development of dry and moist desquamation (Figure 4.2A). The time-course for development of skin reactions is related to the cycle time of the basal cells (about 4.5 days) and to the number of cell layers in the mouse skin (~3–4). When the cycle time of the proliferating basal cells is reduced to 2 days by plucking of the hair, the skin reactions develop much faster, as is shown in Figure 4.2B. Recovery after the high doses is slower than after low doses since more cells have been killed and a greater number of cell divisions is required to restore the original cell number.

Tissues or organs with a slow rate of renewal (*e.g.* kidney, lung, spinal cord) are usually not as well characterized as the mucosal tissues but it is clear that both vascular and parenchymal components may contribute to the development of organ failure after irradiation. In general, however, the rate of development of radiation reactions in these tissues depends on dose, as would be predicted in a situation in which the target cells are responsible for a specific radiation effect. In this case, the rate of functional cell loss increases with radiation dose, since the mature cells are also the proliferating cells which die in mitosis.

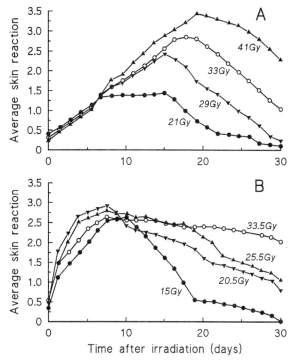

Figure 4.2 The time-course of radiation-induced desquamation in the dorsal skin of the mouse foot. **A**: normal skin; **B**: plucked skin, showing the faster development of reactions due to a shorter cell cycle time. From Hegazy and Fowler (1973), with permission.

Late vascular effects in irradiated skin (*e.g.* telangiectasia) are examples of a flexible-type tissue response. These changes develop progressively, over a period of many years, and there is a clear relationship between the rate of development of damage and total dose (Figure 4.3) in contrast to the development of acute epithelial desquamation after irradiation (Figure 4.2). One important consequence of the progressive dose-dependent development of such reactions is that the follow-up time must carefully be taken into account when comparing the extent of damage after different irradiation doses or schedules (Section 11.4).

It is important to note that the *latent period* before the onset of radiation-induced functional damage depends on the rate of cell turnover of the tissue in question and *not* on its radiation sensitivity. Thus, tissues with a high proliferative activity (*i.e.* rapid cell turnover), such as intestinal and skin epithelia, express their radiation damage within a

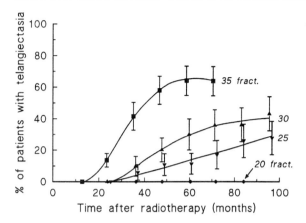

Figure 4.3 The cumulative proportion of patients with distinct telangiectasia as a function of time after treatment with daily 2 Gy fractions. Note that telangiectasia develops more quickly after higher total doses. From Turesson and Notter (1986), with permission.

few days, whereas in slow turnover tissues, such as lung and kidney, there is a considerable latency before any functional damage is apparent (Figure 4.4, Table 4.1).

4.3 Changes in cell proliferation after irradiation

Normal tissues respond to radiation-induced cell death by increasing their rate of proliferation. This is a homeostatic response to cell loss and is not

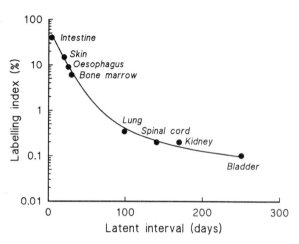

Figure 4.4 Correlation between latent interval before the development of moderate to severe radiation-induced functional damage and the rate of proliferation (indicated by the labelling index) in different rodent tissues. Tissues with a rapid cell turnover express damage much earlier than slowly dividing tissues. Data derived from various published sources.

unique to radiation damage. However, since radiation-induced cell death mainly occurs as cells attempt to divide at mitosis, the time of onset of accelerated proliferation varies widely in different tissues and is related to the unstimulated cell turnover times of proliferating cells in the tissue (Table 4.1). For example, the cell turnover time of basal epithelial cells in unirradiated mouse skin is approximately 4 days. Accelerated proliferation starts at about day 8 after irradiation when the cell turnover time is reduced to about 1 day. By con-

Table 4.1 Cell turnover times and time of onset of compensatory proliferation in normal tissues

Tissue	Cell turnover time (days)		Time of onset (days)	Authors
	Control	Stimulated		
Jejunum	0.6	0.4	3	Dewit et al (1986)
Skin	4	1	8	Denekamp et al (1976)
Lung	82	24	20/120	Coggle (1987)
Spinal cord	144	3	150	Zeman et al (1964)
			20/120	Hornsey et al (1981)
Kidney	144	11	120	Soranson and Denekamp (1986)
			90	Otsuka and Meistrich (1991)
Bladder	200	9	90–180	Stewart (1986)

The data in this table are approximate.

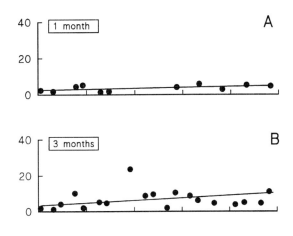
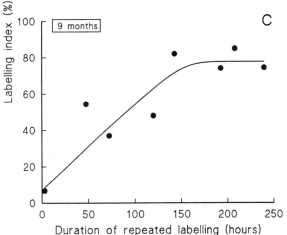
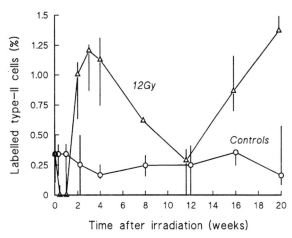

Figure 4.5 Cell proliferation in the epithelium of the mouse bladder at (A) 1, (B) 3 and (C) 9 months after irradiation with 25 Gy. The proliferation rate increased significantly from 3 months after irradiation but was not maximal until 9 months, coinciding with the histological appearance of hyperplasia and the occurrence of functional deficit. From Stewart (1986), with permission.

trast, accelerated proliferation in bladder urothelium does not begin until 3 months after irradiation and is not maximal until 9 months (Figure 4.5). At 9 months after irradiation the cell turnover time is reduced from over 200 days in unstimulated bladders to 9 days.

For both skin and bladder there is a clear correlation between the time of onset of functional damage and accelerated, compensatory cell proliferation. Such a relationship is not always so obvi-

Figure 4.6 Changes in the rate of proliferation in type II pneumonocytes after 12 Gy thoracic irradiation of mice (△) compared with unirradiated mice (○). From Coggle (1987), with permission.

ous in other organs, especially those with a flexible tissue organization. Irradiated mouse lung has two waves of accelerated proliferation in the type II pneumonocytes (Figure 4.6). The early wave, at 20–40 days after irradiation, precedes the onset of functional damage (*i.e.* pneumonitis), but coincides with the release of surfactant from type II cells in response to increased microvascular permeability and leakage of plasma proteins on the alveolar surface. The second wave of proliferation coincides with the onset of pneumonitis (16–20 weeks after irradiation) and is probably a homeostatic response to cell depletion.

Two waves of accelerated proliferation are also seen in irradiated rat spinal cord (Figure 4.7). The early wave, at about 20 days, precedes functional damage but may be related to a proliferative response identified in glial progenitor cells (Figure 5.3). The late wave of proliferation beyond 120 days coincides with the onset of white matter necrosis and paralysis after high doses to the cord.

Thus, proliferative responses do occur in slowly renewing tissues, but are often not so clearly related to the expression of and recovery from functional injury as in acutely responding tissues. It should be noted that early chemical stimulation of proliferation in slow turnover tissues (for instance following cyclophosphamide treatment of the bladder or lung) tends to precipitate latent radiation damage and does not lead to a more rapid recovery from damage.

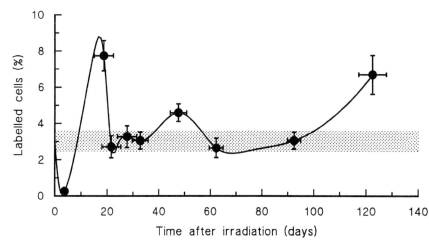

Figure 4.7 Kinetics of glial-cell proliferation in the cervical spinal cord after a single dose of 20 Gy. In spite of the late onset of functional damage (5–6 months) an early wave of proliferation is observed. From Hornsey *et al* (1981), with permission.

Key points

1. The time of appearance and dose-dependence of radiation damage in a normal tissue depends upon its proliferative organization.
2. Rapid renewal systems are organized in a hierarchical way with separate stem cells, a proliferating amplification compartment and mature functional cells. The latent interval before expression of functional radiation damage is dose-independent and is related to the lifespan of the functional cells. The rate of recovery is dose-dependent and related to the number of surviving stem cells.
3. Slowly renewing cell systems have been termed 'flexible' because they have no clear separation of proliferative and functional compartments. The time of onset of functional damage is radiation dose-dependent.
4. Both rapid-renewal and slow-renewal normal tissues are capable of an accelerated rate of proliferation in response to radiation-induced cell depletion. The time of onset of accelerated proliferation ranges from a few days to many months.

BIBLIOGRAPHY

Coggle JE (1987). Proliferation of type II pneumonocytes after x-irradiation on cell proliferation in mouse lung. *Int J Radiat Biol* 51: 393–99.

Denekamp J, Stewart FA and Douglas BG (1976). Changes in the proliferation rate of mouse epidermis after irradiation: continuous labelling studies. *Cell Tissue Kinet* 9:19–29.

Dewit L, Oussoren Y and Bartelink H (1986). The effects of cis-diamminedichloroplatinum (II) and radiation on the proliferative kinetics of mouse duodenal crypt cells and on a partially synchronized crypt cell population. *Int J Radiat Oncol Biol Phys* 12:1977–85.

Hegazy MAH and Fowler JF (1973). Cell population kinetics and desquamation skin reactions in plucked and unplucked mouse skin. II. Irradiated skin. *Cell Tissue Kinet* 6:587–602.

Hornsey S, Myers R, Coultas PG, Rogers MA and White A (1981). Turnover of proliferative cells in the spinal cord after X-irradiation and its relation to time-dependent repair of radiation damage. *Br J Radiol* 54:1081–85.

Otsuka M and Meistrich ML (1991). Acceleration of late radiation damage of the kidney by unilateral nephrectomy. *Int J Radiat Oncol Biol Phys* 22:71–8.

Soranson J and Denekamp J (1986). Precipitation of latent renal radiation injury by unilateral nephrectomy. *Br J Cancer* 53: 268–72.

Stewart FA (1986). Mechanisms of bladder damage and repair after treatment with radiation and cytostatic drugs. *Br J Cancer* 53 (Suppl VII): 280–91.

Turesson I and Notter G (1986). Dose-response and dose-latency relationship for human skin after various fractionation schedules. *Br J Cancer* 53:67–72.

Zeman W, Carsten A and Biondo S. (1964). Cytochemistry of delayed radionecrosis of the murine spinal cord. In *Response of the Nervous System to Ionizing Radiation*. (Eds) Haley TJ and Snider RS Academic Press; New York, p 105.

5

Radiation response and tolerance of normal tissues

Albert J. van der Kogel

5.1 The concept of normal-tissue tolerance

Whenever radiation therapy is given with curative intent there is the risk of serious damage to normal tissues. This risk increases with radiation dose, as does the probability of local tumour control. As described in Section 1.6, the achievable tumour control rate depends on the radiation *tolerance* of the normal tissues.

Tolerance is a complex concept. In experimental studies in the clinic or in experimental animals it can be defined in relation to a particular end-point such as 50% moist desquamation of skin, 5% pneumonitis in lung or 1% paralysis following spinal cord irradiation. In clinical practice it is necessary to consider not only the measurable or life-threatening effects but also the patient's perception of morbidity. Often the therapist is aware of a variety of risks associated with the proposed treatment and these must all be evaluated in deciding how intense the treatment schedule should be. *Clinical tolerance* depends on the medical status of the patient, on medication (especially with cytotoxic agents) and, very importantly, on the treatment volume (Sections 5.4, 11.5).

5.2 Pathogenesis of early and late effects

Radiation effects on normal tissues are commonly divided into two categories: 'early' and 'late' reactions. The difference between these is important for the clinical presentation of treatment-related morbidity (Section 11.2). The need to distinguish between them was stressed by the observation (as described in Section 14.2) that early and late reactions show different patterns of response to fractionated radiotherapy, reflected in a difference in α/β ratio.

The development of early effects in rapidly renewing tissues such as skin, gastrointestinal tract and the haemopoietic system is largely determined by a hierarchical cell lineage, composed of stem cells and their differentiating offspring (Section 4.1). The time of onset of early radiation reactions correlates with the lifespan of the differentiated functional cells, and the intensity of reactions reflects the balance between the rate of stem cell killing and the rate of regeneration of surviving clonogens. For these early effects of radiation the identity of the *target cells* is usually clear, in contrast to the late effects that have a long latent period. Late effects occur predominantly in slowly proliferating tissues, such as lung, kidney, heart, liver and central nervous system.

Late effects are not necessarily restricted to these slowly renewing cell systems. For instance in the skin, in addition to the early epidermal reactions, several later waves of injury occur such as fibrosis, atrophy and telangiectasia (Figure 5.1). Thus, different types of injury may develop sequentially in one organ, with different underlying mechanisms and different target cells. This obviously also complicates the issue of normal-tissue tolerance, as the various end-points are associated with different tolerance doses.

The distinction between early and late effects has important clinical implications. Since early

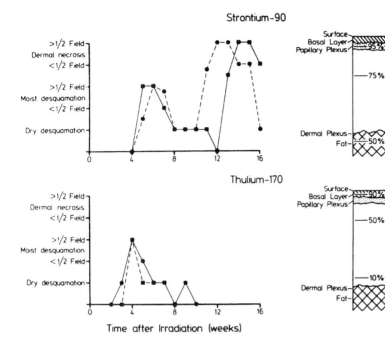

Figure 5.1 Time-course of epithelial and dermal reactions in pig skin irradiated with single surface doses of 40 Gy ^{90}Sr or 130 Gy ^{170}Tm (solid and broken lines show results on two individual skin fields). The marked difference in radiation response is due to the greater penetration of ^{90}Sr emission into the dermis. From Hopewell (1986), with permission.

reactions are usually observed during the course of a conventionally fractionated radiotherapy schedule (1.8–2 Gy per fraction, five times a week) it is possible to adjust the dose in the event of unexpectedly severe reactions, allowing a sufficient number of stem cells to survive. These surviving stem cells will repopulate and restore the integrity of the rapidly proliferating tissues. If, however, the overall treatment time is reduced, the early reactions may not reach the maximal intensity before completion of treatment. This precludes the possibility of adjusting the dose regimen to the severity of reactions. If intensive fractionation schedules reduce the number of surviving stem cells to below the level needed for effective tissue restoration, early reactions may persist as chronic injury, also called *consequential* late complications.

5.3 Normal-tissue tolerance: early and late effects

In this section the response and tolerance of some clinically important dose-limiting normal tissues will be summarized. As a guideline to tolerance doses as used in the clinic, Table 5.1 provides approximate dose levels, incidence estimates, and α/β values. The data from this table should be used with great caution, as they are influenced by many variables such as irradiated volume, energy of the radiation and dosimetry. The documentation and measurement of normal-tissue damage are dealt with in Chapter 11.

SKIN

In the early days of radiotherapy, when orthovoltage X-rays were used, skin was often a dose-limiting structure as the full dose was deposited near the surface of the body. Depending on the dose, erythema develops in the second to third week of a fractionated course of radiotherapy, followed by dry or moist desquamation due to the depletion of the basal stem cell population. When severe, moist desquamation may lead to a secondary ulceration. With modern megavoltage equipment the maximum dose deposition is 0.5–4 cm below the surface and epidermal reactions are usually limited to dry desquamation and increased pigmentation. This moderate skin response is seen after conventional irradiations at a total dose of 60–66 Gy (about 40–50 Gy in the basal layer of the epidermis). Such

Table 5.1 Normal-tissue tolerance doses and α/β values

Tissue	End-point	Risk (%)	Dose* (Gy)	α/β (Gy)
Skin	Desquamation	50	55–60	10–12
	Fibrosis	50	60–65	2–3
	Telangiectasia	50		3–4
Oral mucosa	Confluent mucositis	50	65–70	~10
Small intestine	Late fibrosis, fistulae	5	50	3–4
Colon, rectum	Late fibrosis, fistulae	5	60	3–4
Brain, spinal cord	Necrosis	<1	50	2
		5	60	
Nerve plexus	Demyelination, fibrosis	5	65	3–4
Lung	Pneumonitis	5	20**	2–4
Heart	Pericarditis	<5	35–40	2–3
		50	50–60	
	Cardiomyopathy	15–20	30–36	
Kidney	Glomerulosclerosis	<5	20**	2–3
Liver	Hepatitis, VOD***	5	20–25**	3
Bladder	Cystitis, ulcers	<5	60–65	5–10

* The tolerance doses and associated risks should be considered as approximations. They only apply for conventional daily fractionation with fraction doses of 2 Gy of high-energy photons. Values generally apply to large irradiated volumes and tolerance doses may increase with volume reduction, but the volume effect is widely different for different organs (see text).
** Whole organ irradiated, partial volume tolerance is usually much higher.
*** Veno-occlusive disease.

a dose is tolerated by the skin because a substantial part of the recovery takes place by stem cell proliferation during the 6–8 weeks of the course of radiotherapy. For skin, as well as for the oral or gastrointestinal mucosa, the total tolerated dose depends to a much smaller extent on fraction size than is the case with most late-responding critical organs. This is explained in Section 14.2. In contrast, variations in overall time can have a large but less predictable influence on the tolerance of early-responding tissues. As an approximation, skin tolerance doses decrease by about 3–4 Gy/week when treatment duration is shortened from the standard 6–8 weeks.

With high-energy X-rays a higher dose is deposited below the surface and late damage may therefore occur in the dermis in the absence of early reactions. These are characterized by atrophy leading to contraction of the irradiated area and a clinical appearance of 'radiation fibrosis'. The development of telangiectasia beyond 1 year (Figure 11.2) shows the progression of vascular injury in the late phases of radiation injury to the dermis.

As described in Chapter 12, deficiencies in the Nominal Standard Dose (NSD) approach to fractionation largely resulted from the fact that this did not take account of the differing responses of early and late reactions, such as those seen in the skin. In the LQ approach, different time–dose relationships can be applied to early and late effects. A clear demonstration of the importance of this in human skin was in patients with breast cancer receiving radiotherapy with 200-kV X-rays or 12- to 13-MeV electrons (Turesson and Thames, 1989). Spotty to confluent desquamation of the skin started to occur at epidermal doses of 35–40 Gy, with a 50% incidence at about 45–50 Gy in 2-Gy fractions. The total isoeffective dose for a 10-fraction regimen (45.9 Gy) was only about 7% lower than that of a 25-fraction schedule (49.2 Gy) delivered over the same overall time of 33 days. In contrast, for late telangiectasia at 5 years in 50 % of the patients, the difference in total dose between the two fractionation schemes was 23% (38.7 vs 50.2 Gy).

ORAL MUCOSA

The intensity of early mucous membrane reactions is a major factor limiting the daily and weekly dose accumulation with skin-sparing megavoltage equipment. Early studies by G.H. Fletcher, using a total dose of 55 Gy in 6–6½ weeks to the oropharynx resulted in mucosal reactions ranging from marked erythema to patchy mucositis. When the time was reduced by 1 week for the same total dose, a few patients developed confluent mucositis but the majority experienced spotted mucositis lasting about 4 weeks. With a further reduction of overall treatment time to about 4 weeks all patients developed confluent mucositis starting in the third week of treatment and lasting 3–6 weeks before recovery set in. One of the commonly used radiation schedules in the curative management of head and neck cancers is administered as five fractions per week, 2 Gy per fraction, up to 68–70 Gy over 7 weeks. This regimen induces spotted-confluent mucositis in a large number of patients (Figure 5.2). When reducing the time in a modestly accelerated schedule to 5.5 weeks (concomitant boost), clearly maximal tolerance levels are reached.

Gastrointestinal Tract

As with the skin and oral mucosa, both early and late complications are observed in the GI tract. Acute mucositis frequently occurs, with clinical symptoms such as diarrhoea or gastritis depending on the treatment field. When the total dose is limited to about 50–54 Gy in 2-Gy fractions, early reactions are usually not dose-limiting and if they do occur a few days' interruption of treatment is usually sufficient to overcome acute problems.

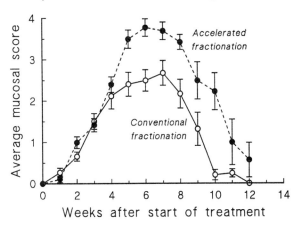

Figure 5.2 Time-course of mucosal reactions in patients treated with conventional daily fractionation or an accelerated schedule, concomitant-boost type (fractions/day for last 7–8 treatment days) to a dose of 68–70 Gy. Score 4 is fully confluent mucositis. From Kaanders et al (1992), with permission.

More serious are the various types of chronic injury, which may either be the result of persistent severe early reactions such as ulcerations, or develop independently of acute damage in the submucosal, muscular and serosal layers. Lesions vary but include mucosal atrophy, submucosal fibrosis and cystica profunda, perhaps with progression to severe ulceration, necrosis and fistulae.

The Central and Peripheral Nervous System

The nervous system is less sensitive to radiation injury than some other late-responding tissues such as the lung or kidney. Damage to this organ results in severe consequences such as paralysis. Although tolerance doses are often quoted at the 5% complication level (*i.e.* TD_5), they generally include a wide margin of safety.

Brain

The most important types of injury in the central nervous system are all late syndromes, developing a few months to several years after therapy. The often-used separation into *early* or *late* delayed injury is not very useful, as different types of lesions with overlapping time distributions occur. Thus, some reactions occurring within the first 6 months comprise transient demyelination (somnolence-syndrome) or the much more severe leukoencephalopathy. The more typical radiation necrosis may also occur by 6 months, but even after as long as 2–3 years. Histopathologically, changes that occur within the first year are mostly restricted to the white matter. For times beyond 6–12 months the grey matter usually also shows changes along with more pronounced vascular lesions (telangiectasia and focal haemorrhages). Radionecrosis of the brain with latent times between 1 and 2 years usually shows a mixture of histological characteristics. A schematic outline of the development of various delayed lesions in the CNS as studied in animals is given in Figure 5.3.

Spinal Cord

Radiation-induced changes in the spinal cord are similar to those in the brain in terms of latent period, histology and tolerance dose. Among the

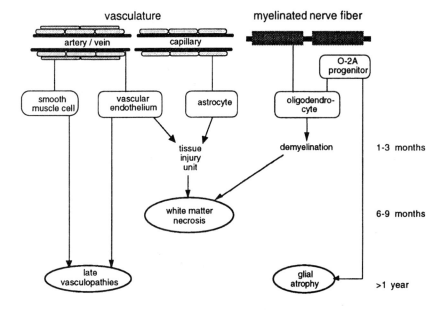

Figure 5.3 Schematic outline of the pathogenesis of early and late delayed types of injury in the central nervous system. Adapted from van der Kogel (1991), with permission.

relatively early syndromes, the Lhermitte sign is a frequently occurring, usually reversible type of demyelinating injury, which develops several months after completion of treatment and lasts for a few months to more than a year. It may occur at doses as low as 35 Gy, well below the tolerance for permanent radiation myelopathy when long segments of cord are irradiated, and does not predict for later development of permanent myelopathy.

As in the brain, the later types of myelopathy include two main syndromes. The first, occurring from about 6 to 18 months, is mostly limited to demyelination and necrosis of the white matter, whereas the second (with a latency of 1–4 years) is mostly a vasculopathy. The tolerance dose of the spinal cord largely depends on the size of the dose per fraction; variations in overall treatment time up to 10–12 weeks have a negligible effect in conventional irradiations with one fraction per day. For longer times or intervals, recovery occurs but with a different time-scale for the two types of injury (white-matter necrosis and late vascular damage). This has been studied in the rat spinal cord after split-dose irradiation using a range of time intervals (Figure 5.4).

Peripheral Nerves

Radiation effects in peripheral nerves, mainly plexuses and nerve roots, are probably more common than effects in the spinal cord but are less well documented. Peripheral nerves are often quoted as being more resistant to radiation than the cord or the brain, but this view is not well supported by clinical data. As is the case for all nervous tissues, a dose of 60 Gy in 2-Gy fractions is associated with a less than 5% probability of injury, but this rises steeply with increasing radiation dose.

The brachial plexus is often included in treatments of the axillary and supraclavicular nodes in breast cancer patients. Clinically, plexopathy is

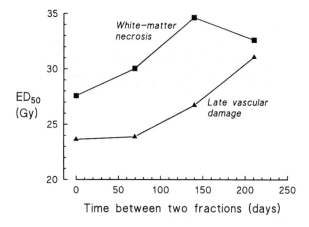

Figure 5.4 Recovery characteristics of two types of injury in the rat spinal cord (early white-matter necrosis and late vascular damage) after split-dose irradiation with different time intervals. Adapted from van der Kogel et al (1982), with permission.

characterized by mixed sensory and motor deficits, developing after a latent time ranging from 6 months to several years. The pathogenesis involves progressive vascular degeneration, fibrosis and demyelination with loss of nerve fibres.

Lung

The lung is an intermediate-to-late responding tissue, in which at least two separate radiation-induced syndromes are recognized: acute pneumonitis at 2–6 months after treatment, and fibrosis which develops slowly over a period of several months to years. The lung is among the most sensitive of late-responding organs, but because of the structural organization of the functional units it only becomes dose-limiting when large volumes of the chest are irradiated, and when the remaining lung is not capable of providing a minimal supportive function. Thus, except in patients with severe emphysema, lung tolerance is usually only dose-limiting when both lungs are irradiated. Clinical signs or symptoms of acute radiation pneumonitis are a reduced pulmonary compliance, progressive dyspnoea, decreased gas exchange and dry cough. When there is insufficient functional reserve, cardiorespiratory failure may occur in a short time-span.

Travis and colleagues (1997) have shown in mouse lung the existence of two to three separate waves of injury and demonstrated that late fibrosis can be distinguished from early pneumonitis (Figure 5.5). The type II pneumocytes and/or endothelial cells are considered as most directly involved in pneumonitis. Fibrosis is not merely the formation of scar tissue developing after acute injury, but the result of an imbalance in collagen production and degradation by fibroblasts. These changes are likely to be mediated through cascades of cellular interactions, including vascular endothelium, alveolar type II cells and macrophages.

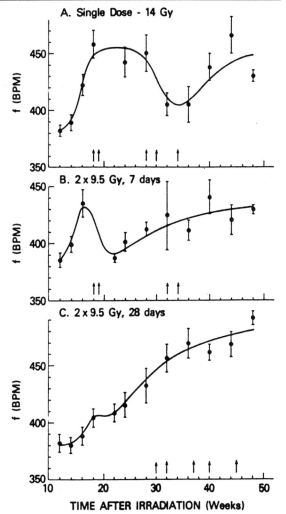

Figure 5.5 Evidence for two phases of radiation response in the mouse lung. Using breathing rate in breaths per minute (f_{BPM}) as a measure of damage, two waves of response are seen after single doses (panel **A**). Split-dose irradiation, especially with a gap of 28 days, reduced the severity of the first phase (acute pneumonitis), but had little effect on the second phase (fibrosis). Arrows indicate the death of individual mice. From Travis and Down (1981), with permission.

Kidney

The kidney, together with the lung, is among the most sensitive late-responding critical organs. Radiation damage develops very slowly (Section 22.8) and may take years to be recognized. Radiation nephropathy usually manifests as proteinuria, hypertension and impairment in urine concentration. Anaemia is usually present, and has been attributed both to haemolysis and to a decreased production of erythropoietin. A mild form of nephritis, presenting as a sustained proteinuria only, may be observed over a period of many years. Parts of one or both kidneys can receive much higher doses without affecting excretory function. However, after partial kidney irradiation hypertension may develop after a latent period of up to or beyond 10 years.

The pathogenesis of radiation nephropathy is complex, but most current studies suggest glomerular endothelial injury as the start of a cascade leading to glomerulosclerosis. Dietary protein restriction and ACE-inhibitors (captopril or aspirin) can reduce the severity and progression of radiation nephropathy.

In contrast to many other organs and tissues, the dose tolerated by the kidney does not increase with increasing treatment time but even declines due to a continuous progression of damage, even after doses well below the threshold for induction of functional deficits (Section 22.8).

Heart

The tolerance of the heart is intermediate between kidney/lung and CNS. The most common type of radiation-induced heart injury is pericarditis with variable degrees of pericardial effusions. This complication has a relatively early onset (~50% occurrence within the first 6 months, remainder within 2 years). It is asymptomatic and clears spontaneously in the majority of patients.

Radiation-induced cardiomyopathy is another form of complication that presents as either reduced ventricular ejection or conduction blocks and develops slowly over a period of 10–20 years. A significantly reduced cardiac function occurred in about 15–20% of patients treated for Hodgkin's disease after receiving a dose of 30–36 Gy to most of the heart (Schultz-Hector, 1991). Histopathologically, this damage is characterized predominantly by diffuse interstitial fibrosis.

Liver

The liver ranks in radiosensitivity immediately below kidney and lung for whole-organ irradiation. Because all these organs have their functional units organized in parallel structures, the tolerance to partial-organ irradiation can be much higher than that of the whole organ, depending on the functional reserve. Thus, liver tolerance is only dose-limiting when the whole organ is irradiated. An example is total body irradiation preceding bone marrow transplantation. In this situation the lung is well known as a dose-limiting organ, but also liver and kidney are at risk, especially after regimes equivalent to single doses of 10 Gy or higher.

Two phases of radiation hepatopathy are recognized, with acute radiation hepatitis being the more dominant. The acute phase develops approximately 2–6 weeks after irradiation, with signs of liver enlargement and ascites. Liver function tests are abnormal, especially alkaline phosphatase. Acute hepatitis usually presents as veno-occlusive disease, characterized by central vein haemostasis whereby occlusion of the centrilobular veins causes atrophy and loss of the surrounding hepatocytes. Chronic hepatopathy has a variable latency ranging from 6 months to more than a year post-irradiation, and shows progressive fibrotic changes in both centrilobular and periportal areas. These alterations are accompanied by blood-flow redistribution through recanalization or newly formed veins, and regenerative proliferation of hepatocytes and bile ducts.

Bladder

In contrast to most other epithelia, the mucosal lining of bladder and ureters shows a very slow cellular turnover (Table 4.1). As a result, radiation-induced desquamation takes a long time to develop, unless precipitated by other injury such as infection or drug-related toxicity. In clinical studies, three phases of damage have been described. An acute phase occurs 4–6 weeks after irradiation and is characterized by hyperaemia and mucosal oedema. Infection may complicate this early damage which then may progress to desquamation and ulceration. A chronic phase develops from about 6 months to 2 years and presents as vascular ischaemia accompanied by progressive mucosal breakdown, ranging from superficial denudation to ulceration and even formation of fistulae. Fibrosis of the bladder wall with a reduced urine capacity may occur up to 10 years after irradiation.

5.4 Normal-tissue tolerance: volume effects

Volume effects differ among the various organs of the body, depending on their structural organization (compartmentalization) as well as on the migration characteristics (rate and distance) of the target cells. The volume of tissue irradiated is an important factor determining the *clinical tolerance* of an organ, which can be quite different from the tissue sensitivity *per se* (Chapter 11). Examples of this are the kidney or lung, both of which are among the most sensitive organs when totally irradiated ($TD_5 \sim 20$ Gy in 2-Gy fractions), but quite large partial volumes can be treated to far higher radiation doses. This is because there is considerable reserve capacity in these organs and only one quarter to one third of the functional organ volume is required to sustain life under normal physiological conditions.

The radiosensitivity of a tissue depends largely on its organization into separate functional units, as well as on the possibility of surviving clonogenic target cells migrating into and repopulating the irradiated tissue. Tissues with a high migratory capacity include skin, mucosae and the intestinal tract, and in these tissues relatively small volumes can be treated to high doses, since repopulation takes place rapidly from the surrounding unirradiated tissue. However, once a critical migration distance is passed, central necrosis develops in the absence of adequate repopulation. Also, above a certain dose level, structural damage to connective tissue and the vasculature may prevent regeneration. The high sensitivity of the kidney and the lung may largely be related to their high degree of compartmentalization into functional subunits, *i.e.* nephrons and the alveoli. Once all stem cells in such a subunit are sterilized it is unlikely that repopulation from neighbouring subunits will occur. The large reserve capacity and higher tolerance to partial-volume irradiations of these tissues is due to a parallel organization of functional units.

In contrast to this, other more tube-like structures such as the spinal cord are often assumed to have a serial organization of functional subunits (Withers et al, 1988). In this theoretical model, inactivation of any one subunit causes loss of function of the whole organ. With the same radiation dose to a larger volume, the probability of inactivation of any single unit increases. In such a statistical model, the probability of inactivation is only related to the total irradiated volume, regardless of the spatial distribution of functional subunits. This model does not take into account the possibility that the tolerance of small volumes may rise steeply due to extensive cell migration in tissues such as skin and intestine and probably also in spinal cord.

Intermediate between the parallel or serial organization are tissues composed of specialized elements each carrying out very specific functions, such as the brain. Radiation-induced damage even to a small area of this tissue leads to a permanent deficit for that function since the uninjured components are not able to take over the functions of the affected elements. As a consequence, the tolerance dose of such tissues is only to a small extent influenced by the volume of the tissue irradiated. Irradiating a larger volume may result in a more severe functional failure because of the elimination of a larger number of functional elements.

The modelling of volume effects on the basis of serial and parallel organization supports the paradox of normal-tissue tolerance: a whole organ with a high tolerance may be lost by inactivation of a small part (*e.g.* spinal cord), while an organ with a very low tolerance (*e.g.* kidney, lung) may sustain the loss of more than half of its functional mass without serious repercussions.

5.5 Volume effects in specific tissues

Skin

In this tissue the migration of surviving basal cells from the periphery of the irradiated field is thought to be an important factor in the volume effect for small irradiated fields. A large field-size effect in pig skin has been demonstrated after irradiation with circular ^{90}Sr sources, showing a steep increase in isoeffective dose for the induction of moist desquamation with decreasing diameter of the plaques (Figure 5.6). Field sizes larger than 22.5 mm did not show a further change in the dose–response curve.

Spinal Cord

A similar volume dependence is seen in the rat spinal cord (Figure 5.7). The rate of cell renewal is much slower than in skin and also the development of injury occurs much later (5–7 months). A steep volume effect for the induction of white matter necrosis is observed, suggesting that cell migration is an important factor. The target cells may be the glial progenitor cells, vascular endothelial cells, or even co-operative units of glial and endothelial cells. For

38 Radiation response and tolerance of normal tissues

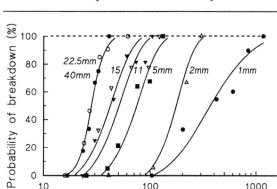

Figure 5.6 The volume effect in pig skin. The skin was irradiated with circular ^{90}Sr sources ranging in diameter from 1.0 to 40 mm. The percentage of skin fields showing moist desquamation is shown as a function of the skin surface dose. From Hopewell *et al* (1987), with permission.

lengths of cord larger than about 1–2 cm, tolerance doses decrease only slightly with increasing volumes.

LUNG

As mentioned earlier, this organ has a 'parallel' tissue architecture and irradiation of small volumes

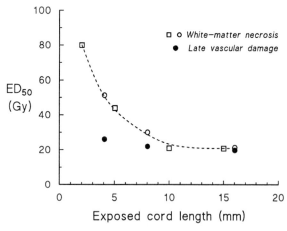

Figure 5.7 The volume effect in rat spinal cord. Data are shown on the ED_{50} for induction of white-matter necrosis from two studies. Full symbols indicate values for late vascular damage, suggesting less volume dependence. From Hopewell *et al* (1987) and van der Kogel (1991), with permission.

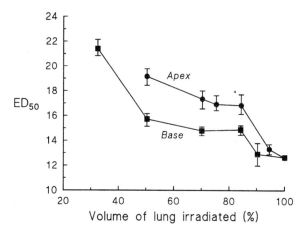

Figure 5.8 The volume effect in mouse lung. ED_{50} values are shown for increased breathing rate as a function of partial volume irradiated in the apex or base of the lung. From Travis *et al* (1997), with permission.

does not greatly impair lung function. An extensive study recently completed in mouse lung showed that functional units are not homogeneously distributed in the lung, and that a measurable decrease in function occurred when 30–40% of the lung including the base was irradiated, but 50–70% when volumes included the apex (Figure 5.8). When more than 80% of the volume was included, tolerance of the lung rapidly dropped. This is the first large study quantifying the inhomogeneity of volume effects within an organ, but similar spatial inhomogeneities are to be expected in other organs.

KIDNEY

The structural and functional organization of the kidney resembles that of the lung and liver. It has a large reserve with regards to its excretory function. Irradiation of a portion of a kidney to high doses will not, therefore, compromise its exocrine function. However, the resultant small area of renal fibrosis and glomerular sclerosis could lead to malignant hypertension by endocrine (*i.e.* renin-angiotensin) perturbations. Hence, the relative tolerance dose for partial renal irradiation is not as high as that of the lung.

> **Key points**
>
> 1. The concept of *normal-tissue tolerance* is basic to considerations of time-dose relationships in radiotherapy.
> 2. Both early and late effects may develop in one organ system, due to injury to different target cell populations or tissue elements.
> 3. Tolerance doses for late effects are more sensitive to changes in dose per fraction (low α/β value) when compared to early effects.
> 4. The *volume paradox* of normal-tissue tolerance is that a tissue with an intrinsically high tolerance may fail as a result of the inactivation of a small segment (as in the spinal cord), while a tissue with an intrinsically low tolerance (kidney, lung) may lose a substantial number of its functional units without impact on clinical tolerance.

BIBLIOGRAPHY

Hopewell JW (1986). Mechanisms of the action of radiation on skin and underlying tissues. *Br J Radiol* (Suppl. 19):39–51.

Hopewell JW, Morris AD and Dixon-Brown A (1987). The influence of field size on the late tolerance of the rat spinal cord to single doses of X-rays. *Br J Radiol* 60:1099–1108.

Kaanders JHAM, van Daal WAJ, Hoogenraad WJ and van der Kogel AJ (1992). Accelerated fractionation radiotherapy for laryngeal cancer, acute and late toxicity. *Int J Radiat Oncol Biol Phys* 24: 497–503.

Schultz-Hector S (1991). Heart. In: *Radiopathology of Organs and Tissues*, pp 347–68. (Eds) Scherer E, Streffer C and Trott KR. Springer-Verlag, Berlin.

Travis EL and Down JD (1981). Repair in mouse lung after split doses of X rays. *Radiat Res* 87:166–74.

Travis EL, Liao Z-X and Tucker SL (1997). Spatial heterogeneity of the volume effect for radiation pneumonitis in mouse lung. *Int J Radiat Oncol Biol Phys*, in press.

Turesson I and Thames HD (1989). Repair capacity and kinetics of human skin during fractionated radiotherapy: erythema, desquamation, and telangiectasia after 3 and 5 years' follow-up. *Radiother Oncol* 15:169–88.

van der Kogel AJ, Sissingh HA and Zoetelief J (1982). Effect of X-rays and neutrons on repair and regeneration in the rat spinal cord. *Int J Radial Oncol Biol Phys* 8:2095–7.

van der Kogel AJ (1991). Central nervous system radiation injury in small animal models. In: *Radiation Injury to the Nervous System*, pp 91–112. (Eds) Gutin PH, Leibel SA and Sheline GE. Raven Press; New York.

van der Kogel AJ and Ang KK (1995). Complications related to radiotherapy. In: *Oxford Textbook of Oncology* Vol.2, pp 2295–306. (Eds: Peckham M, Pinedo H and Veronesi U). Oxford University Press; Oxford

Withers HR, Taylor JMG and Maciejewski B (1988). Treatment volume and tissue tolerance. *Int J Radiat Oncol Biol Phys* 14: 751–9

6
Clonogenic cells and the concept of cell survival

G. Gordon Steel

6.1 Concept of clonogenic cells

As indicated in Chapter 4, the maintenance of tissue size and therefore of tissue function in the normal renewal tissues of the body depends upon the existence of a small number of primitive 'stem cells':

Stem Cells – cells that have the capacity to maintain their numbers whilst at the same time producing cells that can differentiate and proliferate to replace the rest of the functional cell population. Stem cells are at the base of the hierarchy of cells that make up the epithelial and haemopoietic tissues.

Carcinomas are derived from such hierarchical tissues, and our ability to recognize this in histological sections derives from the fact that these tumours often maintain many of the features of differentiation of the tissue within which they arose. Well-differentiated tumours do this to a greater extent than anaplastic tumours. It follows that not all the cells in a tumour are neoplastic stem cells: some have embarked on an irreversible process of differentiation. In addition, carcinomas contain many cells that make up the stroma (fibroblasts, endothelial cells, macrophages, etc.). Stem cells thus may comprise only a small proportion of the cells within a tumour.

When a tumour regrows after non-curative treatment, it does so because some neoplastic stem cells were not killed. Radiobiologists have therefore recognized that the key to understanding tumour response is to ask: How many stem cells are left? It is almost impossible to recognize tumour stem cells *in situ*, and therefore assays have been developed that allow them to be detected after removal from the tumour. These assays generally detect stem cells by their ability to form a colony within some growth environment. We therefore call these 'clonogenic' or 'colony-forming' cells:

Clonogenic Cells – cells that form colonies exceeding about 50 cells within a defined growth environment. The number 50 represents 5–6 generations of proliferation. It is chosen in order to exclude cells that have a limited growth potential as a result of having embarked on differentiation, or having been sublethally damaged by therapeutic treatment.

After exposure to a therapeutic dose of radiation, damaged cells do not die immediately and they may produce a modest family of descendants. This is illustrated in Figure 6.1. The growth of single mouse L-cells was observed under the microscope and one selected colony was irradiated with 200 röntgens of X-rays at the 4-cell stage (Trott, 1972). Subsequent growth was carefully recorded and in the figure each vertical line indicates the lifetime of a cell from birth at mitosis to its subsequent division. The two irradiated cells on the left and the right of this figure produced continuously expanding colonies, although some daughter cells had long intermitotic times. The other two irradiated cells fared badly: they underwent a number of irregular divisions, including a tripolar mitosis. But note that at the end of the experiment cells are present from each of the original four cells: the difference is that two produced expanding colonies and the other two did not. The first two were 'surviving clonogenic

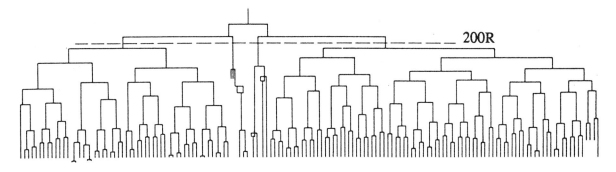

Figure 6.1 Pedigree of a clone of mouse L-cells irradiated with a dose of 200R (*i.e.* röntgens) at the 4-cell stage, illustrating the concept of surviving and non-surviving clonogenic cells. From Trott (1972), with permission.

cells' and the other two are usually described as 'killed' by radiation, since their regrowth is probably unimportant for clinical outcome. It would be more precise to say that two of the cells lost their proliferative ability as a result of irradiation.

Some cells fail to undergo even one division after irradiation. *Interphase cell death* occurs in many cell types at very high radiation doses, and at conventional therapeutic dose levels it is characteristic of lymphoid cells, parotid acinar cells and some cells in the intestinal crypts. Although interphase cell death and apoptosis are related concepts (Sections 8.6, 9.3) they are not synonymous for the same process. But the conventional radiobiological view is that it is *loss of reproductive integrity* that is the critical response to irradiation (either in tumour or normal-tissue cells): this occurs within a few hours of irradiation through damage to the genome, and the subsequent metabolic and death processes are 'downstream' of this event and of little relevance to therapeutic response.

6.2 Clonogenic assays

Clonogenic assays have formed the basis of cellular response studies in tumours, and also in some normal tissues. The basic idea is to remove cells from the tumour, place them in a defined growth environment and test for their ability to produce a defined colony of descendants. Many types of assay have been described; we illustrate the principle by a simple assay in tissue culture that is analogous to a microbiological assay.

A single-cell suspension of tumour cells is prepared and divided into two parts. One is irradiated, the other kept as an unirradiated *control*. The two suspensions are then plated out in tissue culture under identical conditions, except that since we anticipate that irradiation has killed some cells we will have to plate a larger number of the irradiated cells. We here envisage plating 100 control cells and 400 irradiated cells. After a suitable period of incubation the colonies are scored (Figure 6.2). There are 20 control colonies, and we therefore say that the *plating efficiency* was 20/100 = 0.2. The plating efficiency of the treated cells is lower: 8/400 = 0.02. We calculate a *surviving fraction* as the ratio of these plating efficiencies:

$$\text{Surviving fraction} = \frac{\text{PE}_{\text{treated}}}{\text{PE}_{\text{control}}} = \frac{0.02}{0.2} = 0.1$$

thus correcting for the efficiency with which undamaged clonogenic cells are detected and for the different numbers of cells plated. Surviving fraction is often given as a percentage (10% in this case).

The above description started with a suspension of tumour cells. In order to measure *in vivo* cell survival we take two groups of experimental tumours (often subcutaneously implanted tumours in mice),

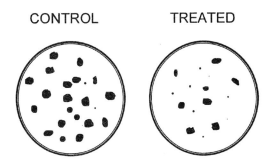

Figure 6.2 Illustrating the principle of measuring a surviving fraction.

irradiate one and keep the other as a control, then at some time after irradiation we make cell suspensions from both groups and plate them out under identical conditions as before. The difference here is that the cells are irradiated under *in vivo* conditions.

The colonies in Figure 6.2 have been drawn also to illustrate a feature of colony assays that was mentioned in the previous section. Irradiation not only reduces the colony numbers; it also increases the *number of small colonies*. Some of these small colonies may represent clones that eventually die out; others may arise from cells that have suffered non-lethal injury that reduces colony growth rate. Unless they reach the usual cut-off of 50 cells they will not be counted, although their implications for the evaluation of radiation effects on tumours may be worthy of greater attention (Seymour and Mothersill, 1989).

6.3 Cell survival curves

A cell survival curve is a plot of surviving fraction against dose (of radiation, cytotoxic drug or other cell killing agent). Figure 6.3A shows that when plotted on *linear* scales the survival curve for cells irradiated in tissue culture is often sigmoid: there is a shoulder followed by a curve that asymptotically approaches zero survival. To indicate the sensitivity of the cells to radiation we could just read off the dose that kills say 90% of the cells. This is sometimes called the ED_{90} (*i.e.* Effect Dose 90%). In doing this we need make no assumptions about the shape of the curve.

There are two reasons why survival curves are always plotted on a *logarithmic* scale of survival:

(i) If cell killing is random then survival will be an exponential function of dose, and this will be a straight line on a semi-log plot.
(ii) A logarithmic scale more easily allows us to see and compare effects at very low survivals.

Such a plot is illustrated in Figure 6.3B. The shapes of radiation survival curves and ways of describing their steepness are dealt with in the next chapter.

Note that for the data shown in Figure 6.3, radiation doses above 5 Gy reduce the survival of clonogenic cells to below 10%. Measurement of radiosensitivity in terms of the parameter D_0 (Section 7.3) is made on the exponential part of the survival curve, which in this case is above 5 Gy. These measurements are therefore made in a dose range where the surviving fraction is very low. Such D_0 values are relevant to the problem of exterminating the last few clonogenic cells, but if the cell

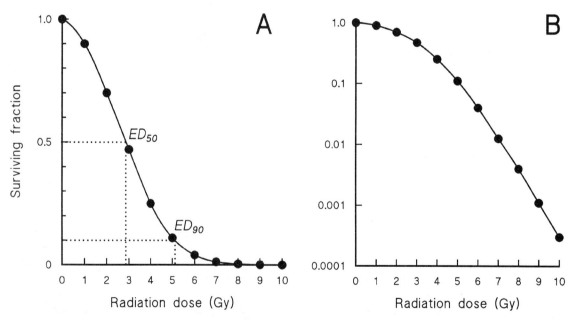

Figure 6.3 A typical cell survival curve for cells irradiated in tissue culture, plotted (**A**) on a linear survival scale. (**B**) shows the same data plotted on a logarithmic scale.

population contains cells of differing radiosensitivity these values may not be typical of the radiosensitivity of the bulk of the tumour cell population.

6.4 Assays for the survival of clonogenic cells

Many techniques have been described for detecting colony formation by tumour cells and thus for measuring cell survival. They almost all require first the production of single-cell suspensions. This is not a simple matter, for tumour tissues differ widely in the ease with which they can be desegregated. Enzymes such as trypsin, collagenase and pronase are often used and some tissues can be disaggregated mechanically.

Such assays can also be used for the assay of colony-forming cells in normal tissues, especially the haemopoietic tissues that can easily be sampled and made into cell suspensions. In addition a variety of *in situ* assays for normal-tissue stem cells have been described (Potten, 1983).

The following are some of the principal assays that have been used for tumour cells.

IN VITRO COLONY ASSAYS

Some tumour cells grow well attached to plastic tissue culture dishes. Others can be encouraged to do so by first laying down a *feeder layer* of lethally irradiated connective tissue or tumour cells. For cells that have been established as an *in vitro* cell line this often works well but for studies on tissue samples taken directly from patients or animals it is commonly observed that fibroblasts grow better than the tumour cells and overgrow the cultures.

An alternative is therefore to thicken the growth medium with agar or methylcellulose. This inhibits the growth of anchorage-requiring cell types, but many epithelial cells will still grow. A widely used assay of this type is the Courtenay–Mills assay for human tumour cells (Courtenay and Mills, 1978). Agar cultures are grown in 15-ml plastic tubes overlayed with liquid medium that can regularly be replenished. The addition of rat red blood cells to the agar has been found to promote the growth of a number of human tumour cell types. An important feature of the Courtenay–Mills assay is the use of a low oxygen tension (a gas phase of 90% nitrogen, 5% oxygen and 5% carbon dioxide) which enhances the plating efficiency of human tumour cells.

SPLEEN COLONY ASSAY

Till and McCulloch (1961) showed that when mouse bone marrow cells were injected intravenously into syngeneic recipients that had received sufficient whole-body irradiation to suppress endogenous haemopoiesis, colonies were produced in the spleen which derived from the stem cells in the graft. The colonies varied in morphology (erythroid, granulocyte or mixed) and these stem cells are therefore termed *pluripotent*. Their precise identity was not known and they are therefore often called *colony-forming units (CFU)*. Using this assay, Till and McCulloch obtained the first survival curve for bone marrow cells and found it to be very steep. The spleen colony assay has also been used for some types of mouse lymphoma cells.

LUNG COLONY ASSAY

This is analogous to the spleen colony assay and is applicable to any transplanted mouse tumour that readily forms colonies in the lung following intravenous injection of a single-cell suspension. The cloning efficiency can often be increased by mixing the test cells with an excess ($\sim 10^6$) of lethally irradiated tumour cells or plastic microspheres. Not all the tumour cells grow: a few colonies per thousand tumour cells injected would be regarded as satisfactory. Although colonies are formed throughout the lung, they are usually scored only on the lung surface. The method was used by Hill and Stanley (1975) on two experimental tumours and they give further experimental details. See also Steel (1977).

LIMITING-DILUTION ASSAY

This is a non-cloning assay that was used in early radiation cell survival studies and which for some experimental tumours has the advantage of high sensitivity (see Steel, 1977, for detailed description). The principle of the method is to prepare a suspension of tumour cells and to make a large number of subcutaneous implants into syngeneic animals, covering a range of inoculum sizes and if possible spanning the level of 50% tumour takes. The animals, usually mice, are then observed for a long enough period to record nearly every tumour that can grow from a single-cell implant. Take-rate is then plotted against inoculum size and the point of 50% takes is interpolated; this is usually called the 'TD_{50}' cell number. The experiment is performed simultaneously on treated cells and control cells and the surviving fraction is given by the ratio of the TD_{50} values. The addition of an excess of lethally irradiated cells improves the take-rate; using this manoeuvre Steel and Adams (1975) found a TD_{50} of 1–3 cells for the Lewis lung tumour and were thus able to measure survival down to 10^{-6}. The method only works well in the absence of an immune response against the tumour grafts, a relatively uncommon situation.

Short-Term in Vitro Assays

The need to develop *in vitro* assays that yield a quicker result than a true clonogenic assay arises from the current interest in prediction of tumour response to treatment (Chapter 25). A variety of assays have been proposed but their reliability has yet to be fully established. The failure of the Salmon-Hamburger 'Human Tumour Stem Cell Assay' that was widely publicized 15 years ago may serve as a warning, both to the developers and users of new assays. There are three pitfalls that must be avoided:

(i) Biopsy samples of human tumours contain both tumour cells and normal connective tissue cells; both may grow under the assay conditions and it may be difficult to distinguish colony formation by tumour cells.
(ii) If the method requires the production of single-cell suspensions, great care must be taken to exclude cell clumps, for these may preferentially give rise to scorable colonies.
(iii) Radiation-killed cells take time to die and in a short-term assay they may be confused with surviving tumour cells; if this is not done, the method may not distinguish between *radiosensitive* cells and cells that *die rapidly* after irradiation.

Many aspects of the prediction of tumour response are dealt with in the book edited by Chapman *et al* (1989). Non-clonogenic assays for tumour cells have been reviewed by Mitchell (1988) and they include the following.

The Micronucleus Test. Tumour cells are cultured in the presence of cytochalasin-B which blocks cytokinesis and creates binucleate cells, and thus allows nuclei that have undergone one post-treatment division to be identified. Micronuclei can be scored as small extranuclear bodies. Their frequency is linearly related to radiation dose and gives a measure of radiation sensitivity (Streffer *et al*, 1986). The reliability of the method is limited by the fact that diploid, polyploid and aneuploid cells may differ in their tolerance of genetic loss and therefore of micronucleus formation.

The Adhesive Tumour Cell Culture System. Human tumour biopsy specimens are disaggregated into a single-cell suspension and plated out in 24-well culture plates that have been coated with a commercially patented adhesive coating (Baker *et al*, 1986). The medium is supplemented with added growth factors. Two weeks after irradiation or drug treatment the cultures are stained with crystal violet and the absorbance is measured using an image analyser.

Growth Assays. A variety of methods have been used to measure the growth of cultures derived from treated and control tumour specimens, thus to derive a measure of radiosensitivity. Incorporation of radioisotopes such as ^{3}H-thymidine has been widely used. MTT is a tetrazolium salt that can be used to stain cell cultures and thus by a colorimetric assay to estimate the extent of growth (Carmichael *et al*, 1987; Wasserman and Twentyman, 1988). It can be used to evaluate growth in microtitre plates and with careful attention to technical factors it can yield a measure of radiosensitivity. Such methods are vulnerable to the variable growth of fibroblasts and for studies on leukaemic cells it may be preferable to stain the cells differentially and analyse the cultures microscopically (Bosanquet, 1991).

6.5 Comparison of assays

Intercomparison of the results of assays of cell survival provides an important check on their validity, yet it has seldom been done. The information can be valuable both at a practical and a fundamental level. At the practical level, it is logical to check a rapid short-term assay against the results of a more laborious but more reliable clonogenic assay.

The more general question is whether assay of cell survival in two different growth environments does actually identify the same population of surviving tumour cells. It is cell survival *in situ* in the patient or in the experimental animal that we seek to determine, and to subject tumour cells to additional stresses and to artificial growth environments might well produce artefacts. It is therefore reassuring that some careful comparisons between clonogenic assays, *in vitro*, in the mouse lung, and by subcutaneous transplantation, have demonstrated good agreement for mouse tumours (Steel and Stephens, 1983).

6.6 The relationship between cell survival and gross tumour response

The objective of studies of clonogenic cell survival is to be able to understand, or to make predictions about, the main features of tumour response to therapy: growth delay and tumour cure.

TUMOUR GROWTH DELAY

Inadequate treatment of a tumour leads to a temporary phase of tumour regression that is subsequently followed by tumour recurrence. This pattern is illustrated in Figure 2.7. Regression is due to the death and disappearance of cells killed by radiation, also to the loss of those differentiated cells of limited lifespan that would have been produced by the killed stem cells. The rate of tumour regression differs widely from one tumour to another, as illustrated in Figure 2.6.

The regrowth component in Figure 2.7 is due to repopulation by surviving clonogenic cells. The speed of regrowth probably varies considerably from one tumour to another, and the broken lines in the figure illustrate the possibilities of a lag period before repopulation gets fully under way, or repopulation at the speed of a small untreated tumour. Fowler (1991) has expressed the view that rapid repopulation may be the norm. He has postulated that irradiation will arrest the proliferation of tumour cells and that if this is the driving force behind the loss of cells to necrosis (Section 15.2) then this loss will temporarily be interrupted. The result will be a period of tumour cell repopulation at a rate that is close to the potential doubling time (Section 3.2) and faster than would have been the case in the absence of treatment. Fowler describes this as the 'unmasking' of rapid tumour cell proliferation.

Growth delay is defined as the difference in time for treated and untreated tumours to reach a fixed multiple of the size at the time of treatment (Figure 17.2A). This has been widely used as a measure of tumour response in transplanted animal tumours.

What is the relationship between growth delay and cell survival following treatment? For a particular radiation dose, Figure 17.5 shows that the colony-forming ability of tumour cells is depressed immediately after irradiation but that the tumour volume response takes time to appear. In that study, a radiation dose that reduced survival to 0.01 gave only partial remission of tumour growth. Greater cell kill would be expected to lead to longer growth delay and the results of a representative dose–response study (for chemotherapy) are shown in Figure 6.4. Transplanted B16 mouse melanomas were treated either with cyclophosphamide or the nitrosourea CCNU and both parameters were measured over a range of drug doses. Following cyclophosphamide treatment there was an almost linear relationship between the time the tumours took to reach four times the treatment size and the log(survival). The curve for CCNU was much flatter: reducing survival to 10^{-2} gave a growth delay of roughly 4 days, as compared with around 14 days for cyclophosphamide. The explanation appeared to lie in a reduced *rate of repopulation* after cyclophosphamide. The conclusion is that growth delay is a function both of the level of clonogenic cell survival and of the rate of regeneration.

LOCAL TUMOUR CONTROL

The eradication of every clonogenic tumour cell must lead to tumour cure; this is, however, a daunting objective. Every gram of tumour may contain 10^9 cells of which perhaps 1% might be clonogenic. A human tumour at presentation could weigh some tens or hundreds of grams and the total num-

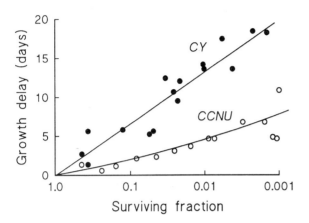

Figure 6.4 Relationship between tumour growth delay and surviving fraction, for B16 mouse melanomas treated either with cyclophosphamide (CY) or CCNU. For a particular level of cell killing, greater growth delay was observed after CY than CCNU. From Steel and Stephens (1983), with permission.

ber of clonogenic cells might therefore exceed 10^9. Cell kill by radiotherapy or chemotherapy is roughly exponential with dose. Thus, when treatment has reduced survival to 10^{-2} (which could lead to complete disappearance of all visible tumour) we would expect that four or five times this dose will be needed to reach a curative level. This is illustrated in Figure 2.1.

Whether it is always necessary to eradicate the last clonogenic tumour cell in order to achieve local tumour control has been a matter of intense debate. During the 1970s there was a widespread belief that the immune responses of patients against their tumours may be strong and this gave rise to many attempts at immunotherapy. Unfortunately, it has subsequently been realized that the animal tumours used to support the optimistic claims for immunotherapy were not truly syngeneic and were misleading (Alexander, 1977; Hewitt, 1979). Thus, whilst it is possible that weak tumour-directed immune responses may exist within cancer patients, sufficient perhaps to eradicate the last decade or so of surviving tumour cells, this cannot be relied upon.

6.7 Repair and recovery

The majority of damage induced in cells by radiation is satisfactorily repaired. Evidence for this comes from studies of strand breaks in DNA, most of which disappear during the few hours after irradiation (Section 8.6). Further evidence for repair comes from the wide variety of recovery experiments that have been done, both on *in vitro* cell lines and on normal and tumour tissues *in vivo*. It is useful to draw a distinction between these two sources of evidence:

Repair – refers to the process by which the function of macromolecules is restored. *Rejoining* of DNA strand breaks provides some evidence for this, although the rejoining of a break does not necessarily mean that gene function is restored. *Repair fidelity* is a significant additional factor. 'Repair' has often loosely been used for cellular or tissue recovery.

Recovery – refers to an increase in cell survival or a reduction in the extent of radiation damage to a tissue when time is allowed for this to occur.

There are a number of experimental sources of evidence for recovery, including the following.

Split-Dose Experiments. The effect of a given dose of radiation is less if it is split into two fractions, delivered a few hours apart. This effect has been termed recovery from *sublethal damage (SLD)*, or *Elkind recovery* (Elkind and Sutton, 1960). SLD recovery can be observed using various experimental endpoints: for instance using cell survival (Figure 6.5A), tumour growth delay (Figure 6.5C) or mouse lethality after irradiating a vital normal tissue (Figure 6.5D). The typical timing of split-dose recovery is shown in Figure 6.5A. Considerable recovery occurs within 15 min to 1 hour, and recovery often seems to be complete by roughly 4 hours. Recovery seems to be slower in some normal tissues such as the spinal cord (Table 13.4). When the split-dose technique is applied to cycling cells there is usually a wave in the data caused by cell cycle progression effects (Section 6.8).

Delayed-Plating Experiments. If cells are irradiated in a non-growing state and left for increasing periods of time before assaying for survival, an increase in survival is often observed (Figure 6.5B). During the delay the cells are recovering the ability to divide when called upon to do so. This has been

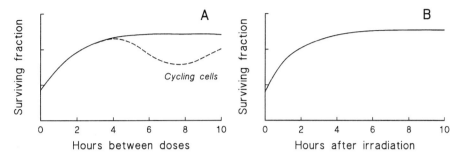

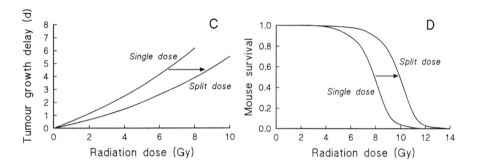

Figure 6.5 Illustrating four ways of measuring recovery from radiation damage (see text). Panels **A**, **C** and **D** show three types of split-dose experiment; panel **B** shows the results of a 'delayed-plating experiment'. The arrows in panels **C** and **D** indicate the measurement of $(D_2 - D_1)$ values.

termed recovery from *potentially lethal damage (PLD)*. The kinetics of PLD recovery and SLD recovery are similar.

Dose Rate Effect. The sparing of radiation damage as dose rate is reduced to around 1 Gy/hour is primarily due to cellular recovery (Sections 18.1, 18.2).

Fractionation. The sparing effect of fractionating radiation treatment within a relatively short overall time is primarily due to recovery. This is therefore the main reason why isoeffect curves slope upwards as the fraction number is increased (Sections 12.1, 12.2).

What is the relationship between these various ways of detecting recovery? The damage induced in cells by ionizing radiation is complex, as are the enzymatic processes that immediately begin to repair it. The various types of 'recovery experiment' listed above *evaluate* this complex repair process in slightly different ways. For instance, the evaluation based on giving a second dose (*i.e.* SLD recovery) is different from that obtained by asking irradiated non-dividing cells to divide (*i.e.* PLD recovery). There is only one complex entity, but a variety of ways of detecting it.

6.8 Variation of cell killing through the cell cycle

The radiosensitivity of cells varies considerably as they pass through the cell cycle. Although this has not been studied in a large number of cell lines, there seems to be a general tendency for cells in the S-phase (in particular the latter part of the S-phase) to be the most resistant and for cells in G2 and mitosis to be the most sensitive. The reason for the resistance in S may be related to the conformation of DNA at that time; the sensitivity in G2 probably results from the fact that those cells have little time to repair radiation damage before the cell is called upon to divide.

The classic results of Sinclair and Morton (1965) are illustrated in Figure 6.6. They synchronized Chinese hamster cells at five different points in the cell cycle and performed cell survival experiments. The survival curves showed that it was mainly the *shoulder* of the curve that changed: there was little shoulder for cells in G2 or mitosis and the shoulder was greatest for cells in S. The right-hand panel shows the profile of variation in cell killing through the cell cycle, constructed from these data.

The effect of this phenomenon is that it must create a degree of synchrony in the cells that survive irradiation. Immediately after a dose of X-rays, all the cells will still be at precisely the same point in the cell cycle as they were before irradiation, but some will have *lost their reproductive integrity* and it is the number that retain this which will tend to be greatest in the S phase.

MITOTIC DELAY AND CELL CYCLE PROGRESSION

Important effects become apparent during the course of the next cell cycle following irradiation. Mitotic delay (*i.e.* delay in the entry of cells into mitosis from G2) is very commonly observed, as also is a delay from G1 into the S phase. The genetic control mechanisms involved in these processes are receiving much research attention (Section 9.3).

The phenomena of induced cell synchrony and *cell cycle progression* are illustrated in Figure 6.7 for cells treated with an S-phase-specific cytotoxic drug. Untreated cells in exponential growth have a distribution through the cell cycle shown by (a). Immediately after treatment the distribution of *surviving clonogenic cells* will be as shown by (b). S-phase killing agents also tend temporarily to block the movement of cells into the S phase, and a few hours later there may be a pile-up of survivors in the latter part of the G1 phase (c). Later, these cells will begin to recover and move on in a semi-synchronous wave through the S phase (d), when the effect of a second dose of the drug would be greatest. Similar effects occur after irradiation and this is the basis of *reassortment*, as described in the next section.

In the 1970s there was much interest in *synchronization therapy*. This was the attempt to exploit cell cycle progression phenomena by treating with a second agent (usually a cytotoxic drug) at the optimum time interval after a priming treatment with drug or radiation. In the example illustrated in Figure 6.7, a second treatment with an S-phase-specific agent would be expected to be most effective at the time of panel (d). Although this approach to improving tumour therapy was thoroughly researched it proved in most cases to be disappointing. One possible reason for this is that tumours tend to be very heterogeneous from a kinetic point of view: cells move at very different speeds through the phases of the cell cycle and

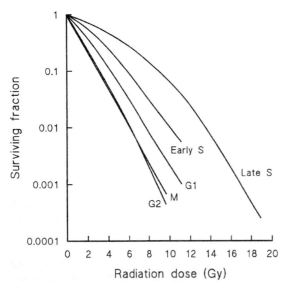
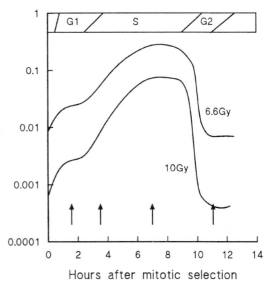

Figure 6.6 Variation of radiosensitivity through the cell cycle of Chinese hamster cells. Adapted from Sinclair and Morton (1965), with permission.

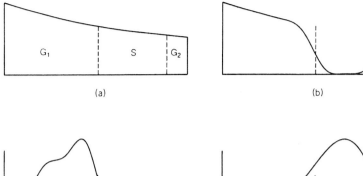

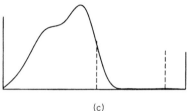
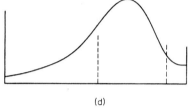

Figure 6.7 Cell-cycle progression (*i.e.* reassortment) induced in an exponentially growing cell population by an S-phase specific cytotoxic agent. From Steel (1977), with permission.

induced cell synchrony is therefore quickly lost (chapter 8 in Steel, 1977).

6.9 The 5 Rs of radiotherapy

The biological factors that influence the response of normal and neoplastic tissues to fractionated radiotherapy were summarized by Withers (1975) in the *4 Rs of Radiotherapy*:

Repair – As evidenced by cellular recovery during the few hours after exposure (Section 6.7).

Reassortment – Cell cycle progression effects, otherwise known as 'redistribution'. Cells that survive a first dose of radiation will tend to be in a resistant phase of the cell cycle and within a few hours they may progress into a more sensitive phase (Section 6.8).

Repopulation – During an extended course of radiotherapy, cells that survive irradiation may proliferate and thus increase the number of cells that must be killed (Section 14.3).

Reoxygenation – In a tumour, cells that survive a first dose of radiation will tend to be hypoxic but thereafter their oxygen supply may improve, leading to an increase in radiosensitivity (Section 15.5).

Note that two of these processes (repair and repopulation) will tend to make the tissue *more resistant* to a second dose of radiation; the other two (reassortment and reoxygenation) tend to make it *more sensitive*. These four factors modify the response of a tissue to repeated doses of radiation and are responsible for the slope of an isoeffect curve (Chapter 12). The overall radiosensitivity of the tissue (*i.e.* the height of the isoeffect curve on the page) depends on a fifth 'R': *Radiosensitivity*. Thus for a given fractionation course (or for single-dose irradiation) the haemopoietic system shows a greater response than the kidney, even allowing for the different timing of response. Similarly, tumours such as lymphomas are more radioresponsive than others and this is largely due to differences in radiosensitivity (Steel *et al*, 1989).

6.10 Remission and relapse

The concept of clonogenic cells provides a basic understanding of the main features of the time-course of cancer in response to treatment (Figure 6.8). At presentation there may well be in the region of 10^{10} to 10^{12} clonogenic cells present. During a course of treatment with radiation or cytotoxic drugs this number will be reduced. If the total body burden of tumour cells is significantly reduced, this is often described as a 'partial remission'. If it falls below the limit of detection by physical and imaging techniques then such a response is described as 'complete remission'. If treatment leads to the elimination of all clono-

Key points

1. Tumour recurrence after treatment depends upon the survival of clonogenic cells, which may comprise only a small proportion of the total cells within the tumour.
2. Evaluation of the survival of clonogenic cells following treatment is an important aspect of experimental cancer therapy. In experimental situations this is relatively simple to perform, but for cells removed directly from human tumours great care is necessary in the selection and performance of the assays.
3. In tumour therapy, as with tumour growth, it is important to *think logarithmically*. Partial or complete tumour remission may involve only perhaps two decades of clonogenic cell reduction, yet over nine decades of cell kill may be necessary before long-term tumour control can be assured.
4. Repair of radiation damage is an important feature of irradiated tissues. It can be assessed by a variety of experimental tests.
5. Variation of cell killing through the cell cycle is considerable and may give rise to cell cycle progression phenomena.
6. The 5 Rs of radiotherapy.

genic cells then the disease will be controlled. If not, then in the majority of cases there will be a period of remission followed by relapse. The duration of remission will depend greatly on the level of cell kill achieved by the treatment, *i.e.* on the depth of the nadir in the curve in Figure 6.8 (see also Figures 2.7 and 17.5).

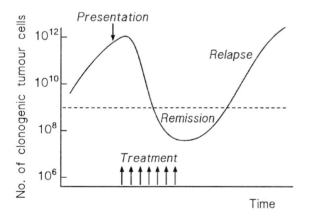

Figure 6.8 Time-course of changes in the body burden of clonogenic cells through the four main phases of uncontrolled cancer: presentation, treatment, remission and relapse.

Bibliography

Alexander P (1977). Back to the drawing board – the need for more realistic model systems for immunotherapy. *Cancer* 40:467–70.

Baker FL, Spitzer G, Ajani JA *et al* (1986). Drug and radiation sensitivity measurements of successful primary monolayer culturing of human tumour cells using cell-adhesive matrix and supplemented medium. *Cancer Research* 46:1263–74.

Bosanquet A (1991). Correlations between therapeutic response of leukaemias and *in vitro* drug-sensitivity assay. *Lancet* 337:711–14.

Carmichael J, DeGraff WG, Gazdar AF *et al* (1987). Evaluation of a tetrazolium-based semiautomated colorimetric assay: assessment of radiosensitivity. *Cancer Research* 47:943–6.

Chapman JD, Peters LJ and Withers HR (1989). *Prediction of Tumour Response to Treatment.* Pergamon; New York.

Courtenay VD and Mills J (1978). An *in vitro* colony assay for human tumours grown in immune-suppressed mice and treated *in vivo* with cytotoxic agents. *Brit J Cancer* 37:261–68.

Elkind MM and Sutton H (1960). Radiation response of mammalian cells grown in culture: I. Repair of X-ray damage in surviving Chinese hamster cells. *Radiat Res* 13:566–93.

Fowler JF (1991). The phantom of tumour treatment – continually rapid proliferation unmasked. *Radiother Oncol* 22:156–8.

Hewitt HB (1979). A critical examination of the foundations of immunotherapy for cancer. *Clin Radiol* 30:361–9.

Hill RP and Stanley JA (1975). The lung colony assay: extension to the Lewis lung tumour and the B16 melanoma. *Int Radiat Biol* 27:377–87.

Mitchell JB (1988). Potential applicability of nonclonogenic assays to clinical oncology. *Radiat Res* 114:401–14.

Seymour CB and Mothersill C (1989). Lethal mutations, the survival curve shoulder and split-dose recovery. *Int J Radiat Biol* 56:999–1010.

Sinclair WK and Morton RA (1965). X-ray and ultraviolet sensitivity of synchronised Chinese hamster cells at various stages of the cell cycle. *Biophys J* 5:1–25.

Steel GG and Adams K (1975). Stem-cell survival and tumour control in the Lewis lung carcinoma. *Cancer Research* 35:1530–5.

Steel GG, McMillan TJ and Peacock JH (1989). The 5 Rs of radiobiology. *Int J Radiat Biol* 56:1045–8.

Steel GG and Stephens TC (1983). Stem cells in tumours. In: *Stem Cells*. (Ed) Potten CS. Churchill Livingstone; Edinburgh.

Streffer C, van Beunigen D, Gross E *et al* (1986). Predictive assays for the therapy of rectum carcinoma. *Radiother Oncol* 5:303–10.

Till JE and McCulloch EA (1961). A direct measurement of the radiation sensitivity of normal mouse bone marrow. *Radiat Res* 14:213–22.

Trott KR (1972). Relation between division delay and damage expressed in later generations. *Curr Topics Radiat Res* 7:336–7.

Wasserman TH and Twentyman P (1988). Use of a colorimetric microtiter (MTT) assay in determining the radiosensitivity of cells from murine tumours. *Int J Radiat Oncol* 15:699–702.

Withers HR (1975). The four R's of radiotherapy. *Adv Radiat Biol* 5:241–7.

Further Reading

Elkind MM and Whitmore GF (1967). *The Radiobiology of Cultured Mammalian Cells.* Gordon and Breach; New York.

Potten CS (Ed) (1983). *Stem Cells, their Identification and Characterisation.* Churchill Livingstone; Edinburgh.

Steel GG (1977). *The Growth Kinetics of Tumours.* Oxford University Press; Oxford.

Tannock IF and Hill RP (1992). *The Basic Science of Oncology.* Second edition. Pergamon; New York.

7
Models of radiation cell killing

Michael C. Joiner

7.1 Introduction

Research in experimental radiobiology covers studies at the cell, animal and human levels. It deals at the fundamental level with the molecular, biochemical and biophysical nature of radiation damage. Models are a necessary part of radiobiology research, to provide a framework in which to analyse and compare data and ultimately to assist in building up a consistent theory of radiation action both *in vitro* and *in vivo*. Models and mathematics are also sometimes necessary to relate experimental studies to clinical cancer treatment with the aim of improving therapy. This chapter describes some models that are used to analyse the relationship between cell survival and radiation dose.

7.2 Molecular targets

Radiation kills cells by producing secondary charged particles and free radicals in the nucleus which in turn produce a variety of types of damage in DNA. Evidence that damage to DNA is the primary cause of radiation cell killing and mutation is set out in Section 8.2. Each 1-Gy dose of low-LET radiation produces about 1000 initial single-strand breaks and 40 initial double-strand breaks. Some lesions are more important than others and radiation lethality correlates most significantly with the number of residual, unrepaired double-strand breaks (DSB) several hours after irradiation. Table 8.2 shows that for a fixed radiation dose, if cell kill is modified by changing LET, oxygen level, thiol concentration or temperature, only the number of DSB follows the change in cell kill. Single-strand breaks, base damage and DNA–protein cross-links do not reflect the change in cell kill for all of these modifiers. The DNA double-strand break is therefore thought to be the most important type of cellular damage. Just one residual DSB (or 'hit') in a vital section of DNA may be sufficient to produce a significant chromosome aberration and thus to sterilize the cell.

7.3 Target theory

One way of explaining the shape of cell survival curves is the idea that there may be regions of the DNA that are important to maintain the reproductive ability of cells. These sensitive regions could be thought of as specific *targets* for radiation damage so that the survival of a cell after radiation exposure will be related to the number of targets inactivated. There are two versions of this idea that have been used commonly. The first version of the theory proposes that just one hit by radiation on a single sensitive target leads to death of the cell. This is called *single-target single-hit inactivation*, and it leads to the form of survival curve shown in Figure 7.1A. The survival curve is exponential (*i.e.* a straight line in a semi-logarithmic plot of survival against dose). To derive an equation for this survival curve, Poisson statistics can be applied, as during irradiation there are a very large number of hits on different cells taking place, but the probability (p) of the next hit occurring in a given cell is very small. Thus for each cell,

$$p(\text{survival}) = p(0 \text{ hits}) = \exp(-D/D_0)$$

where D_0 is defined as the dose that gives an average of one hit per target. A dose of D_0 Gy reduces survival from 1 to 0.37 (*i.e.* to e^{-1}), or from 0.1 to

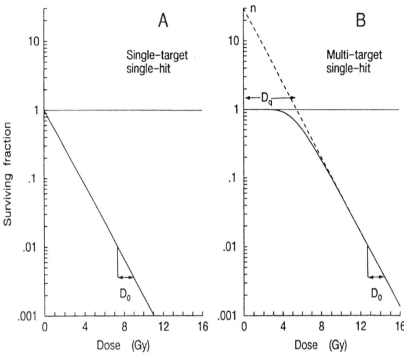

Figure 7.1 The two most common types of target theory. **A**: single-target inactivation; **B**: multi-target inactivation.

0.037, etc. D/D_0 is the average number of hits per target (and in this case per cell). This is the reason why (as in Figure 7.4) a scale of cell survival is sometimes labelled -ln(S): this is a scale of the natural logarithm of surviving fraction and it is also the *equivalent* number of lethal lesions per cell.

In this example (Figure 7.1A) D_0 = 1.6 Gy. Straight survival curves of this sort are usually found for the inactivation of viruses and bacteria. They may also be appropriate in describing the radiation response of some very sensitive human cells (normal and malignant) and also the radiation response at very low dose rates (Section 18.2) and response to high LET radiations (Section 19.3).

A more general version of target theory is called *multi-target single-hit inactivation*. This proposes that just one hit by radiation on each of n sensitive targets in the cell is required for death of the cell. The shape of this survival curve is shown in Figure 7.1B. Again, using Poisson statistics,

p(0 hits on a specific target) = $\exp(-D/D_0)$
Thus p(specific target inactivated) = $1 - \exp(-D/D_0)$
As there are n targets in the cell,
p(all n targets inactivated) = $(1 - \exp(-D/D_0))^n$
Thus p(survival) = p(not all targets inactivated)
  = $1 - (1 - \exp(-D/D))^n$ (Eqn 7.1)

Figure 7.1B shows that multi-target single-hit survival curves have a shoulder whose size can be indicated by the *quasi-threshold dose* (D_q). This is related to n and D_0 by the relation:

$$D_q = D_0 \log_e (n) \qquad \text{(Eqn 7.2)}$$

For the example in Figure 7.1B we have chosen n = 30 and D_0 = 1.6 Gy, giving D_q = 5.4 Gy. Such multitarget survival curves have proved useful for describing the radiation response of mammalian cells at high doses, 'off the shoulder'. They do not describe the survival response well at lower more clinically relevant doses.

7.4 More complex models

The main shortcoming of the multi-target model is that, as shown in Figure 7.1B, it predicts a response that is flat for very low radiation doses. On the contrary, much experimental cell survival data show evidence for a finite initial slope. To take account of this, the multi-target model can be modified by adding an additional single-target component. The resulting equation for the survival curve is called the *two-component model*:

$$p(\text{survival}) = \exp(-D/D_1)(1-(1-\exp(-D(1/D_0-1/D_1)))^n) \quad \text{(Eqn 7.3)}$$

This type of survival curve is illustrated in Figure 7.2A. In addition to the parameters n, D_0 and D_q, this curve also has a parameter D_1 which fixes the initial slope, i.e. the dose required in the low-dose region to reduce survival from 1 to 0.37. In this example, $n = 30$ and $D_0 = 1.6$ Gy, and $D_1 = 4.6$ Gy. This type of curve does predict finite cell killing in the low-dose region but it has the drawback that the change in cell survival over the range 0–D_q occurs almost linearly. This implies that no sparing of damage should occur as dose per fraction is reduced below ~2 Gy, which is usually not found to be the case either experimentally or in clinical radiotherapy (Section 14.2).

7.5 The linear-quadratic model

A better description of radiation response in the low-dose region (0 – 3 Gy) is given by the linear-quadratic (LQ) model shown in Figure 7.2B:

$$p(\text{survival}) = \exp(-\alpha D - \beta D^2) \quad \text{(Eqn 7.4)}$$

This is a continuously bending survival curve with no straight portion at high radiation doses. Its shape (or 'bendiness') is determined by the ratio α/β. Since the dimensions of the parameters are α: Gy^{-1} and β: Gy^{-2}, the dimensions of α/β are Gy: as shown in Figure 7.2B, this is the dose at which the *linear* contribution to damage (αD on the logarithmic scale) equals the quadratic contribution (βD^2). The response of cells to densely-ionizing radiations like neutrons or α-particles is usually a steep and almost exponential survival curve (Section 19.3). As shown in Figure 7.2, this would be explained in the two-component model by the ratio D_1/D_0 being near to 1.0, or in the LQ model by a high α/β ratio.

The LQ model is in widespread use in radiobiology and generally works well in describing responses to radiation *in vitro* and also *in vivo* (Section 13.2). Is there a mechanistic justification for the LQ model? One simple idea is that the linear component [$\exp(-\alpha D)$] might be due to *single-track* events while the quadratic component [$\exp(-\beta D^2)$] might arise from *two-track* events. This interpretation is supported by studies of the dose rate effect (Section 18.2) which show that as dose rate is reduced, cell survival curves become straight and tend to extrapolate the initial slope of the high dose-rate curve: the quadratic component of cell killing disappears, leaving only the linear

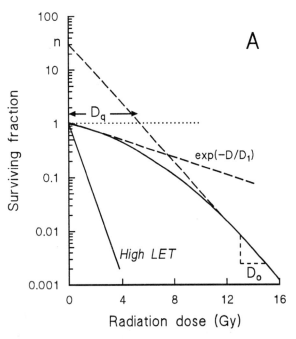
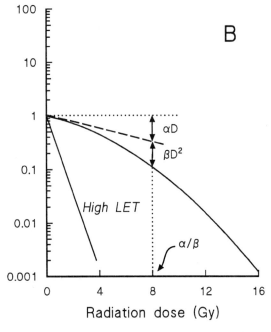

Figure 7.2 Models with non-exponential cell killing but a finite initial slope. **A**: the two-component model and **B**: the linear-quadratic model.

component. This would be expected, for at low dose rate single-track events will occur far apart in time and the probability of interaction between them will be low. Although this interpretation of the LQ equation seems reasonable, the nature of the interactions between separate tracks is still a matter of considerable debate. Chadwick and Leenhouts (1973) postulated that separate tracks might hit opposite strands of the DNA double helix and thus form a double-strand break; this now seems an unlikely mechanism in view of the very low probability of such a close interaction at a dose of a few gray. Interaction between more widely spaced regions of the complex DNA structure, or between DNA in different chromosomes, may be a more plausible mechanism (Chapter 8).

7.6 The lethal, potentially lethal damage (LPL) model

Curtis (1986) proposed this model as a 'unified repair model' of cell killing. Ionizing radiation is considered to produce two different types of lesion: repairable (*i.e.* potentially lethal) lesions and non-repairable (*i.e.* lethal) lesions. The non-repairable lesions produce single-hit lethal effects and therefore give rise to a linear component of cell killing [= $\exp(-\alpha D)$]. The eventual effect of the repairable lesions depends on competing processes of repair and *binary misrepair*. It is this latter process that leads to a quadratic component in cell killing. As shown in Figure 7.3, the model has two sensitivity parameters (η_L determines the number of non-repairable lesions produced per unit dose, and η_{PL} the number of repairable lesions). There are also two rate constants (ϵ_{PL} determines the rate of repair of repairable lesions, and ϵ_{2PL} the rate at which they undergo interaction and thus misrepair).

This model produces almost identical cell survival curves to the LQ equation, down to a survival level of perhaps 10^{-2}. It can therefore be taken to provide one possible mechanistic model for the LQ equation. It predicts that as dose rate is reduced the probability of binary interaction of potentially lethal lesions will fall and parameter values can be found that allow the model accurately to simulate cell survival data on human and animal cells irradiated at various dose rates (Section 18.2).

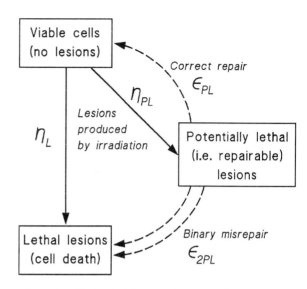

Figure 7.3 The Lethal, Potentially Lethal (LPL) damage model of radiation action.

Curtis' LPL model is an example of a lesion interaction model, which additionally incorporates repair processes. Figure 7.4A shows how this produces the downward-bending cell survival curve: the dashed curve indicates the component of cell killing that is due to single-track non-repairable lesions. It is the extra lethal lesions produced by the binary interaction of potentially lethal lesions which give the downward-bending curve.

7.7 Repair saturation models

Another class of model are the *repair saturation models*, which propose that the shape of the survival curve depends only on a dose-dependent rate of repair. Figures 7.4B and 7.4C demonstrate this idea. Only one type of lesion and single-hit killing are postulated, and in the absence of any repair these lesions produce the steep dashed survival curve in Figure 7.4B. The final survival curve (solid line) results from repair of some of these lesions but if the repair enzymes become saturated (Figure 7.4C), there is not enough repair enzyme to bind to all damaged sites simultaneously and so the reaction velocity of repair no longer increases with increasing damage. Therefore at higher doses (more lesions), there is proportionally less repair

56 Models of radiation cell killing

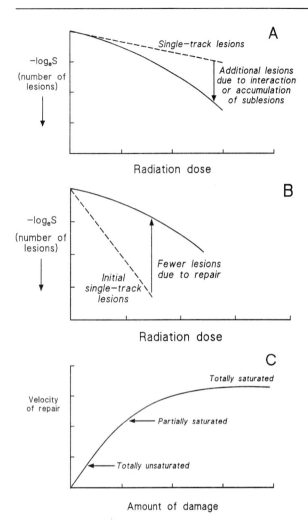

Figure 7.4 The contrast between lesion-interaction models and repair-saturation models. **A**: the LPL model; **B**: the effect of repair becoming less effective at higher radiation doses; **C**: the basic concept of repair saturation. Adapted from Goodhead (1985), with permission.

during the time available before damage becomes fixed; this will lead to more residual damage and to greater cell kill. The mechanisms of fixation of non-repaired damage are not understood but they may be associated with the entry of cells carrying such damage into DNA synthesis or mitosis. It should be noted that an alternative 'saturation' hypothesis, leading to the same consequence, is that the pool of repair enzymes is used up during repair, so that at

Table 7.1 Different interpretations of radiobiological phenomena by lesion-interaction and saturable repair models

Observation	Explanation	
	Lesion interaction	*Repair saturation*
Curved dose–effect relationship	Interaction of sublesions	Saturation of capacity to repair sublesions
Split-dose recovery	Repair of sublesions (sublethal damage repair)	Recovery of capacity to repair sublesions
RBE increase with LET	More non-repairable lesions at high LET	High-LET lesions are less repairable
Low dose rate is less effective	Repair of sublesions during irradiation	Repair system not saturating

Adapted from Goodhead (1985).

Key points

1. A number of different mathematical models adequately simulate the shape of cell-survival curves for mammalian cells. So far it has not been possible strongly to prefer one model to others.
2. Target theory proposes that a specific number of targets or DNA sites must be inactivated or damaged to kill the cell. This approach is only satisfactory if a component of single-hit killing is also introduced. So far it has not been possible to identify the location of these vital 'targets' within the cell nucleus.
3. Lesion interaction models explain downward-bending cell survival curves by postulating two classes of lesion. One class is directly lethal, but the other type is only *potentially* lethal and may be repaired enzymatically or may interact with other potentially lethal lesions to form lethal lesions.
4. Repair saturation models assume that potentially lethal lesions are either repaired successfully or fixed as the cell goes through some critical stage in its cycle (*e.g.* mitosis). Downward-bending survival curves occur because higher radiation doses produce too much damage for the repair system to handle and it saturates; proportionally less damage can therefore be repaired in the time available before fixation than at lower doses where the system is unsaturated.

higher doses the repair system is depleted and is less able to repair all the induced damage.

Table 7.1 illustrates how the basic conceptual difference between the lesion accumulation/interaction models such as Curtis' LPL and the dose-dependent repair models affects the interpretation of some radiobiological phenomena (Goodhead, 1985). Both types of model predict linear-quadratic cell survival curves in the clinically relevant dose region. They also provide good explanations of split-dose recovery (Section 6.7), changing effectiveness with LET (Section 19.3) and the dose-rate effect (Section 18.2). At present, radiation scientists are uncertain whether lesion interaction or repair saturation really exist in cells but it may well be that molecular and microdosimetric studies will eventually determine which explanation (maybe both!) is correct.

Bibliography

Chadwick KH and Leenhouts HP (1973). A molecular theory of cell survival. *Phys Med Biol* 13:78–87.
Curtis SB (1986). Lethal and potentially lethal lesions induced by radiation: a unified repair model. *Radiat Res* 106:252–70.
Frankenberg-Schwager M (1989). Review of repair kinetics for DNA damage induced in eukaryotic cells *in vitro* by ionizing radiation. *Radiother Oncol* 14:307–20.
Goodhead DT (1985). Saturable repair models of radiation action in mammalian cells. *Radiat Res* 104:S58–S67.

Further Reading

Alpen EL (1990). *Radiation Biophysics*. Prentice-Hall; London.
Chapman JD (1980). Biophysical models of mammalian cell inactivation by radiation. In: *Radiation Biology in Cancer Research*, pp.21–32. (Eds) Meyn RE and Withers HR. Raven Press; New York.
Douglas BG and Fowler JF (1976). The effect of multiple small doses of X-rays on skin reactions in the mouse and a basic interpretation. *Radiat Res* 66:401–26.
Elkind MM and Sutton H (1960). Radiation response of mammalian cells grown in culture: I. Repair of X-ray damage in surviving Chinese hamster cells. *Radiat Res* 13:556–93.
Ward JF (1990). The yield of DNA double-strand breaks produced intracellularly by ionizing radiation: a review. *Int J Radiat Biol* 57:1141–50.

8
DNA damage and cell killing

Trevor J. McMillan and G. Gordon Steel

DNA represents the most important target for radiation action on a cell. This chapter examines how this damage is inflicted, the nature of the damage, and how it leads to cell death and mutation.

8.1 Initial processes of radiation damage

Irradiation of a biological system initiates a series of processes that can be classified in terms of the time-scale over which they act (Figure 1.2). The physical, chemical and biological phases of this process have been described in Section 1.3.

An electron with an energy of 1 MeV has a range in soft tissue of a few millimetres. In the early part of its track the particle moves very quickly and its rate of energy deposition is low; the result is a relatively straight track in which the ionizations may be separated by distances of around 0.1 μm on average. We describe this as radiation with a low linear energy transfer (low LET, Section 19.2). As the electron slows down it interacts more strongly with the orbital electrons in the medium. Its rate of energy loss increases, the track becomes more tortuous due to stronger collisions, and the ionization density increases. Figure 8.1A shows a computer simulation of the tracks of 1-keV electrons, representing a very small terminal part of the tracks of 1-MeV electrons. The important feature is the tendency towards *clustering* of the ionization events at the ends of the tracks, each cluster having the size of a few nanometres. Within each electron track there is opportunity for interactions between the products of separate ionization events and it is thought that, particularly at low dose rate or following acute radiation doses up to a few gray, the main biological effects of radiation (cell killing and cell mutation) are predominantly due to damage that is produced by these 'hot spots'.

Within perhaps 10^{-10} seconds of exposure to either photon or particle beams the irradiated volume will contain atoms that have been ionized and a corresponding number of free electrons, all produced by the cascade of atomic reactions just described and with a rather non-uniform spatial distribution. The numbers of ionizations produced at therapeutic dose levels is very large: approximately 10^5 ionizations per cell per gray. One gray gives roughly 1 *lethal* lesion per cell and the ineffectiveness of most of the ionizations is due to three main factors: free radical scavenging processes, the small number of ionizations that are close enough to DNA to damage it, and cellular repair processes.

FREE-RADICAL PROCESSES

Since biological systems consist largely of water, the bulk of the ionizations produced by irradiation occur in water molecules. Negatively charged free electrons (e⁻) that are produced by ionization will rapidly become associated with polar water molecules, greatly reducing their mobility. The configuration of an electron surrounded by water molecules (a 'hydrated electron', e^-_{aq}) has a degree of stability and a lifetime under physiological conditions of a few microseconds. The water molecule that has lost an electron is a highly reactive positively charged ion. It quickly breaks down to produce a hydrogen ion (H⁺) and an (uncharged) OH radical. OH is a molecule that normally does not exist in water. Oxygen has valency 2, hydrogen 1, and the stable configuration is H_2O. The uncharged OH radical has an unpaired electron ('unattached valency') that makes it highly reactive. We desig-

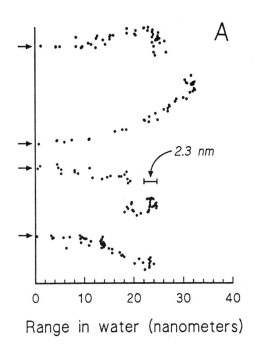
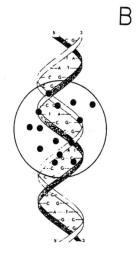

Figure 8.1 Computer-simulated tracks of 1 keV electrons. Note the scale in relation to the 2.3 nm diameter of the DNA double helix (adapted from Chapman and Gillespie, 1981). Illustrating the concept of a Local Multiply-Damaged Site produced by a cluster of ionizations impinging on DNA.

nate it as a free radical thus: OH•. Free radicals are simply fragments of broken molecules. OH• is different from OH⁺ which is a positively charged *ion*: the OH radical has equal numbers of protons and orbital electrons but because of the unpaired electron it is chemically reactive (note that some ions may also be radicals, for example a water molecule that has lost an electron is actually H_2O^+, a radical cation). Similarly, H^+ is a bare proton, positively charged, whilst H• is a proton plus an electron (neutral charge) but again highly reactive because the stable form of hydrogen is H_2.

Around 10^{-10} seconds after irradiation there will be three principal radiolysis products of water:

$$e^-_{aq} \quad H^\bullet \quad OH^\bullet$$

These highly reactive species will go on to take part in further reactions. An important one is

$$OH^\bullet + OH^\bullet \rightarrow H_2O_2$$

the production of hydrogen peroxide. Oxygen, if present, plays an important part in the free radical reactions that follow irradiation. Molecular oxygen has a high affinity for free radicals (R•):

$$R^\bullet + O_2 \rightarrow RO_2^\bullet$$

giving rise to further reactive products and acting to *fix* the free radical damage. The oxygen effect in radiation cell killing has often been explained in terms of this type of process.

In biological systems the free radicals produced in water may react with essential macromolecules. A vast range of reactions take place, most of which are unimportant for the survival and functioning of the cell. The important reactions are those with DNA, because of the uniqueness of the function of many parts of this molecule. Damage to DNA from free radicals produced in water is called the *indirect* effect of radiation; ionization of atoms that are part of the DNA molecule is the *direct* effect.

Compounds containing sulphydryl (–SH) groups have a particular affinity for free radicals. Their presence within the cell may therefore act to 'mop up' a proportion of radicals, thus decreasing radiation effects. The principal non-protein thiols in mammalian cells are:

- Cysteine – a natural amino acid
- Cysteamine – decarboxylated cysteine
- Glutathione – the commonest non-protein thiol

Administration of these compounds to animals or to cells in tissue culture immediately before irradiation reduces the extent of damage. The radiation dose required to produce a given level of damage can go up by as much as a factor of 2.0.

During the first millisecond after radiation exposure a competition takes place between damage-fixing and scavenging reactions for radicals in key target molecules. There is considerable interest in the possibility of manipulating the levels of cellular thiols either to protect normal tissues or, by depressing thiol levels, to increase the radiosensitivity of tumour cells. Thiol depression can be achieved experimentally by blocking the synthesis of glutathione using such agents as diamide or buthionine sulphoximine (BSO).

The development and radiobiological properties of hypoxic cell radiosensitizers are dealt with in Chapter 16. The most important class of sensitizers are the electron-affinic compounds, principally the nitroimidazoles, which act like oxygen in promoting the fixation of free radical damage. Since they are less rapidly metabolized than oxygen, they can diffuse further in tumour tissue and thus reach hypoxic cells.

8.2 Radiation damage to DNA

THE STRUCTURE OF DNA

Deoxyribonucleic acid (DNA) is a large molecule that has a characteristic double-helix structure consisting of two strands, each made up of a sequence of nucleotides (Figure 8.2). A nucleotide is a subunit in which a 'base' is linked through a sugar group to a phosphate group. The sugar is deoxyribose which has a 5-atom ring: 4 carbons and 1 oxygen. The 'backbone' of the molecule consists of alternating sugar–phosphate groups. Note that by the conventional numbering system the connection points of the phosphates to the sugar ring are labelled 5' and 3': this leads to the ends of a sequence of nucleotides also being labelled in this way and it defines the direction in which the sequence is read during transcription. There are four different bases. Two are single-ring groups (pyrimidines): thymine and cytosine. Two are double-ring groups (purines): adenine and guanine. It is the order of these bases along the molecule that specifies the genetic code.

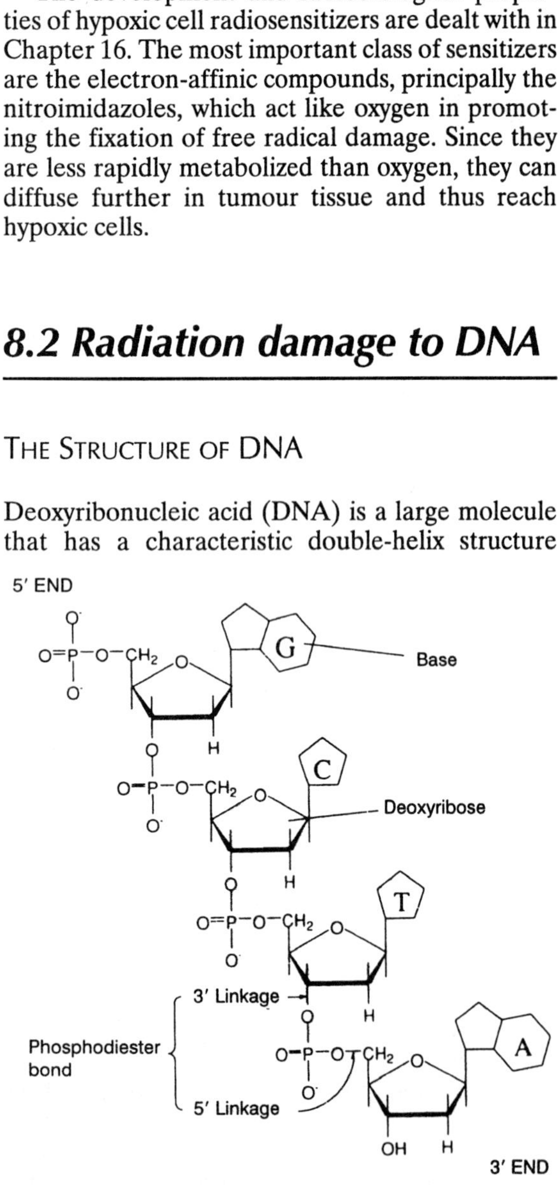

The two strands of the double helix are held together by hydrogen bonding between the bases. These bonds are formed between thymine and adenine, and between cytosine and guanine. The bases are paired in this way along the length of the DNA molecule. During the S-phase of the cell cycle DNA synthesis takes place (the process of *replication*) in which every base pair is accurately duplicated.

The first stage in the manufacture of proteins is the construction by the process of *transcription* of a messenger RNA (labelled mRNA) that has a similar structure to a single strand of DNA except that the sugar groups are ribose in place of deoxyribose, and thymine is replaced by uracil. The decoding is based on the pairing of bases: A-U, C-G, G-C, T-A. Transcription is performed by RNA polymerases which bind to DNA and generate the corresponding RNA. The control of transcription is not yet fully understood but specific DNA sequences at the beginning and ends of genes are known to be involved. Within the *coding region* of the gene there may be several stretches of DNA that are not required for mRNA. These are called *introns* (the required regions are *exons*) and they need to be removed at some point prior to the translation of the mRNA into protein. Protein production occurs at ribosomes where transfer RNA (tRNA) with an amino acid attached recognizes groups of three bases in the mRNA (*codons*) and in this way the amino acids are lined up in the correct sequence for the protein.

The very long DNA double-helix molecule, together with nuclear proteins, is organized within the cell through a number of levels of supercoiling.

Figure 8.2 The structure of DNA, in which the four bases (G, C, T, A) are linked through sugar groups to the sugar phosphate backbone.

The DNA double helix (about 2.5 nm in diameter) is first coiled around protein cores to form a bead-like string of nucleosomes, then coiled into a 25-nm fibre, which is further spiralized and becomes visible in the condensed form of a chromosome at mitosis. During interphase and in chromatin that has been gently extracted from cell nuclei, DNA shows a series of loops or *domains* that are attached to the nuclear matrix. A human chromosome may have around 2600 looped domains, each formed from about 0.4 μm of the 25-nm DNA fibre and containing 20 000 to 80 000 base pairs.

Radiation Damage to DNA

Early experiments showed that irradiation leads to a loss of viscosity in DNA solutions. Subsequently this has been shown to result from DNA strand breaks. There are two categories of DNA strand breaks; single-strand (SSB) and double-strand breaks (DSB). The detection of these depends on a study of the size distribution of fragments of DNA after extraction from irradiated cells (Section 8.3). There are a variety of other types of DNA lesion, some of which are important in the cytotoxic action of chemotherapeutic agents (Figure 8.3).

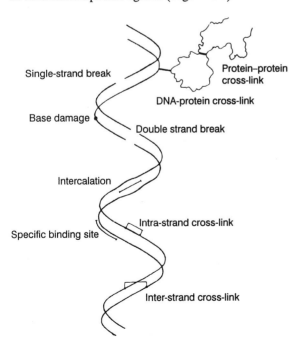

Figure 8.3 Types of damage to DNA produced by radiation and chemical agents.

Why Do We Believe that DNA Damage is the Critical Event in Radiation Cell Killing and Mutation?

There are many sources of evidence, including the following:

(i) Micro-irradiation studies show that to kill cells by irradiation only of the cytoplasm requires far higher radiation dose than irradiation of the nucleus.
(ii) Isotopes with short-range emission (such as ^{3}H, ^{125}I) when incorporated into cellular DNA efficiently produce radiation cell killing and DNA damage (Table 8.1).
(iii) The incidence of chromosomal aberrations following irradiation is closely linked to cell killing.
(iv) Thymidine analogues such as IUdR or BrUdR when specifically incorporated into chromatin modify radiosensitivity.

The number of lesions induced by radiation in DNA is far greater than those that eventually lead to cell killing. A dose of radiation that induces on average one lethal event per cell will kill 63% and leave 37% still viable (this results from Poisson statistics) and we call this the D_0 dose (Section 7.3). D_0 values for oxic mammalian cells are usually in the region of 1–2 Gy. The numbers of DNA lesions per cell that are detected immediately after such a dose have been estimated to be approximately:

	Events per D_0
Base damage	> 1000
Single-strand breaks	~ 1000
Double-strand breaks	~ 40

In addition cross-links between DNA strands and between DNA and nuclear proteins are formed (Figure 8.3). Irradiation at clinically used doses thus induces a vast amount of DNA damage, most of which is successfully repaired by the cell. In a variety of experimental situations it has been found that the incidence of cell killing fails to correlate with the number of SSB induced, but relates better to the incidence of DSB (Table 8.2). Significantly, a dose of hydrogen peroxide that induces many SSB produces little cell killing and few DSB. On this basis it is generally believed that DSB are the critical lesions for radiation cell killing, although as the next paragraph indicates it may be only *some* DSB that are important.

The realization that low-LET radiation produces 'hot spots' in which clusters of ionizations

62 DNA damage and cell killing

Table 8.1 Toxicity of radioisotopes depends upon their subcellular distribution

	Radiation dose to part of the cell* (Gy)		
	To nucleus	To cytoplasm	To membranes
X-ray	3.3	3.3	3.3
^{3}H-thymidine	3.8	0.27	0.01
^{125}I-conconavalin	4.1	24.7	516.7

* For each of these three treatments a dose has been chosen that gives 50% cell killing in CHO cells. The absorbed radiation doses to the nucleus, cytoplasm or membranes have then been calculated. ^{3}H-thymidine is bound to DNA, ^{125}I-conconavalin to cell membranes. It is the *nuclear* dose that is constant and thus correlates with cell killing, not the cytoplasmic or membrane doses.
From Warters *et al* (1977).

may occur within a diameter of a few nanometres (Section 8.1) has led to the notion that such an event may produce a particularly severe lesion if it impinges on the DNA molecule. This lesion might consist of one or more DSB together with a number of SSB, and also DNA base damage, etc. Ward (1986) has termed this a local multiply-damaged site (LMDS) (Figure 8.1B). The importance of these hypothetical lesions derives from the fact mentioned above that the great majority of DNA lesions are repaired. Any difference in repair among DSB will lead to lethality being associated with the rare lesion that fails to repair, which may *inherently* be difficult to repair. An LMDS, which will be recognized by a strand-break assay as merely one DSB, may have a low repair probability and high probability of leading to cell death.

Table 8.2 Double-strand DNA breaks correlate best with cell killing

Modifier	Cell kill	DSB	SSB	Base	DNA-protein crosslinks
High-LET radiation	↑	↑	↓	↓	–
Hypoxia	↓	↓	↓	0	↑
Thiols	↓	↓	↓	0	↓
Hyperthermia	↑	↑	0	0	0
Hydrogen peroxide	0	0	↑	↑	–

↑ increased; ↓ decreased; 0 little or no effect; – not known.
See Frankenburg-Schwager (1989) for further information on these relationships.

The importance of the initial slope of cell survival curves for clinical radiotherapy is described in Sections 17.5 and 18.2. The concept of a linear component of cell killing is also essential to the linear-quadratic and LPL models of cell killing (Sections 7.5 and 7.6). LMDS are a type of radiation-induced lesion that could cause single-hit cell killing and therefore conceivably could give rise to the linear component.

Even strand breaks induced by single events may be heterogeneous. Over 20 products of thymine have been detected using gas chromatography/mass spectrometry. It is therefore highly likely that the chemical residues on the edges of a strand break vary markedly. The biological consequences of this are not yet known.

RADIATION-INDUCED MUTATIONS

Some damage induced by radiation may be insufficient to abolish the colony-forming ability of a cell but enough to lead to mutation. Mutations generally appear as a non-lethal alteration in the base sequence in DNA, although if severe they may be lethal. Mutations result in the expression of an altered protein or in the increase or decrease in the level of a normal protein.

The frequency of radiation-induced mutations usually increases in a dose-dependent manner in the range of doses per fraction commonly used in clinical radiotherapy. At higher radiation doses lethal events may predominate and the frequency of mutations will fall. The biological consequences of mutations depend on where they occur. Cellular resistance to cytotoxic agents can arise by mutation of the enzyme to which the drug is targeted. Cellular transformation can occur if a mutated gene leads to a decrease in the normal control mechanisms of cell proliferation. Each of these examples has obvious implications for the use of radiation in therapy and for the effects of environmental exposure to radiation.

The molecular analysis of non-lethal mutations has allowed a detailed examination of the results of radiation-induced DNA damage. A wide spectrum of types of damage has been detected and an interesting observation is that ionizing radiation tends to produce a higher proportion of large deletions relative to simple base changes when compared with mutations that have arisen spontaneously (Yandell *et al*, 1990).

8.3 Methods for detecting DNA damage

Since it is widely believed that strand breaks are the critical lesion produced by ionizing radiation a large amount of effort has gone into producing methods by which they can be measured. These include the following.

Sucrose Gradient Sedimentation. Sucrose solutions are prepared within plastic centrifuge tubes with a sucrose concentration that varies continuously from around 30% at the bottom to 5% at the top. Cells are exposed to ^{14}C-thymidine for a sufficient period to label most of the DNA. They are then mixed with a solution that induces cell lysis and releases the DNA. Some solutions (such as 0.2M NaOH) release DNA in a single-stranded form whereas at lower (i.e. less strongly alkaline) pH the double-stranded structure is retained. The mixture is floated on the top of the gradient. When the tubes are spun in a high-speed centrifuge at around 30 000 rpm the large DNA fragments travel further and the profile of ^{14}C concentration down the gradient reflects the amount of DNA damage. This is detected by piercing the bottom of the centrifuge tube, collecting fractions of the fluid, and assaying for ^{14}C. The strand-break frequency is then deduced from the distribution of fragment sizes.

Neutral Filter Elution. Cells whose DNA has been made radioactive are lysed on the top of a filter whose pore size is about 2 μm (remember that the DNA fibre has a diameter of ~25 nm, 100 times smaller). A flow of elution buffer washes DNA fragments through the membrane and the rate of elution of DNA is related to the size of the DNA molecules on the membrane. To measure DSB the elution buffer is at pH 7.4 or 9.6. The use of pH 9.6 increases the sensitivity of the technique but also allows the detection of *alkali-labile sites* that are not actual DSB at physiological pH. Comparison of the two buffers has rarely led to different answers to biological questions. Increasing pH even further to 12.3 leads to denaturation of the two strands of the DNA and allows the measurement of SSB.

Nucleoid Sedimentation Technique. The lysis of cells at neutral pH in the presence of high salt concentration and a non-ionic detergent allows the interphase nucleus to open up and reveal the tangled mass of chromatin. The resulting structures, often called *nucleoids*, consist of supercoiled DNA still retaining attachment to residual protein structures. The sedimentation of these structures in a sucrose gradient is influenced by the induction of SSB which allow the domains to relax and therefore enlarge. One adaptation of this technique, the *halo* method, assesses the expansion of nucleoids by incorporating a fluorescent dye (usually ethidium bromide) into the DNA and measuring the size of halos by microscopy. The concentration of the intercalating dye greatly influences the degree of unwinding of the domains, and the relationship between halo size and dye concentration gives information about the chromatin structure in the nucleoid.

Pulsed-Field Gel Electrophoresis. Fragments of DNA carry a net negative charge and when incorporated into an agarose gel they generally migrate under an electric field at a speed that is inversely related to their size. In order to detect the movement of the DNA this is either made radioactive (in the live cells prior to lysis) or is stained with a DNA-specific fluorescent dye after electrophoresis. The separation of fragments is improved by pulsing the electric field and alternating it, for instance between directions at 30° to the axis of migration. This technique overcomes problems of anomalous movement of large DNA molecules in an electric field so that their separation can be translated into a measure of strand breakage produced by small radiation doses. DNA of known molecular weight (for instance intact yeast chromosomes) is used to calibrate the movement of irradiated DNA in the gels.

Single-Cell Gel Electrophoresis. Cells are embedded in low-density agarose on a microscope slide and then lysed to release DNA. The domain structures unwind and when the preparations are subjected to electrophoresis the broken DNA migrates away from the general mass of DNA. The shape of the structure formed by this migration has led to this assay commonly being referred to as the 'comet' assay. Variations in the lysis conditions allow the measurement of different lesions (*e.g.* alkaline conditions assess SSB while neutral conditions allow the assessment of DSB) and the effect of chromatin structure can also be manipulated to some degree in a manner similar to nucleoid sedimentation. This assay has the advantages of high sensitivity to SSB (though not to DSB) and a requirement for only small numbers of cells.

8.4 Chromosome aberrations

One of the most obvious cytological effects of irradiation is the production of damage to chromosomes. Irradiation induces a dose-dependent delay in the entry of cells into mitosis and when cells that were irradiated while in interphase begin to divide, some of them reveal chromosome aberrations. Whilst the most serious of these will lead to early cell death, some aberrations can be carried through many divisions. Some of the main types of aberration are illustrated in Figure 8.4. They consist of a variety of exchanges and deletions: fragments may be exchanged between chromosomes, between the arms of a single chromosome, or even within a single arm. This can lead to chromosomes in which arm-lengths are abnormal, and also to chromosomes sticking together and forming X or O structures, or to dicentric chromosomes containing two centromeres plus a chromosome fragment.

Aberrations may also be classified as chroma*tid* or chromo*some* aberrations. The irradiation of cells in the G2 phase leads mainly to chroma*tid* damage; radiation damage in G1, if unrepaired, will lead to defects involving both chromatids and thus to chromo*some* aberrations. Irradiation of cells in the S-phase can lead to either type, depending on whether the affected chromosome sites themselves had undergone replication.

What is the relationship between chromosome aberrations and cell viability? Cells do seem to be able to tolerate a variety of structural chromosomal changes and in irradiated individuals some changes persist throughout life. Tumours are characterized by chromosomal instability (both in terms of chromosome structure and number of chromosomes) and irradiation increases the extent of this. But some chromosome changes tend to be lethal. In general terms, the lethal events are those that eventually lead to the loss of a substantial part of the genome. Any rearrangement that leads to a portion of a chromosome lacking a centromere (an *acentric fragment*) will usually lead to its eventual loss from the cell. This may be seen in a subsequent interphase as a *micronucleus*. In diploid cells the formation of a micronucleus signals cell death; in polyploid cells the presence of multiple copies of chromosomes makes such genetic loss less serious. The significance of this manifestation of chromosomal damage has been exploited in the *micronucleus assay* for cell survival (Section 6.4).

THE TECHNIQUE OF PREMATURE CHROMOSOME CONDENSATION

The morphological signs of chromosome damage are only visible when cells undergo mitosis. Cells irradiated early in the cell cycle take some hours to enter mitosis and therefore the chromosome

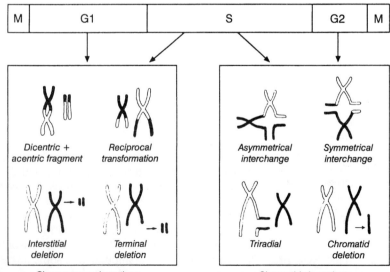

Figure 8.4 Significant types of radiation-induced chromosome aberration. The cartoons show some of the products of damage to a chromosome, or interaction between two chromosomes (one shown black, the other white). Redrawn from Bedford (1991), with permission.

breaks that are observed are those that have failed to rejoin during this period. However, if an interphase cell is fused with a mitotic cell it is found that the interphase cell undergoes a process of *premature chromosome condensation (PCC)* in which its chromosomes become visible. The mitotic cell can be of a different cell type and its chromatin can be labelled with BrUdR so that within the binucleate fusion product it is possible to identify the chromosomes of the target cell. This technique enables breaks in chromatin to be scored within 10–15 minutes of irradiation and the speed of their rejoining can also be determined (Cornforth and Bedford, 1983; Bedford, 1991).

The analysis of both chromosome aberrations and PCC has been greatly facilitated by the development of chromosome-specific lengths of DNA (*i.e.* probes) that can be used in *fluorescence in situ hybridization (FISH)*. In this technique the chromosomes are spread and fixed on microscope slides, then heated so that much of their DNA becomes single-stranded. The specimens are incubated with labelled probe DNA and the probe binds to those regions of the chromosomal DNA with which it is homologous. The bound probe is detected with a fluorescent ligand that binds to the probe and may be seen under fluorescence microscopy. This technique has made the identification of individual chromosomes and translocations between chromosomes much easier.

8.5 Mechanisms by which DNA damage is 'processed'

The use of the word 'processed' reflects the current view that damage induced in DNA is modified by a series of enzymic processes that may lead either to successful repair or to fixation of damage (*i.e.* misrepair). The outcome depends not only on the speed and efficiency of repair but also on the competition between these two processes. The nature of repair processes depends to a large degree on the type of lesion that is being repaired. Where damage involves simply a change in one of the DNA bases, it can be removed by a simple *excision-repair* process. This involves the nicking of the DNA on either side of the lesion, the removal of a few bases around the point of damage, synthesis of new DNA within the damaged region, and finally ligation of the newly synthesized DNA to the orig-

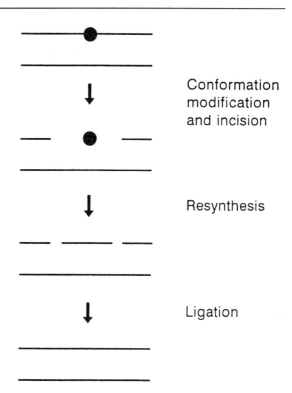

Figure 8.5 Excision-repair of damage in a single strand of DNA. The damage is recognized, the DNA unwound and the DNA strand nicked around the damage site. Following removal of the damaged bases, DNA is resynthesized using the opposite strand as a template and finally the new piece is ligated back into the DNA molecule.

inal DNA strand (Figure 8.5). This process has been well characterized in the repair of damage induced by ultraviolet light and a number of the genes that are involved have been identified.

Repair of other types of lesion is based essentially on this simple scenario but with some variations. Cytotoxic drugs often produce specific lesions (Figure 8.3) and unique enzymes may be required to recognize and remove their particular adducts. Where both strands of the DNA are involved there may be a loss of information either directly as a result of the damage or during the excision stage. In this situation a strand of DNA with a homologous sequence can be used as a template in what is known as *recombination repair*. After these processes are over there may still be some loss of sequence information and this may be one way in which misrepair occurs. The extent of damage may not be sufficient to compromise the viability of the cell but the genetic code has been disrupted and the resulting message translated into RNA and protein may be erroneous (*i.e.* a mutation has occurred). The genetics of these processes are described in the next chapter.

8.6 Cell death

The amount of chromosome damage observed at the first mitosis after irradiation is perhaps the sub-cellular end-point that correlates best with cell killing for most cell types. This has encouraged the view that the major cause of cell death after irradiation is physically aberrant mitosis leading to uneven distribution of chromosomes or loss of chromosome fragments. Such large-scale loss of DNA probably leads to metabolic imbalance that is incompatible with further proliferation.

It has been recognized for many years that some irradiated cells do not enter mitosis before they begin to degenerate. Normal and leukaemic lymphocytes die rapidly in what has been called 'interphase death' and this leads to clinical consequences of a rapid fall in cell numbers after drug or radiation treatment. The phenomenon of *apoptosis* is closely related to interphase cell death. It is characterized by distinctive morphological features that differ from necrosis (Table 8.3). Apoptosis is important in the embryological development of tissues, where it can truly be described as being a 'programmed' process. In recent years apoptosis has been described and studied in a wide range of normal and neoplastic tissues of the adult organism and it has been hypothesized that here too this process may be programmed and therefore potentially controllable for the benefit of cancer therapy. A large number of genes influence the incidence of apoptosis and the associated morphological changes have been identified in cells that have been treated with a number of different cytotoxic agents. However, the link between the incidence of apoptosis and tumour response is somewhat controversial as it is still not clear whether induction of apoptosis significantly affects the reproductive integrity of cells that otherwise would survive treatment.

8.7 The sequence of events that determine radiosensitivity

The complex sequence of processes that follow the initial induction of free radical damage to DNA and may eventually lead to cell death or mutation are illustrated in Figure 8.6. This sequence may be divided into three main sections: *induction*, *processing* and *manifestation*, each of which has been described in the foregoing sections of this chapter.

At what point in this sequence is radiosensitivity determined? There is no single answer to this question. Mammalian cells that are more than usually sensitive to ionizing radiation fall into a number of different categories, including:

(i) stem cells of certain radiosensitive normal tissues (*e.g.* lymphocytes, spermatocytes);
(ii) cells from patients with inherited hypersensitive syndromes (*e.g.* ataxia telangiectasia, A-T);
(iii) radiosensitive tumour types (*e.g.* lymphomas, neuroblastomas);
(iv) radiosensitive mutants of established cell lines (*e.g. xrs* mutants, L5178Y-S).

Two of these categories are described further in Section 9.1. Current evidence suggests that the mechanisms of sensitivity differ within and between these categories.

Table 8.3 Characteristic features of necrosis and apoptosis

Necrosis	Apoptosis
Increase in cell volume	Cell shrinkage
Less stainable nucleus	Condensation of chromatin
General DNA degradation	'Laddering' in electrophoresis gels*
Increased plasma membrane permeability	Plasma membrane blebbing
Decline in protein synthesis	Compaction of organelles and nuclear pores

* *i.e.* endonuclease-mediated DNA digestion producing fragments whose lengths are multiples of a regular unit, derived from cutting in inter-nucleosomal regions.

INITIAL DNA DAMAGE

The radiosensitivity of mammalian cells may to a considerable extent be determined during the *induction* phase (Figure 8.6). Radford (1985, 1986)

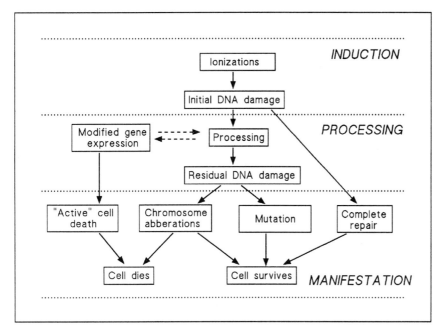

Figure 8.6 The sequence of processes that take place in cells following exposure to ionizing radiation and which may lead eventually to cell death.

measured initial DNA damage in V79 and other cell lines irradiated under a variety of conditions; the cells were irradiated at 4°C and assayed immediately for DSB, thus preventing enzymatic processing. He found a good correlation between initial damage and cellular radiosensitivity. Whitaker et al (1995) performed a similar study on a range of human tumour cell lines (category (iii) above) and the level of damage induction, as detected by pulsed-field gel electrophoresis, was found to relate well to radiosensitivity.

THE RATE, EXTENT, AND FIDELITY OF REPAIR

Many studies have detected a correlation between the rejoining of DNA strand breaks and cell survival. In some cases it is the *speed* of rejoining that appears to be important (Schwartz et al, 1988), in others it is the *residual level* of unrepaired DNA damage. These two parameters may of course be linked. It has also been appreciated that the rejoining of a DSB does not necessarily mean that the function of damaged genes has been restored. Using endonuclease-induced strand breaks in plasmid DNA, it has been possible to probe for the ability of mammalian cells to restore gene function and thus to gain a measure of the *fidelity* of repair of these lesions (Powell and McMillan, 1994). There was evidence that fidelity of repair was lower in more radiosensitive human tumour cell lines.

In addition to the processes directly involved in the induction and repair of damage following irradiation, the following factors can indirectly alter these processes.

Chromatin Structure. DNA in the cell is associated with a variety of proteins, together making up the chromatin. The nature of DNA–protein interactions may influence both the enzymatic repair of DNA lesions and the chemical modification of the radiation-induced free radicals which do the damage. This has been well characterized following UV irradiation where it has been shown that transcribing regions of DNA are repaired more readily than non-transcribing regions. Indeed, it has also been shown that the *transcribing strand* may be more repairable than its opposite DNA strand. With ionizing radiation the primary influence of chromatin structure seems to be at the level of damage induction, where DNA-associated histone proteins have been demonstrated to have a dramatic protective effect against radiation-induced strand breaks.

Membrane Structure. It was suggested some years ago by Dr Tikvah Alper that not all of the effects of radiation can be explained by a simple direct action on DNA itself. She and others have postulated that damage to cell membranes, perhaps at the point of attachment between DNA and the nuclear membrane, may also be important in some situations.

68 DNA damage and cell killing

Inducible Responses to Irradiation. Bacteria sometimes respond to DNA-damaging agents by increasing the expression of proteins that subsequently promote DNA repair. In mammalian cells there is also a limited amount of evidence for this process of *inducible repair*. For example, if a mammalian cell is irradiated and then exposed to irradiated virus, the virus may be repaired to a greater degree than if the mammalian cell had not been pre-irradiated (Jeeves and Rainbow, 1979). Pre-irradiation appears to have induced the repair of the viral DNA damage. The significance of these observations for radiation therapy is still a matter of debate. There is, however, increasing evidence for gene activation by exposure to low doses of ionizing radiation. The genes that are switched on are diverse and include those for growth factors such as the fibroblast growth factor (FGF), nuclear signal transducers and natural cytotoxins like tumour necrosis factor (TNF-α). These genes are potentially involved in many biological responses to radiation, including repopulation, fibrosis, repair and indeed cell killing (see Chapter 9). The process of apoptosis (Section 8.6) may also be classed as an inducible response.

Chromosomal Instability. The principal effects of radiation on cells are seen soon after irradiation as mitotic delay, loss of colony-forming ability and cell death. Recently recognized is the tendency for cells that survive radiation exposure to have a persistently raised level of chromosomal aberrations (Khadim *et al*, 1992), which indicates that radiation can produce long-lasting sublethal effects that are manifested as 'chromosomal instability'. This phenomenon is predominantly seen after high-LET irradiation and it has important implications for the long-term effects of environmental radiation exposure. It may also be linked to the observation that cells irradiated *in vitro* often exhibit a low plating efficiency for many cell generations after irradiation, well beyond the time of initial cell death (Seymour and Mothersill, 1989).

Key points

1. A therapeutic dose of low-LET radiation produces in every cell a large number of ionizations. Some of these directly damage DNA; others generate free radicals that react with DNA. Free radicals are extensively 'scavenged' by SH-compounds. Chemical lesions in DNA are very effectively repaired by enzymatic reactions. Some of them fail to repair and become fixed, and it is these that lead to cell lethality.
2. The death of most cells is associated with certain types of chromosomal aberrations, in particular those aberrations that lead to the loss of a substantial chromosome fragment in the form of a micronucleus.
3. The relative radiosensitivity of a given cell type can be determined at various stages in the post-irradiation sequence of cellular events.

Bibliography

Chapman JD and Gillespie CJ (1981). Radiation-induced events and their time-scale in mammalian cells. *Adv Radiat Biol* 9:143–98.

Cornforth MN and Bedford JS (1983). High-resolution measurement of breaks in prematurely condensed chromosomes by differential staining. *Chromosoma* (Berlin) 88:315–18.

Friedberg EC (1995). *DNA Repair and Mutagenesis.* ASM Press; Washington.

Jeeves WP and Rainbow AJ (1979). Gamma-ray enhanced reactivation of gamma-irradiated adenovirus in human cells. *Biochem Biophys Res Commun* 90:567–74.

Kadhim MA, MacDonald DA, Goodhead DT *et al* (1992). Transmission of chromosomal instability after plutonium α-particle irradiation. *Nature* 355:738–40.

Powell SN and McMillan TJ (1994). The repair fidelity of restriction-enzyme-induced double strand breaks in plasmid DNA correlates with radioresistance in human tumor cell lines. *Int J Rad Oncol Biol Phys* 29:1035–40.

Radford IR (1985). The level of induced DNA double-strand breakage correlates with cell killing after X-irradiation. *Int J Radiat Biol* 48:45–54.

Radford IR (1986). Evidence for a general relationship between the induced level of DNA double-strand breakage and cell killing after X-irradiation of mammalian cells. *Int J Radiat Biol* 49:611–20.

Schwartz JL, Rotmensch J, Giovanazzi S *et al* (1988). Faster repair of DNA double-strand breaks in radioresistant human tumor cells. *Int J Rad Oncol Biol Phys* 15:907–12.

Seymour CB and Mothersill C (1989). Lethal mutations, the survival curve shoulder and split-dose recovery. *Int J Radiat Biol* 56:999–1010.

Ward JF (1986). Mechanisms of DNA repair and their potential modification for radiotherapy. *Int J Radiat Oncol Biol Phys* 12:1027–32.

Warters RL, Hofer KG, Harris CR and Smith JM (1977). Radionuclide toxicity in cultured mammalian cells: elucidation of the primary site of radiation damage. *Curr Top Radiat Res Q* 12:389–407.

Whitaker SJ, Ung YC and McMillan TJ (1995). DNA double strand break induction and rejoining as determinants of human tumour cell radiosensitivity. A pulsed field gel electrophoresis study. *Int J Radiat Biol* 67: 7–18.

Yandell DW, Dryja TP and Little JB (1990). Molecular genetic analysis of recessive mutations at a heterozygous autosomal locus in human cells. *Mutation Research* 229:89–102.

Further Reading

Bedford JS (1991). Sublethal damage, potentially lethal damage, and chromosomal aberrations in mammalian cells exposed to ionising radiations. *Int J Radiat Oncol Biol Phys* 21:1457–69.

Frankenburg-Schwager M (1989). Review of repair kinetics for DNA damage induced in eukaryotic cells in vitro by ionising radiation. *Radiother Oncol* 14:307–20.

Lehmann AR, Hoeijmakers JHJ, van Zeeland AA *et al* (1992). Workshop on DNA repair. *Mutat Res* 273:1–28.

Powell SN and McMillan TJ (1990). DNA damage and repair following treatment with ionizing radiation. *Radiother Oncol* 19:95–108.

Von Sonntag C (1987). *The Chemical Basis of Radiation Biology*. Taylor & Francis; London.

Whitaker SJ, Powell SN and McMillan TJ (1991). Molecular assays of radiation-induced DNA damage. *Eur J Cancer* 27:922–8.

9

Genetic control of the cellular response to ionizing radiation

Trevor J. McMillan

Understanding of the pathways that are important in control of the processing of cellular damage by radiation and the genes involved has expanded greatly during the 1990s. This chapter provides a brief overview of this field.

9.1 Categories of radiation-sensitive cells

Cells that have an unusual response to radiation are an important resource in the study of the genetic control of radiosensitivity. The most productive studies have been those involving sensitive mutants derived from cells established in tissue culture in the laboratory. In addition, cells from human syndromes that demonstrate an abnormal response to radiation have an important role in mechanistic studies as well as in the translation of this research to humans.

LABORATORY-DERIVED SENSITIVE CELL MUTANTS

A large number of radiosensitive mutants have been isolated from cell lines maintained in the laboratory, especially from rodent cell systems (Table 9.1). These have been well characterized phenotypically and our understanding of the genetic changes in these cells has increased greatly in recent years. Among the best characterized are the *xrs* mutants that were isolated by Jeggo and Kemp (1983). These are very sensitive to the killing effects of radiation and this has been related to their much reduced ability to rejoin DNA double-strand breaks (DSB, Figure 9.1). Not all sensitive mutants have such a clear defect in DSB rejoining and in most of them the critical defect has not yet been identified.

HUMAN SYNDROMES WITH AN ABNORMAL RADIATION RESPONSE

Abnormally severe normal-tissue reactions to radiotherapy are not only a practical problem to the radiotherapist but in those cases where this has a genetic basis they are also of great interest from the point of view of the molecular nature of the underlying genetic abnormality (Section 25.3). Some patients who show such a response have been identified as exhibiting the traits of specific inherited syndromes (Table 9.2). It should be stressed, however, that it is only in the case of *ataxia telangiectasia* (A-T) that cells taken from the patient have frequently been found to be more radiosensitive than normal, and even in this case there are examples of clinically diagnosed A-T patients who do not exhibit the laboratory radiosensitivity test of increased chromosomal damage following treatment with ionizing radiation. A-T is an autosomal recessive syndrome which presents clinically as oculocutaneous telangiectasia and progressive cerebellar ataxia. Immunodeficiency and a high frequency of neopla-

Table 9.1 Genetic complementation groups in rodent cell lines

Complementation group	Gene	Examples of cells	Identified defect and gene function
IR1	XRCC1	EM9	Reduced SSB rejoining, protein associated with DNA ligase III function
IR2		irs1	Reduced fidelity of DSB repair
IR3		irsSF	Slow base-damage repair
IR4		XR1	Defective V(D)J recombination
IR5	XRCC5, Ku80	xrs1–6, XR-15B	Reduced DSB rejoining, defective V(D)J recombination, protein binds to DNA ends
IR6	XRCC6, Ku70	sxi-1	Reduced DSB rejoining, defective V(D)J recombination, protein binds to DNA ends
IR7	XRCC7, DNA-PKcs	V3, SCID	Reduced DSB rejoining, defective V(D)J recombination, protein binds to Ku molecules at site of DNA end, triggers signal transduction pathway (?)

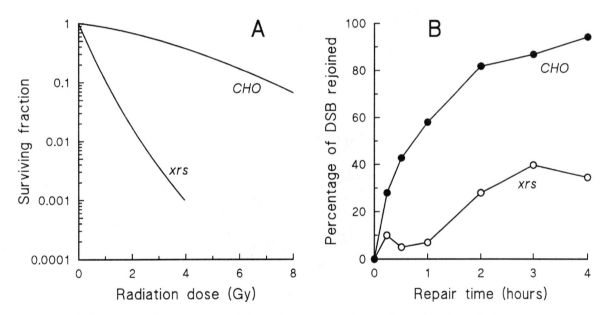

Figure 9.1 Cellular sensitivity and DSB rejoining efficiency of CHO-*xrs* cells. The highly radiation-sensitive *xrs* cells have a clear defect in their ability to rejoin DNA double-strand breaks.

sia are also associated with this disease. An excessive degree of normal-tissue reaction following radiotherapy was the first indication that these patients may have an increased sensitivity to ionizing radiation. This has been confirmed subsequently in the laboratory through experiments on lymphocytes and fibroblasts from A-T patients (Figure 9.2).

The way in which the genetic defect in A-T cells leads to radiosensitivity has yet to be identified conclusively. There is one report of a decrease in DSB rejoining; other studies have indicated that the cells may have a reduced *fidelity* of repair leading to an increased incidence of *misrepair* of double-strand breaks. A-T cells have been found to lack the usual radiation-induced delay in the entry of cells into and through the S-phase of the cell cycle, the so-called G1 delay. This may allow replication of damaged DNA to take place, damage that normally would have had time to repair during the period of

Table 9.2 Some human syndromes in which there have been reports of increased sensitivity to ionizing radiation

Ataxia telangiectasia
Basal cell nevoid syndrome
Cockayne's syndrome
Gardner's syndrome
Fanconi's anaemia
Down's syndrome
Nijmegen breakage syndrome
Usher's syndrome

arrest. The recent cloning of the gene defective in A-T (the *ATM* gene) has given further insight into the function of this gene (see below). Interest in A-T extends beyond the response of these individuals to radiation, as it has been reported that those who are heterozygous at the A-T locus may have an increased susceptibility to cancer. It has been estimated that as many as 15% of breast cancer patients may be A-T heterozygotes.

9.2 Complementation analysis

In view of the many factors involved in the determination of radiosensitivity it is likely that many genes are involved. A first estimate of this number can be obtained by complementation analysis in which two sensitive cells are fused together (Figure 9.3). If the resulting hybrid is also sensitive, then this suggests that the two cell lines are deficient in the same gene. On the other hand, if the hybrid is resistant, then this implies that the two cell lines have different defective genes and therefore that at least two genes are involved in radiation sensitivity. Among rodent cell lines that are sensitive to ionizing radiation several complementation groups have been identified (Table 9.1); three of these groups have been characterized as having a reduced ability to rejoin DNA double-strand breaks. Multiple complementation groups have also been identified in cells from patients with radiosensitive syndromes. The UV-sensitive cells from xeroderma pigmentosum fall into seven complementation groups. Complementation analysis involving cells from individuals with ataxia telangiectasia suggests the presence of four complementation groups, but this is inconsistent with the recent finding of a single gene that is mutated in all A-T individuals tested so far. The reason for this inconsistency is not yet known.

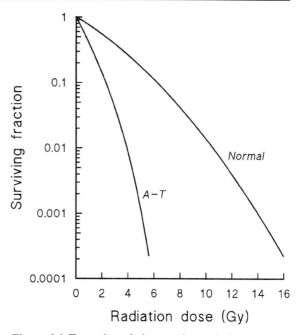

Figure 9.2 Examples of clonogenic survival curves for transformed fibroblasts taken from a normal individual and a patient with *ataxia-telangiectasia* (A-T).

9.3 Stages in the genetic control of radiosensitivity

The scheme of processes that are known to influence radiosensitivity (Figure 8.6) points us towards genes that are important to consider in this context. Each stage is controlled by several genes so that the total number of genes that can have an impact on the radiation response is large, although some may have a more significant influence than others.

CONTROL OF THE INITIAL LEVEL OF DAMAGE INDUCTION

The level of induced damage is markedly influenced by the ability of cells to scavenge the reactive molecules that are produced by ionizing radiation (Section 8.1). This in turn is dictated by the activity of a series of enzymes that can catalyse the conversion or neutralization of reactive species (*e.g.* glutathione-s-transferases, superoxide dismutases, glutathione peroxidase) and by the presence of

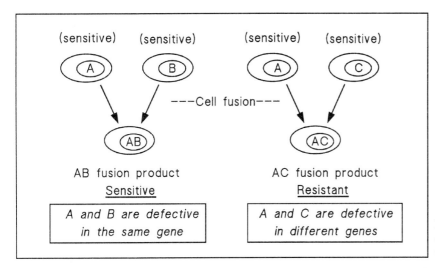

Figure 9.3 The principle of complementation analysis (see text).

thiol-containing compounds that can neutralize these reactive species. Of the enzymes involved, the glutathione-s-transferases are among the most significant since they facilitate the activity of glutathione which is the most abundant non-protein thiol compound in the cell. Chemical manipulation of glutathione has been shown to alter the incidence of DNA damage and also cell survival after irradiation, especially under hypoxic conditions. Whether differences in the sensitivity *between* cells can be explained on the basis of changes in expression of these genes is still not clear.

DNA Repair Genes

A number of genes have been isolated that are involved in the repair of radiation-induced DNA damage. These include glycosylases and endonucleases which can excise a wide range of different modified nucleotide derivatives. However, since it is the DNA double-strand break which appears to be most closely linked to the cell-killing effects of radiation, it is perhaps not surprising that it is the identification of genes that influence the repair of DSB that has led to great interest in this area.

The *xrs* radiosensitive mutants mentioned in Section 9.1 are sensitive because of a defect in DSB rejoining. Through a combination of gene mapping and gene transfer methods the gene defective in these cells has been identified as the one encoding the Ku80 subunit of a DNA-dependent kinase (DNA-PK) that binds to the free DNA ends at the site of a DSB (Taccioli *et al*, 1994). Defects in either of the other two components of this complex also lead to a greatly increased sensitivity to ionizing radiation (Jackson and Jeggo, 1995). One of these, the catalytic subunit of the DNA-PK (DNA-PKcs) is of particular importance because it is the protein that is defective in mice with the 'severe combined immune deficiency' syndrome (SCID). These mice had been recognized as being sensitive to radiation and having a defect in their ability to rejoin radiation-induced DSB. The immunodeficiency in these mice is due to a defect in the process of V(D)J recombination which is the genomic rearrangement process that produces variability in immunoglobulins and T-cell receptor proteins. Hence there is now a strong link between DSB rejoining and V(D)J recombination, although the precise mechanics of that link are not yet defined.

The Control of Apoptosis

Apoptosis is a mode of radiation-induced cell death which is highly cell-type dependent (Section 8.6). Cells from the haemopoietic and lymphoid systems are particularly prone to rapid radiation-induced apoptotic cell death. In most tumour cells mitotic cell death and necrosis are observed at least as frequently as apoptosis. For many radiosensitive cells (*e.g.* stem cells at the base of the crypts of the small intestine) apoptosis appears to be a major mechanism of cell death while other cells, such as the differentiated cells higher in the crypt, do not appear to undergo radiation-induced apoptosis. The high sensitivity of crypt stem cells to apoptosis may be part of an efficient self-screening mechanism against carcinogenesis.

A large number of genes are known to influence the onset of apoptosis and there has been much research into their role in the cellular response to DNA damage. Of particular interest is the gene *TP53* which encodes the tumour suppressor protein, p53. Thymocytes from *p53* null mice (whose cells have no functional copy of *p53*) do not undergo apoptosis after treatment with ionizing radiation, unlike those from mice with normal *p53*. Thus apoptosis after ionizing radiation seems commonly to be a *p53*-dependent process. Consistent with this is a small increase in radioresistance in fibroblasts in which the p53 protein has been made non-functional; however, this is not a totally general finding and the association between *p53* status, apoptosis and radiosensitivity still requires clarification.

There is also a family of genes that includes the *bcl-2* gene that are intimately involved in the control of apoptosis. *Bcl-2* is a suppressor of apoptosis and this is thought to be an important route through which overexpression of *bcl-2* leads to cellular transformation, for instance in the t(14;18) abnormality in non-Hodgkin's lymphomas. There is evidence that the level of *bcl-2* expression can influence the growth of a tumour cell population after irradiation and that this is related to the incidence of apoptosis.

While it is clear that apoptosis is an important mode of cell death after irradiation of some cell types there is still debate as to whether in other systems it may simply be a description of the way a cell dies rather than a significant determinant of *whether* it dies. Evidence from murine tumour systems suggests that in contrast to other forms of rapid apoptosis, apoptosis after ionizing radiation can be somewhat delayed (occurring beyond 24–48 hours). In some cases apoptosis can even be a postmitotic event and this makes the distinction between apoptotic and mitotic cell death very difficult. In a group of experimental cell systems it has been suggested that it is the timing of apoptosis that is the critical factor in radiosensitivity, with sensitive cells showing earlier evidence of apoptosis than resistant cells, but that the genetic manipulation of the susceptibility of cells to undergo apoptosis may not necessarily affect the ultimate clonogenic potential of the cells (Aldridge *et al*, 1995). The signalling pathways leading to apoptosis and in particular the initial apoptotic trigger after irradiation clearly require further investigation.

Cell Cycle Control and Radiosensitivity

Treatment of mammalian cells with ionizing radiation causes delays in the Gl, S and G2 phases of the cell cycle. Early theories suggested that these cell cycle delays allow time for cells to repair DNA damage, thus making them more resistant than they otherwise would be. While this may still be true to some degree, the relationship between cell cycle delay and radiosensitivity is turning out to be more complex as we learn more about the control of the cell cycle. The Gl cell cycle delay seems to be highly dependent on the normal function of the tumour suppressor protein p53. This has been most clearly demonstrated with *p53* null mice which lack a Gl delay. These mice are more prone to radiation-induced carcinogenesis than mice with an active *p53* gene but, as mentioned above, data regarding effects on radiosensitivity are equivocal. At the molecular level cells appear to block in Gl because of a damage-induced increase in p53 protein levels, leading to the activation of downstream cell cycle control proteins, principally the cyclin-dependent kinase inhibitor (CDKI) p21 (alternatively known as cdi-1, sdi-1 or waf-1). CDKI p21 then interacts with other proteins to block replication.

In yeast cells the importance of the G2 delay is demonstrated by the isolation of mutants, such as *rad9*, that are defective in their ionizing-radiation-induced G2 delay. Such mutants are radiosensitive and have been shown to be defective in gene products involved in the monitoring of DNA damage prior to mitosis. In mammalian cells much less is known about the precise control mechanisms of the G2 delay. It has been shown that the addition of low levels of caffeine to cells can reduce the G2 delay, sensitize cells to radiation and increase the levels of cyclin B_1 mRNA, but the direct relationship between these endpoints is unclear. An important case that involves the modification of a G2 delay involves transformation of cells with the oncogenes *H-ras* and *V-myc* (McKenna *et al*, 1992). Transformation with both oncogenes produces cells that are significantly more radioresistant than their parental cell lines and they show a greater G2 delay after ionizing radiation. The protein products of these oncogenes are important components of signal transduction pathways that control cell growth. How such pathways are initiated in response to DNA damage remains to be determined.

9.4 Conservation of DNA repair enzymes in evolution

The molecular biology of repair processes in lower organisms such as yeast and bacteria has been very extensively researched. As well as giving important pointers towards the general principles that are likely to be relevant to mammalian cell repair systems this may prove to be useful in the indentification of mammalian DNA repair genes. DNA repair is such a fundamental aspect of cellular metabolism that there are significant homologies between repair genes in different organisms. This is well demonstrated in genes involved in the repair of UV-induced damage. For example, the protein encoded by the human *ERCC1* gene has significant amino acid similarities to the RAD*10* gene of yeast and the *uvrA* gene from bacteria. Thus the approach of using isolated bacterial and yeast genes to identify homologous human genes is likely to be very productive.

9.5 The ATM gene

Genes associated with ataxia telangiectasia (A-T) were mapped several years ago to chromosome 11q22–23 (Gatti *et al*, 1988) but finer mapping was required to identify a gene that is in that region of chromosome 11 that is mutated in A-T patients (Savitsky *et al*, 1995). This gene has been called the *ATM* (A-T mutated) gene and it appears to be mutated in members of all A-T complementation groups previously identified on the basis of the abnormal suppression of DNA synthesis. Thus it now looks as though there is only a single A-T gene rather than four as previously thought.

It has been found that there is considerable similarity between the predicted amino acid sequence of the ATM protein and a number of other proteins (Table 9.3) and the picture is emerging that the ATM protein may be part of signal transduction pathways involved in many physiological processes. Of particular relevance to radiosensitivity are the similarities to cell cycle controlling genes, a gene that can influence apoptosis, and the catalytic subunit of the DNA-dependent protein kinase (DNA-PKcs) that has been shown to be part of a protein complex that binds to double-strand breaks in DNA (see Section 9.3).

9.6 Epigenetic modulation of radiation sensitivity

Epigenetic processes are those that alter gene expression without modifying the sequence of bases in DNA. Differentiation, for example, is largely an epigenetic process because it involves the switching on and off of specific genes at given times during cellular development. Within an individual not all cells have the same sensitivity. Bone marrow stem cells, for example, are more radiosensitive than fibroblasts or the stem cells of some epithelia. Little is known of what determines these differences but epigenetic processes no doubt play an important role. Some of the determinants of radiosensitivity in normal cells probably

Table 9.3 The predicted ATM protein has amino acid sequences that are similar to those of many other proteins

Gene/protein	Organism	Protein is involved in:
MEI-41	*Drosophila melanogaster*	G2/M cell cycle transition
MEC1	*Saccharomyces cerevisiae*	Cell cycle checkpoint
TEL1	*Saccharomyces cerevisiae*	Controls telomere length
rad3	*S. pombe*	Cell cycle checkpoint
DNA-PKcs	Human	V(D)J recombination, rejoining of DNA strand breaks
TOR1	*Saccharomyces cerevisiae*	Progression from G1
TOR2	*Saccharomyces cerevisiae*	Progression from G1
RAFT1	Rat	Progression from G1
FRAP	Cow	Progression from G1

carry over into the neoplastic cells that derive from them, as evidenced by the radiosensitivity of leukaemias and some lymphomas.

One example in the laboratory where switching off a critical gene leads to an increase in radiosensitivity is in *xrs* mutants of Chinese hamster ovary cells. Treatment of these radiosensitive mutants with 5-azacytidine tends to abolish their increased radiosensitivity (Jeggo and Holliday, 1986). This drug is known to increase gene expression by reducing the methylation status of DNA, and the observed increase in resistance is believed to be due to the switching back on of the *XRCC5 (Ku80)* gene that is defective in these cells.

9.7 Inherited predisposition to cancer and the cellular response to ionizing radiation

It has been recognized for some years that human syndromes which exhibit abnormal sensitivity to cytotoxic agents sometimes have a predisposition to certain types of cancer. For example, patients with xeroderma pigmentosum, who carry a severe sensitivity to UV light, have a high susceptibility to skin cancers. The ionizing-radiation-sensitive syndrome ataxia telangiectasia has an associated high frequency of lymphomas in the homozygotes and A-T heterozygotes are also believed to have a high cancer susceptibility, especially to breast cancer. In some cases the sensitivity to ionizing radiation may be the direct cause of the tumours (Swift, 1994) but it is more likely that the biochemical defect leading to the increased cancer incidence also alters the sensitivity to the cytotoxic agent. Some intriguing data relevant to this have come from the study of ionizing-radiation-induced chromosome aberrations in cells irradiated in the G2 phase of the cell cycle. It has been reported that even when no abnormal sensitivity to the killing effects of ionising radiation is known, a large number of cancer-prone syndromes can be detected by this so-called G2 assay (Parshad *et al*, 1983). These syndromes include xeroderma pigmentosum, Gardner's syndrome and ataxia telangiectasia. Thus the possibility of a close relationship between chromosomal sensitivity to ionizing radiation and cancer predisposition is raised and this deserves considerable experimental attention.

Key points

1. A number of genes have now been identified that can influence the cellular response to radiation. Many of them are genes that are involved in the response of cells to other damaging agents.
2. Various stages in the response of cells to irradiation can be affected by one or more of these radiosensitivity genes.
3. There may be a link between the cellular response to ionizing radiation and cancer predisposition.

Bibliography

Aldridge DR, Arends MJ and Radford IR (1995). Increasing the susceptibility of the rat 208F fibroblast cell line to radiation-induced apoptosis does not alter its clonogenic survival dose-response. *Br J Cancer* 71:571–7

Gatti RA, Berkel L, Boder E *et al* (1988). Localisation of an ataxia telangiectasia gene to chromosome 11q22–23. *Nature* 336:577–80.

Jackson SP and Jeggo PA (1995). DNA double-strand break repair and V(D)J recombination: involvement of DNA-PK. *TIBS* 20:412–15.

Jeggo PA and Holliday R (1986). Azacytidine-induced reactivation of a DNA repair gene in Chinese hamster ovary cells. *Mol Cell Biol* 6:2944–9.

Jeggo, PA and Kemp, LM (1983). X-ray sensitive mutants of Chinese hamster ovary cell line. Isolation and cross sensitivity to other DNA-damaging agents. *Mutation Research* 112:313–27.

McKenna WG, Iliakis G and Muschel RJ (1992). Mechanism of radioresistance in oncogene transfected cell lines. In: *Radiation Research: A Twentieth Century Perspective*, pp. 392–7. (Eds) Dewey WC,

Eddington M, Fry RJM, Hall EJ and Whitmore GF. Academic Press; San Diego.

Parshad R, Sanford KK and Jones CM (1983). Chromatid damage after G2 phase X-irradiation of cells from cancer prone individuals implicates deficiency in DNA repair. *Proc Natl Acad Sci USA* 80:5612–16.

Savitsky K, Bar-Shira A, Gilad S *et al* (1995). A single ataxia telangiectasia gene with a product similar to PI-3 kinase. *Science* 268:1749–53.

Swift M (1994). Ionizing radiation, breast cancer and ataxia-telangectasia. *J Natl Cancer Inst* 21:1571–2.

Taccioli CE, Cottlieb TM, Blunt T *et al* (1994). Ku8O: product of the *XRCCS* gene and its role in DNA repair and V(D)J recombination. *Science* 265:1442–5.

FURTHER READING

Friedberg EC (1995). *DNA Repair and Mutagenesis*. ASM Press; Washington.

Yarnold JR, Stratton M and McMillan TJ (1996). *Molecular Biology for Oncologists*. Second edition. Elsevier Science Publishers; Amsterdam.

10

Dose–response relationships in radiotherapy

Søren M. Bentzen

10.1 Dose–response relationships

Clinical radiobiology is concerned with the relationship between a given physical absorbed dose and the resulting biological response and with the factors that influence this relationship. Although the term tolerance is frequently (mis-)used when discussing radiotherapy toxicity, it is important to realize that there is no dose below which the complication rate is zero: there is no clear-cut limit of tolerance. What is seen in clinical practice is a broad range of doses where the risk of a specific type of radiation reaction increases from 0% towards 100% with increasing dose, i.e. a dose–response relationship.

An *end-point* is a specific event that may or may not have occurred at a given time after irradiation. The idea of dose–response is almost built into our definition of a radiation end-point: to classify a specific biological phenomenon as a radiation effect we would require that this phenomenon be never or rarely seen after zero dose and seen in nearly all cases after very high doses. Exceptions from this rule are certain surrogate end-points that occur at earlier times because of some underlying radiation effect. In experimental animals, an example could be death after thoracic irradiation, which might be used as a surrogate for functional lung damage if it is seen early compared with the normal lifespan of the animal. Reduced kidney function is another example, where increasing doses of radiation shorten the latent time rather than increasing the incidence of the end-point. The concept of dose–response relationships was introduced in Section 1.5 and various ways to characterize normal-tissue end-points are discussed in Chapter 11.

With increasing radiation dose, radiation effects may increase in severity (*i.e.* grade), in frequency (*i.e.* incidence) or both. A plot of, say, stimulated growth hormone secretion after graded doses of prophylactic cranial irradiation in children may reveal a dose dependence, an example of severity increasing with dose. Here we will concentrate on the other type of dose–response relationship: dose–incidence curves. In that example we mean a curve showing the proportion of children with growth hormone secretion below a certain threshold as a function of dose. Thus, the dependent variable in a dose–response plot is an incidence or a probability of response as a function of dose (Figure 10.1).

In this chapter we will introduce some key concepts in the quantitative description of dose–response relationships. Many of these ideas are important in understanding the general principles of radiotherapy. Furthermore, they form the basis of most of the more theoretical considerations in radiotherapy. We will keep the mathematics to a minimum but a few formulae are needed to substantiate the presentation.

Empirical attempts to establish dose–response relationships in the clinic date back to the first decade of radiotherapy. In 1936 the great clinical scientist Holthusen was the first to present a theo-

Shape of the dose–response curve 79

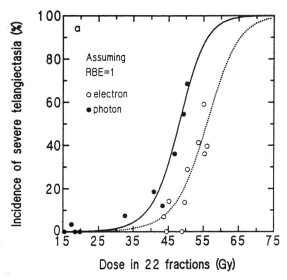

Figure 10.1 Examples of dose–response relationships in clinical radiotherapy. Data are shown on the incidence of severe telangiectasia following electron or photon irradiation. From Bentzen and Overgaard (1991), with permission.

retical analysis of dose–response relationships and this has had a major impact on the conceptual development of radiotherapy optimization. Holthusen demonstrated the sigmoid shape of dose–response curves both for normal-tissue reactions (*i.e.* skin telangiectasia) and local control of skin cancer. He noted the resemblance between these curves and the cumulative distribution functions known from statistics, and this led him to the idea that the dose–response curve simply reflected the variability in clinical radioresponsiveness of individual patients. This remains one of the main hypotheses on the origin of dose–response relationships and this has had a renaissance in recent years with the growing interest in patient-to-patient variability in response to radiotherapy.

10.2 Shape of the dose–response curve

Radiation dose–response curves have a sigmoid (*i.e.* S-) shape, with the incidence of radiation effects tending to zero as dose tends to zero and tending to 100% at very large doses. Many mathematical functions could be devised with these properties, but three standard formulations are used: the *Poisson*, the *logistic* and the *probit* dose–response models. Among these, the first two are most frequently used and we will concentrate on these. The choice between these models for describing and analysing dose–response relationships in radiotherapy is not an easy one. In principle, it is an empirical problem to decide whether one model fits observed data better than the other. In reality, both clinical and experimental dose–response data are generally too noisy to allow statistical discrimination between these models.

THE POISSON DOSE–RESPONSE MODEL

Munro and Gilbert published a landmark paper in 1961 in which they formulated what we could call the *clonogen hypothesis* of tumour control: 'The object of treating a tumour by radiotherapy is to damage every single potentially malignant cell to such an extent that it cannot continue to proliferate.' From this idea and the random nature of cell killing by radiation they derived a mathematical formula for the probability of tumour cure after irradiation of 'a number of tumours each composed of N identical cells'. More precisely, they showed that this probability depends only on the *average* number of clonogens surviving per tumour.

Figure 10.2 shows a Monte Carlo (*i.e.* random number) simulation of the number of surviving clonogens per tumour in a hypothetical sample of 100 tumours with an average number of 0.5 surviving clonogens per tumour. In panel A each tumour is represented by one of the squares in which the figure indicates the actual number of surviving clonogens, these numbers having been generated at random. The cured tumours are those with zero surviving clonogens. In this simulation, there were 62 cured tumours. The relative frequencies of tumours with 0, 1, 2, ... surviving clonogens follow closely a statistical distribution known as the Poisson distribution, as shown in panel B. Many processes involving the counting of random events are (approximately) Poisson distributed, for example the number of decaying atoms per second in a radioactive sample or the number of tumour cells forming colonies in a Petri dish.

When describing tumour cure probability (TCP) it is the probability of zero surviving clonogens in a tumour that is of interest. This is the zero-order term of the Poisson distribution and if λ

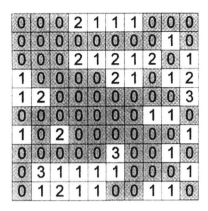
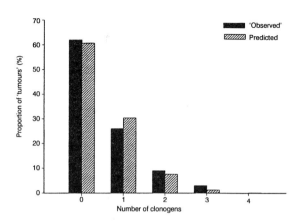

Figure 10.2 Simulation of a Poisson distribution, showing the number of clonogens surviving per tumour in a hypothetical sample of 100 tumours. The average number was 0.5 surviving clonogens/tumour. The histogram shows the proportion of tumours with a given number of surviving clonogens (black bars) and this is compared with the prediction from a Poisson distribution with the same average number of surviving clonogens (hatched bars).

denotes the average number of clonogens per tumour after irradiation this is simply

$$TCP = e^{-\lambda} \quad \text{(Eqn 10.1)}$$

Munro and Gilbert went one step further: they assumed that the average number of surviving clonogenic cells per tumour was a (negative) exponential function of dose. Under these assumptions they obtained the characteristic sigmoid

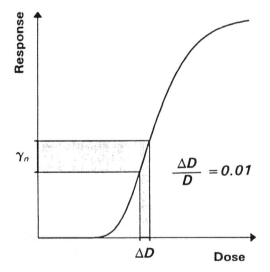

Figure 10.3 Geometrical interpretation of γ. A 1% dose increment (ΔD) from a reference dose D yields an increase in response equal to γ percentage points. From Bentzen (1994), with permission.

dose–response curve (Figure 10.3). Thus the shape of this curve could be explained solely from the random nature of cell killing (or clonogen survival) after irradiation: there was no need to assume variability of sensitivity between tumours.

The Poisson dose–response model derived by Munro and Gilbert has had a strong influence on theoretical radiobiology. The simple exponential dose–survival curve was later replaced by the linear-quadratic model (Section 13.4) and thus we arrive at what could be called the standard model of tumour control

$$TCP = exp[-N_0 \, exp(-\alpha D - \beta D d)] \quad \text{(Eqn 10.2)}$$

Here, N_0 is the number of clonogens per tumour before irradiation and the second exponential is simply the surviving fraction after a dose D given with dose per fraction d according to the linear-quadratic model. Thus when we multiply these two quantities we obtain the (average) number of surviving clonogens per tumour and this is inserted into the Poisson expression in Equation 10.1. N_0 can easily be expressed as a function of tumour volume and the clonogenic cell density (i.e. clonogens per cm³ of tumour tissue). Unfortunately, the clonogenic cell density varies with histopathological differentiation and other patient and tumour characteristics and is seldom known with useful precision. In spite of this, the parameters of the Poisson model have an immediate biological or mechanistic interpretation, and this of course is a great attraction. One drawback is that the parameter estimates will be influenced by biological and dosimetric heterogeneity and therefore cannot usually be regarded as realistic measures of some intrinsic biological property of the tumour (Bentzen et al, 1991; Bentzen, 1992).

The mechanistic interpretation of the Poisson model is most obvious when modelling *tumour* dose–response relationships. For most *normal-tissue* end-points, the biological interpretation of N_0 is not clear. Hendry and Thames (1986) have suggested that the Poisson dose–response model should describe the survival of some hypothetical Tissue Rescuing Units (TRU) and N_0 is then interpreted as the initial number of TRU per volume of normal tissue. Tissue breakdown is assumed to occur when the number of TRU falls below a critical level. However, the biological reality of the TRU is questionable in many organs and tissues and a mechanistic interpretation of the fitted dose–response parameters may therefore be misleading.

The Logistic Dose–Response Model

The logistic model is often introduced and used with more pragmatism than the Poisson model. This model has no simple mechanistic background and consequently the estimated parameters have no simple biological interpretation. Yet it is a convenient and flexible tool for estimating response probabilities after various exposures and is widely used in areas of biology other than radiobiology. The idea of the model is to write the probability of an event (P) as:

$$P = \frac{exp(u)}{1 + exp(u)} \qquad \text{(Eqn 10.3)}$$

where, when analysing data from fractionated radiotherapy, u has the form

$$u = a_0 + a_1 D + a_2 Dd + \ldots \qquad \text{(Eqn 10.4)}$$

Here, D is total dose and d is dose per fraction, and the representation of the effect of dose-fractionation in this way is of course inspired by the linear-quadratic model. Additional terms, representing other patient or treatment characteristics, may be included in the model to see if they have a significant influence on the probability of effect. The coefficients a_0, a_1, . . . are estimated by *logistic regression*, a method that is available in many standard statistical software packages. The parameters a_1 and a_2 play a role similar to the coefficients α and β of the linear-quadratic dose–effect relation-

ship. But note that the mechanistic interpretation is not valid: a_1 is not an estimate of α and a_2 is not an estimate of β. What is preserved is the ratio a_1/a_2, which is an estimate of α/β.

Rearrangement of equation 10.3 yields the expression

$$u = ln\left(\frac{P}{1-P}\right) \qquad \text{(Eqn 10.5)}$$

The ratio of P to $(1 - P)$ is called the *odds* of a response, and the natural logarithm of this is called the *logit* of P. Therefore, logistic regression is sometimes called logit analysis.

10.3 Position of the dose–response curve

Several descriptors are used for the position of the dose–response curve on the radiation dose scale. They all have the unit of dose (Gy) and they specify the dose required for a given level of tumour control or normal-tissue complications. For tumours, the most frequently used position parameter is the TCD$_{50}$, *i.e.* the radiation dose for 50% tumour control. For normal-tissue reactions, the analogous parameter is the radiation dose for 50% response (ED$_{50}$) or in case of rare (severe) complications ED$_5$, that is the dose producing a 5% incidence of complications.

10.4 Quantifying the steepness of dose–response curves

The most convenient way to quantify the steepness of the dose–response curve is by means of the 'γ-value' or, more precisely, the *normalized dose–response gradient* (Brahme, 1984). This measure has a very simple interpretation, namely the increase in response in percentage points for a 1% increase in dose. (Note: an increase in response from, say, 10% to 15% is an increase of 5 percentage points, but a 50% *relative* increase). Figure 10.3 illustrates the definition of γ geometrically.

A more precise definition of γ requires a little mathematics. Let $p(D)$ denote the dose–response function and ΔD a small increment in dose, then the 'loose definition' above may be written

$$\gamma = \frac{p(D+\Delta D) - p(D)}{\Delta D/D \times 100\%} \times 100\% = D\frac{p(D+\Delta D) - p(D)}{\Delta D}$$

The second term on the right-hand side is recognized as a difference-quotient and in the limit where ΔD tends to zero, we arrive at the formal definition of:

$$\gamma = D p'(D)$$

where p' is the derivative of $p(D)$ with respect to dose.

Clinical dose–response curves generally originate from studies where the dose has been changed while keeping either the *dose per fraction* or the *number of fractions* fixed. In the latter case, increasing the dose leads to a simultaneous increase in dose per fraction, and this is associated with an increased biological effect per gray. This is a straightforward consequence of the linear-quadratic model (Section 7.5). Withers (1992) realized that this occurs with dosimetric hot spots in radiotherapy. He called this phenomenon '*double trouble*' because a hot spot would receive not only an increased total dose but also receive that dose with a larger dose per fraction and therefore a greater biological effectiveness than planned (Section 13.8).

10.5 Clinical estimates of the steepness of dose–response curves

Clearly, the value of γ depends on the response level at which it is evaluated: at the bottom or top of the dose–response curve a 1% increase in dose will produce a smaller increment in response than on the steep part of the curve. A compact and convenient way to report the steepness of a dose–response curve is by stating the γ-value at the level of response where the curve attains its maximum steepness: at the 37% response level for the Poisson curve and at the 50% response level for the logistic model. This is a compact measure of steepness because from this single value and a measure of the position of the dose–response curve; the steepness at any other dose can be calculated. Also, it is a very convenient measure in the sense that it is independent of the dose-fractionation details in the case of a dose–response curve generated using a fixed dose per fraction (Bentzen, 1994; Brahme, 1984). In the absence of other sources of variation the maximum steepness of a tumour control curve is fixed by the statistics of survival of clonogenic cells (Figure 10.2), a value of roughly $\gamma_{37} = 7$ (Suit *et al*, 1992). The principal reason why clinical data show shallower dose–response curves than this is inter-patient variability.

Figure 10.4 shows estimates of γ_{37} for head and neck tumours estimated under the assumption of a fixed dose per fraction (Bentzen, 1994). Typical values range from 1.5 to 2.5 with a tendency for vocal cord tumours to be at the upper end of this interval. This probably reflects a lower dosimetric and biological heterogeneity in laryngeal carcinomas. Steepness estimates from dose–response curves for other tumour histologies have been reviewed recently (Okunieff *et al*, 1995), but it should be noted that data for other histologies are generally more sparse than for the head and neck tumours. Also, some estimated values are obviously outliers that cannot be taken as a serious estimate of the steepness of the clinical dose–response curve. These extreme values must be explained by patient selection bias or errors in dosimetry.

Other patient and treatment characteristics will influence both the position and the steepness of the dose–response curve. It can be shown (Brahme, 1984) that the γ_{37} of a Poisson dose–response curve

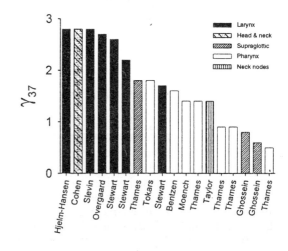

Figure 10.4 Estimated γ_{37}-values from a number of studies on dose–response relationships for squamous cell carcinoma in various sites of the head and neck. From Bentzen (1994), where the original references may be found.

for a fixed dose per fraction depends only on the number of clonogens that have to be sterilized to cure the tumour. As mentioned in Section 10.2, many tumour and treatment variables, for example tumour volume and overall treatment time, are thought to affect the (effective) number of clonogens to be sterilized. Therefore, in a multivariate analysis, γ_{37} will depend on all the significant patient and treatment characteristics.

Figure 10.5 shows a selection of γ_{50}-values for normal-tissue end-points. Estimates are given both for treatment with a fixed dose per fraction and, where possible, also for treatment in a fixed number of fractions, namely 22. The estimates in the latter situation are considerably higher, which is the effect of 'double trouble'. Again, a spectrum of values is seen for the various end-points. The dose–response curves for many late end-points are steeper than for head and neck cancer. An exception is rectosigmoid complications after combined external-beam and intracavitary brachytherapy where a large dose–volume heterogeneity is present due to the steep gradients in the dose-distribution from the intracavitary sources. Also, the lung data arise from a treatment technique where the dose to the lung tissue was heterogeneous. Thus it is likely that dosimetric rather than intrinsic biological factors are the main cause of the relatively low steepness seen for these end-points.

10.6 The therapeutic window

As with any other medical procedure, prescription of a course of radiotherapy must represent a balance between risks and benefits (Section 1.6). The relative position and shape of the dose–response curves for tumour control and a given radiotherapy complication determine the possibility of delivering a sufficient dose with an acceptable level of side-effects. This was nicely illustrated by Holthusen, who plotted dose–response curves for tumour control and complications in the same co-ordinate system for two hypothetical situations: one favourable,

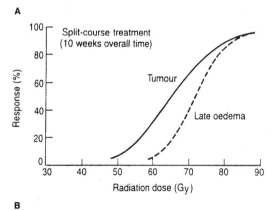

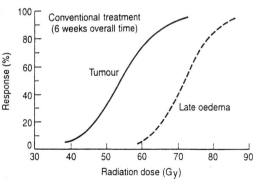

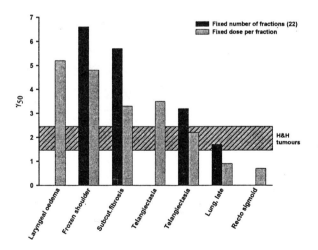

Figure 10.5 Estimated γ_{50}-values for various late normal-tissue end-points. Estimates are shown for treatment with a fixed dose per fraction and a fixed number of fractions. The shaded horizontal band corresponds to the typical γ-values at the point of maximum steepness for dose–response curves in head and neck tumours. Compare with Figure 10.4. Data from Bentzen (1994) and Bentzen and Overgaard (1996), where the original references may be found.

Figure 10.6 Dose–response curves for local control of laryngeal carcinoma and late laryngeal oedema as estimated from the data by Overgaard *et al* (1988). Protraction of overall treatment time narrowed the therapeutic window. From Bentzen and Overgaard (1996), with permission.

that is with a wide *therapeutic window* between the two curves, and the other one less favourable. Figure 10.6 shows an example of how changing treatment parameters may affect the therapeutic window. For split-course treatment (A) the tumour and oedema curves are closer together than for conventional treatment, and the therapeutic window is therefore narrower. In practice, there will be several sequelae of clinical concern and each of these will have its characteristic dose–response curve and will respond differently to treatment modifications. This complicates the simple strategy for optimization suggested by Figure 10.6.

Several parameters are found in the literature for quantifying the effect of treatment modifications on the therapeutic window. Holthusen's proposal was to calculate the probability of *uncomplicated cure*, and this is still used frequently in the literature. The difficulty with this measure is that it gives equal weight to the complication in question and to tumour recurrence, which may often be fatal, and this is against common sense. A simple alternative, which is easy to interpret, is to specify the tumour control probability at isotoxicity with respect to a specific end-point, as illustrated in Figure 1.4.

10.7 Methodological problems in estimating dose–response relationships from clinical data

An increasing number of publications are concerned with the quantitative analysis of clinical radiobiological data. Many methodological problems must be addressed in such a study and these may roughly be grouped as clinical, dosimetric, and statistical.

Clinical aspects include the evaluation of well-defined end-points for tumour and normal-tissue effects. End-points requiring prolonged observation of the patients, such as local tumour control or late complications, should be analysed using actuarial statistical methods. Special concerns exist for evaluation, grading and reporting of normal-tissue injury and these are discussed in some detail in Chapter 11. For dose–response data obtained from non-randomized studies, the reasons for variability in dose should be carefully considered. Subsets of patients treated with low/high doses may not be comparable in terms of other patient characteristics influencing the outcome. An example is where patients receive a lower total dose than prescribed because of their poor general condition, perhaps in combination with severe early reactions, or because of progressive disease during treatment.

Dosimetric aspects involve a detailed account of treatment technique and quality assurance procedures employed. Furthermore, the identification of biologically relevant dosimetric reference points and a proper evaluation of the doses to these points are required.

Statistical aspects include the choice of valid statistical methods that are appropriate for the data type in question and which use the available information in an optimal way. Statistical tests for significance or, preferably, confidence limits on estimated parameters should be specified. When negative findings are reported, an assessment of the *statistical power* of the study should be given. Finally, the censoring (*i.e.* incomplete follow-up) and latency should be allowed for.

For an overview of the quantitative analysis of clinical data, see Bentzen (1993), Bentzen *et al* (1988).

10.8 Clinical implications: modifying the steepness of dose–response curves

The γ-value may be used as a multiplier in converting from a dose change to a change in response. For example, increasing the dose from 64 to 66 Gy in a schedule employing 2 Gy dose per fraction corresponds to a 3.1% increase in dose. If we assume that the γ-value is 1.8, this yields an estimated improvement in local control of $1.8 \times 3.1 \approx 5.6$ percentage points. Because of the sigmoid shape of the dose–response curve this is only a good approximation in a relatively narrow interval around the dose at which γ is specified. For large dose changes this procedure will overestimate the associated change in response.

Similarly, the γ-value may be used as a multiplier for converting an uncertainty in dose into an uncertainty in response. If the standard deviation of the absorbed-dose distribution in a population of patients is ± 5%, a γ-value of 3 would yield an estimated ± 15% standard deviation on the response-probability distribution. Note that in this situation it is generally the γ for a fixed number of fractions that applies. Figure 10.5 shows that the high γ-values at the maximum steepness of the dose–response curve for normal tissues would yield a large variability in response probability for a ± 5% variability in absorbed dose. This provides an indication of the precision required in treatment planning and delivery in radiotherapy.

Another field where the steepness of the dose–response curves for tumours and normal-tissue reactions plays a crucial role is in the design of clinical trials. For a discussion of this topic see Bentzen (1994).

A final issue in this chapter is the prospect for modifying the steepness of the clinical dose–response curve. It was shown by Suit et al (1992) that patient-to-patient variability in tumour biological parameters could strongly affect the steepness of the dose-response curve. A direct illustration of this is obtained from an analysis of local tumour control in patients with oropharyngeal cancers (Bentzen, 1992). Analysing the data with the Poisson model yielded $\gamma_{37} = 1.8$. An analysis taking an assumed variability in tumour cell radiosensitivity into account allowed the dose–response curve to be broken down into a series of very steep curves, each of which would apply to a subpopulation of patients stratified according to intrinsic radiosensitivity (Figure 10.7). Viewing these curves in relation to Figure 10.6, it is clear that some of these sub-

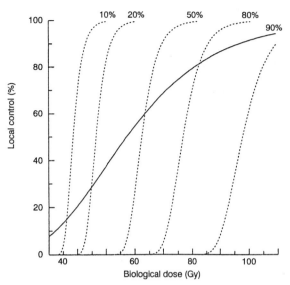

Figure 10.7 Local control of oro-pharyngeal carcinoma as a function of the biologic dose in 2 Gy fractions. Dotted lines are theoretical dose–response curves after stratification for intrinsic radiosensitivity. These represent dose–response relationships from five homogeneous patient populations with radiosensitivity equal to selected percentiles of the radiosensitivity distribution in the total population. From Bentzen (1994), with permission

groups could be expected to have a greater therapeutic window than others. If, by means of a reliable predictive assay, these subgroups could be identified prior to starting therapy, a substantial therapeutic benefit could be realized.

Key points

1. There is no well-defined 'tolerance dose' for radiation complications or 'tumoricidal dose' for local tumour control: rather, the probability of a biological effect rises from 0% to 100% over a range of doses.
2. The steepness of a dose–response curve may be quantified by the γ-value, that is the increase in response in percentage points for a 1% increase in dose.
3. Dose–response curves for late normal-tissue end-points tend to be steeper than dose–response curves for local control of squamous cell carcinomas of the head and neck. For normal tissues, the γ-value at the steepest part of the curve is typically between 2 and 6. For tumours, it ranges from 1.5 to 2.5.
4. The steepness of a dose–response curve is higher if the data are generated by varying the dose while keeping the number of fractions constant ('double trouble') than if the dose per fraction is fixed.
5. Dosimetric and biological heterogeneity cause dose–response curves to be more shallow.

Bibliography

Bentzen SM (1992). Steepness of the clinical dose–control curve and variation in the in vitro radiosensitivity of head and neck squamous cell carcinoma. *Int J Radiat Biol* 61:417–23.

Bentzen SM, Christensen JJ, Overgaard J and Overgaard M (1988). Some methodological problems in estimating radiobiological parameters from clinical data. Alpha/beta ratios and electron RBE for cutaneous reactions in patients treated with postmastectomy radiotherapy. *Acta Oncol* 27:105–16.

Bentzen SM, Johansen LV, Overgaard J and Thames HD (1991). Clinical radiobiology of squamous cell carcinoma of the oropharynx. *Int J Radiat Oncol Biol Phys* 20:1197–206.

Bentzen SM and Overgaard M (1991). Relationship between early and late normal-tissue injury after postmastectomy radiotherapy. *Radiother Oncol* 20:159–65.

Hendry JH and Thames HD (1986). The tissue-rescuing unit. *Br J Radiol* 59: 628–30.

Holthusen H (1936). Erfahrungen über die Verträglichkeitsgrenze für Röntgenstrahlen und deren Nutzanwendung zur Verhütung von Schäden. *Strahlentherapie* 57: 254–69.

Munro TR and Gilbert CW (1961). The relation between tumour lethal doses and the radiosensitivity of tumour cells. *Br J Radiol* 34:246–51.

Okunieff P, Morgan D, Niemierko A and Suit HD (1995). Radiation dose–response of human tumors. *Int J Radiat Oncol Biol Phys* 32:1227–37.

Overgaard J, Hjelm-Hansen M, Johansen LV and Andersen AP (1988). Comparison of conventional and split-course radiotherapy as primary treatment in carcinoma of the larynx. *Acta Oncol* 27:147–52.

Suit HD, Skates S, Taghian A, Okunieff P and Efird JT (1992). Clinical implications of heterogeneity of tumor response to radiation therapy. *Radiother Oncol* 25: 251–60.

Thames HD, Rozell ME, Tucker SL, Ang KK, Fischer DR and Travis EL (1986). Direct analysis of quantal radiation response data. *Int J Radiat Biol* 49:999–1009.

Further Reading

Bentzen SM (1993). Quantitative clinical radiobiology. *Acta Oncol* 32:259–75.

Bentzen SM (1994). Radiobiological considerations in the design of clinical trials. *Radiother Oncol* 32:1–11.

Bentzen SM and Overgaard J (1996). 'Clinical normal-tissue radiobiology'. In: *Current radiation oncology*, pp 37–67. (Eds) Tobias JS and Thomas PRM. Arnold; London.

Brahme A (1984). Dosimetric precision requirements in radiation therapy. *Acta Radiol Oncol* 23:379–91.

Withers HR (1992). Biologic basis of radiation therapy. In: *Principles and practice of radiation oncology* pp 64–96 (Eds) Perez CA and Brady LW. Lippincott; Philadelphia.

11

Clinical manifestations of normal-tissue damage

Søren M. Bentzen and Jens Overgaard

11.1 Documentation of normal-tissue injury is an essential component of radiotherapy research

Radiotherapy is associated with a broad spectrum of normal-tissue reactions and no reporting of the outcome of radiotherapy is satisfactory without a thorough description of the treatment-related morbidity. With an increasing number of long-term survivors after cancer treatment, particularly in childhood and adolescence, long-term side-effects are increasingly important. Systems for tumour staging and tumour response evaluation are generally agreed upon, but there is as yet no widely accepted comprehensive system for classification and grading of normal-tissue injury after radiotherapy. Dische *et al* (1989) surveyed the reporting of radiotherapy trials in leading international journals for the years 1985 and 1988. Over half (in 1985) and one-third (in 1988) of the papers gave no or only anecdotal accounts of treatment-related morbidity. Few papers employed a previously published scoring system. Sismondi *et al* (1989) analysed the reporting of complications of radiotherapy for carcinoma of the uterine cervix in 96 papers published between 1938 and 1986. Only 37 papers used some kind of classification of the complications, and 30 of these employed a grading system. In these 30 papers, 22 different systems were used! This makes a comparison of published complication frequencies virtually impossible.

11.2 Classification of clinical normal-tissue end-points

There are several ways in which normal-tissue end-points may be classified in clinical practice. The choice between them is influenced by clinical and/or biological relevance and by the choice of methods of data recording and analysis. The main classifications are as follows.

SUBJECTIVE VS. OBJECTIVE END-POINTS

Objective end-points are usually preferable in clinical science but they do not provide a complete picture of the patient's situation after therapy. Table 11.1 shows an example where the physician's evaluation of impairment of shoulder movement was compared with the patient's perception of pain.

Table 11.1 Severity of impaired shoulder movement compared with frequency of subjective symptoms

Grade	Pain at movement(%)	Pain* at rest (%)	Reduced working ability (%)
0	6	6	8
1	9	4	28
2	27**	16***	62
3	31**	20***	99

* Defined as moderate or severe (grade 3 only) pain on a scale including no pain and light pain.
** 5% had severe pain.
*** 1.5% had severe pain.
From Bentzen *et al* (1989a).

Whether or not a given grade of objective reaction limits the patient's ability to live a normal life is very important in evaluating treatment-related morbidity. For example, the fact that 27% of patients with a 'moderate' (*i.e.* grade 2) impairment of shoulder movement had pain associated with movement of the shoulder is important information characterizing the patient's condition.

Quality-of-life measures have been used in a few studies as a surrogate end-point for treatment toxicity. This cannot be justified: quality of life is not toxicity and toxicity is not quality of life. Obviously, quality of life end-points are of interest as a characterization of the patient's perception of his or her own situation. However, such measures are not necessarily related in any simple way with objective clinicobiological changes. How the patient actually copes with treatment sequelae is influenced by several cultural and psychosocial factors. Minimization of treatment-related morbidity will always be a goal even if the patient can cope with the sequelae.

The objective evaluation of normal-tissue damage may depend to a considerable extent on the observer. Physical measurements often have minimal observer or operator dependency and should usually be preferred, but the results of such measurements may not be superior to an overall judgement by a physician, taking into account many different aspects of response to radiotherapy.

COSMETIC VS. FUNCTIONAL END-POINTS

Cosmetic changes may constitute a major problem for the patient and must therefore be a concern of the radiotherapist as well. Serious cosmetic changes after radiotherapy may arise from retardation of the growth of muscle and bone in children and adolescents, or visible skin changes on the face or hands. Changes in body image may also lead to secondary problems like sexual dysfunction. Cosmetic end-points have an obvious importance when cosmesis is one of the motivations for changing a treatment schedule, for example in the evaluation of combined tumorectomy and radiotherapy for early breast cancer. Some cosmetic effects rarely trouble the patient, such as telangiectasia on parts of the body normally covered by clothes. In contrast, functional end-points of injury almost inevitably influence the patient's quality of life.

EARLY VS. LATE END-POINTS

A consistent distinction between early responses (*Frühreaktion*) and late responses (*Spätwirkungen*) is found in the German literature of the 1930s. The irreversibility of the late reactions was recognized in the term *Dauerveränderungen* and Holthusen (1936) recommended that late reactions be assessed after a minimum observation time of 2 years.

The distinction between early and late responses came to play a key role with the 1982 paper by Thames and colleagues in which the observation was made that early and late reactions could be classified according to the α/β ratio of the linear-quadratic model (Section 14.2). They interpreted this as reflecting a difference in the shape of the survival curves for the (in some cases hypothetical) underlying target cell populations. Biologically, as well as clinically, this classification is very important as the reactions of early- and late-responding tissues differ in a number of important respects (Table 11.2). These two types of clinical response are defined operationally: any treatment-related morbidity that occurs within the first 90 days of the end of treatment is usually regarded as an early response. The clinical course of the two types of reaction is different, early reactions tending to be transient whereas late reactions are often irreversible. The pathogenesis of radiation effects in various normal tissues is described in Sections 5.2 and 5.3.

GENUINE VS. CONSEQUENTIAL LATE REACTIONS

In general, there is no correlation between the expression of early and late normal-tissue injury in individual patients (for a review see Bentzen and Overgaard, 1996). However, there is accumulating evidence that late reactions in some tissues and organs may occur either as genuine (*i.e.* specific) late reactions or as a consequence of severe early reactions (Bentzen and Overgaard, 1996). Clinical examples of consequential late reactions are the occurrence of bone and soft-tissue necrosis after severe mucosal denudation or the increased risk of developing telangiectasia after moist desquamation. Experimental studies have shown that also in the bladder and rectum severe early reactions increase the risk of subsequent late reactions. Deciding whether a specific late reaction is consequential is generally not possible in individual

Table 11.2 Typical clinical and biological characteristics of early and late radiation reactions

	Early reactions	Late reactions
Latency	<90 days, typically 3–9 weeks. No strong dependence on extent of damage, but greater damage may lead to slower healing of injury	>90 days, typically 0.5–5 years. Shorter latent period with greater damage
Fractionation sensitivity	Low, $\alpha/\beta \approx 10$ Gy	High, $\alpha/\beta \approx 1$–5 Gy
Overall treatment time	Shorter overall time leads to more injury	No significant influence
Clinical course	Transient, but consequential late reactions may occur	Irreversible. Compensatory mechanisms may occur; rehabilitation or treatment for complications may relieve

patients. However, it is important to realize that treatment regimens expected to increase early toxicity may also lead to an increased incidence of late sequelae. One example of this is in accelerated fractionation protocols where preliminary results show an increase in late mucosal reactions after treatment of head and neck tumours; the primary effect may be in early reactions but these may also have a consequential effect on late reactions (Section 14.4).

From a radiobiological point of view, consequential late reactions will 'inherit' some radiobiological characteristics of early reactions: for example sensitivity to changed overall treatment time. Similarly, some sparing from the use of low dose per fraction or low dose rate may be offset in situations where late reactions are consequential.

Binary vs. Graded Responses

Statistically as well as clinically, there is yet another important classification of normal-tissue end-points (Table 11.3). Binary end-points are *all-or-nothing* events, and no grading is possible for these. An example is radiation-induced neoplasia. Binary responses to radiotherapy are often called *stochastic* because it is the *frequency* that varies with radiation dose rather than the *intensity* of the response. A further example is radiation myelopathy, which is a binary-threshold end-point (Withers *et al*, 1988) where damage beyond a critical limit results in an all-or-nothing response.

In contrast, many functional tests produce a *continuum* of responses and these are sometimes called *non-stochastic responses*. In between are the graded response end-points where different outcomes can be ranked in order of increasing severity. However, no numerical relationship can be assumed between the various grades and the reporting of average scores should be avoided as this practice mixes up the severity and the incidence of reactions.

By a process of data reduction it is possible to convert a *continuous* response into a *graded* or even a *binary* response. Continuous responses may be divided into grades, either on the basis of fixed levels of measured response or (as in the case of skin reactions) on the basis of a qualitative judgement of severity (Table 11.5). Graded responses may be converted to an all-or-nothing end-point by defining a binary variable as a response that is above or below a specific threshold value. This is illustrated in Figure 11.1.

11.3 Systems for reporting normal-tissue injury

Several attempts have been made to devise a comprehensive system for grading and reporting of normal-tissue complications (Table 11.4). None of them has so far gained general acceptance, although the RTOG/EORTC Late Morbidity Scoring Criteria and Acute Radiation Morbidity Scoring Criteria have been used in quite a few studies. Recently, the RTOG/EORTC working groups on late effects of normal tissues have jointly proposed a new system for assessment and recording of radiotherapy-related morbidity (Rubin *et al*, 1995; Pavy *et al* 1995), yet at the time of writing this system still awaits validation in clinical studies. One strong warning is needed here. In

90 Clinical manifestations of normal-tissue damage

Table 11.3 Three types of clinical toxicity data

Type of end-point	Statistical data type	Scoring system	Examples
Binary	Categorical; all-or-nothing response	Yes/no	Radiation-induced second tumours
Graded	Ordinal; ranking of severity	e.g. None/mild/moderate/severe	Telangiectasia; subcutaneous fibrosis
Continuous	Continuous	'Laboratory value'	Kidney ^{51}Cr-EDTA clearance; CT density of pulmonary fibrosis

the original publications it is suggested to score several specific toxicity items for an organ and afterwards calculate the average score as an overall measure of toxicity. This could obviously be a

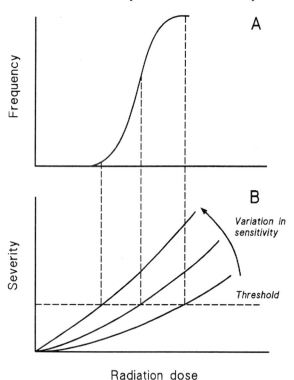

Figure 11.1 The proportion of patients with a reaction above a certain threshold may be treated as a binary variable. The lower panel shows *continuous* dose–response curves corresponding to three different degrees of sensitivity. We can then calculate the frequency of reponses above or below a defined threshold (upper panel). If tissue sensitivity varies among a group of patients, this will lead to a cumulative frequency response that is sigmoid in shape. From Field and Upton (1985), with permission.

very misleading procedure as it might dilute the score of even very serious complications.

The ideal system should be complete, reproducible and sensitive. *Complete* means that it should be possible to record and score any relevant adverse effect of radiotherapy. *Reproducible* means that the inter- and intra-observer variability should be low compared with the variation between patients, *i.e.* that the system can be applied with a reasonable agreement between different observers and that individuals can reasonably reproduce their own scoring. *Sensitive* means that the system should be able to detect changes in treatment intensity, for example in detecting the change in toxicity resulting from dose escalation. Once these aspects of the system are established, the ideal system would be easy to use, would be clinically relevant and would ensure that as much information as possible is available for radiobiological analysis. For example, both the RTOG/EORTC system and the French–Italian Glossary are clinically rather than radiobiologically oriented. In the former, the grading of late skin reactions involves hair loss, telangiectasia and atrophy/ulceration, which biologically are three different end-points, most likely resulting from the depletion of different target-cell populations. An example of part of a system that has proven feasible in practice is given in Table 11.5; it has been incorporated into the European system (Dische *et al*, 1989) and the LENT/SOMA system.

In spite of the problems with these multiple systems, the use of a standard system of reporting is strongly encouraged (Overgaard and Bartelink, 1995). This should not prevent the reporting of supplementary information on toxicity whenever possible. It is to be hoped that the current efforts in developing improved systems for the reporting of toxicity will lead to the realization of a comprehensive system which is widely acceptable both to radiotherapists and biologists.

Table 11.4 Some comprehensive systems for reporting and grading of treatment-related morbidity after radiotherapy

System	Basic features	Limitations	Reference
RTOG/EORTC	Very comprehensive; available for all major organs that may be injured by radiotherapy; proved to be feasible	Mixes various end-points for the same organ; requires judgement by an experienced radiotherapist	Cox et al (1995)
LENT/SOMA	Very comprehensive; scores subjective symptoms, objective signs, and laboratory test results	Not clear how various expressions of damage should be combined into a single grade; needs validation	Rubin et al (1995) Pavy et al (1995)
European	Focuses on end-points rather than organs; attempts to break down scores in specific symptoms, thus allowing retrospective rescoring of grades	Based on previously published systems; still under evaluation	Dische et al (1989)
WHO	Derived from a system used in medical oncology	Focuses on early reactions; not well suited to radiotherapy	Miller et al (1981)
French/Italian	Aimed at treatments for gynaecological cancer; also suitable for surgical complications	Mixes various end-points for the same organ; requires judgement by an experienced therapist	Chassagne et al (1993)
AADK	Aimed at treatments for gynaecological cancer	Mainly based on medical interventions to relieve treatment-related morbidity	Pedersen et al (1993)

11.4 Latent period

Data Recording and Analysis

Late radiation responses in humans occur after latent periods of between 3 months and many years. To obtain a reliable estimate of the incidence of late complications therefore requires an extended period of follow-up (Figure 11.2). Typically a 5-year follow-up is preferable. Patients who die of their disease or from an unrelated cause early after therapy are clearly less likely to have developed late normal-tissue damage than patients who live longer. Similarly, some patients may have incomplete follow-up because they were entered late into the study and only a short period before the data analysis. Such data are said to be *censored*, because the status of the patients after a certain time is unknown. To restrict the analysis to patients who have completed a specified minimum observation time, say 2 years or 5 years, may involve a considerable loss of information. But even more seriously, such a procedure may lead to selection bias, as it is not necessarily a random subset of all patients who would be included

Table 11.5 Definition of early and late cutaneous and subcutaneous reactions

	Erythema	Moist desquamation	Telangiectasia	Subcutaneous fibrosis
Grade 0	None	None	None	None
Grade 1	Mild	<10% of field	<1 per cm^2	Mild (just palpably increased firmness)
Grade 2	Moderate	10–49%	1–4 per cm^2	Moderate (definite, with fixation)
Grade 3	Severe	>50%	>4 per cm^2	Severe (marked, with retraction)

From M. Overgaard et al (1987).

92 Clinical manifestations of normal-tissue damage

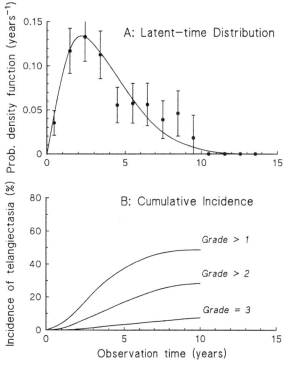

Figure 11.2 A: The latent-time distribution for any grade of telangiectasia as observed in 174 treatment fields with an intermediate probability of developing this reaction (grade ≥ 1). The probability density function may be interpreted as the fraction of patients who express the radiation reaction within a specific year after treatment. Even 9 years after treatment, about 2% of the patients showed the mildest grade of telangiectasia for the first time. **B**: The cumulative incidence of telangiectasia as a function of time for various grades of reaction following 44.4 Gy in 25 fractions. Model calculation based on observation in a total of 401 treatment fields. See also Figure 4.3. From Bentzen *et al* (1990), with permission.

in the analysis. In statistical terminology, data of this type are called *failure-time data*: for each patient we record the time of reaching the end-point or, for patients who were alive without complications when last seen, the time of their last follow-up.

There is an extensive statistical literature on the analysis of failure-time data. Popularly, these methods are often referred to as *survival statistics*, analysed by *actuarial* methods. To provide a full description of this field is far beyond the scope of this chapter but a few key methods should be mentioned. Estimates of the complication rate at a given time may be obtained by the life-table method or the Kaplan-Meier estimate (Machin and Gardner, 1989). In statistical literature the latter is called the product-limit estimate. From these estimates the median latent time can be estimated (Bentzen *et al*, 1990).

Testing for a statistically significant difference between the time-course of complications in two groups may be done by one of the versions of the log-rank test. Finally, multivariate methods have been used for analysing failure-time data in radiobiology: the Cox Proportional Hazards Model (Taylor *et al*, 1987) and the Mixture Model (Bentzen *et al*, 1989c). The former has become very popular in the analysis of survival data and is readily available in many standard statistical computer programs. It has been shown that these methods may provide virtually identical estimates of the incidence of radiation reactions (Taylor and Kim, 1993). However, the mixture model provides an attractive framework for more biologically oriented analyses. For an (admittedly technical!) discussion see Bentzen *et al* (1989c).

CLINICAL IMPORTANCE

Even if normal-tissue reactions are recorded, many studies fail to use proper statistical methods for reporting these data. This is not just a statistical technicality, as two reports using different statistical methods cannot be compared in any meaningful way. Toxicity data are often given in terms of the crude proportion of patients with complications, *i.e.* the number of patients with a specific type and grade of complication divided by the total number of patients treated. As discussed above, this is not a useful characterization of toxicity as it is influenced not only by treatment intensity but also by the life expectancy of the group of patients under study. Table 11.6 shows the difference between crude and actuarial estimates in selected clinical examples. If life expectancy is short, as with the lung cancer study by Hatlevoll *et al*, a treatment regimen may look safe even when it produces a high risk of severe complications among the long-term survivors. Similarly, an increase in treatment intensity may be overlooked if the two groups studied differ in the number of person-years at risk because of different follow-up or a difference in prognosis. Actuarial methods are required to get round these problems. Some authors have advocated the use of alternative statistical methods for the estimation of late effects of cancer treatment.

Table 11.6 Crude and actuarial estimates of complication frequencies

End-point	Primary tumour	Crude estimate	Actuarial	Remarks	Reference
Radiation myelopathy	Lung	4 ± 1% at 3 years	30 ± 15%	Median survival 9 months	Hatlevoll
Marked telangiectasia	Breast	39 ± 6%	62%	Follow-up 1.5–6 years; long latent period	Bentzen
Severe* rectosigmoid complications	Uterine cervix: FIGO IIb FIGO IIIb+IVa	9 ± 5% 15 ± 3%	10 ± 6% at 5 y 39 ± 8% at 5 y	Patients with IIIb and IVa disease received a higher dose but had shorter life expectancy	Unpublished

* ±1 standard error of the estimate.
** requiring treatment.
From Bentzen *et al* (1995), where original references may be found.

For a discussion of these see Bentzen *et al* (1995) and references in that paper. It is strongly recommended that any report on late treatment-related toxicity should include actuarial estimates as a minimum requirement. Any other statistics may be given as a supplement to this.

Biologically, it has been established that higher grades of reactions are on average seen at later times than lower grades, and that increased treatment toxicity may considerably shorten the latent period. Thus, a follow-up at a fixed time after treatment is usually not sufficient. A complete description of the latent-time distribution is required.

11.5 Additional factors influencing normal-tissue damage

Apart from the details of dose fractionation, many other factors can influence the incidence and severity of normal-tissue reactions. These include concomitant treatment with other modalities, irradiated volume and several patient-related factors.

INTERACTION WITH OTHER TREATMENT MODALITIES

Figure 11.3 illustrates the influence of adjuvant chemotherapy on the development of subcutaneous fibrosis following post-mastectomy radiotherapy.

The data have been corrected for latency, and error bars show binomial standard errors on the frequency of responders. Radiotherapy and the first cycles of chemotherapy were given concomitantly. After 44 Gy in 22 fractions the risk of ultimately developing marked subcutaneous fibrosis increased from 19% to 50% when the CMF combination (cyclophosphamide, methotrexate and 5-fluorouracil) was given as an adjuvant to radiotherapy. No such change was seen with cyclophosphamide as a single agent. Figure 20.7C illustrates the time-dependence of gastrointestinal damage in testicular teratoma patients treated with radiotherapy and combination chemotherapy.

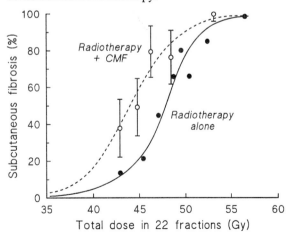

Figure 11.3 Effect of adjuvant chemotherapy on the incidence of radiation-induced subcutaneous fibrosis. (•) post-mastectomy radiotherapy alone; (○) radiotherapy plus adjuvant CMF (cyclophosphamide + methotrexate + 5-fluorouracil). From Bentzen *et al* (1989b), with permission.

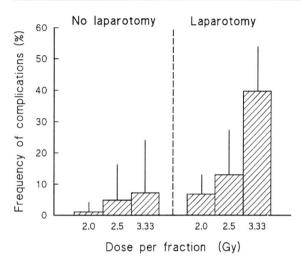

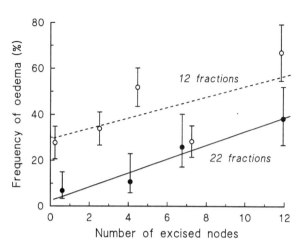

Figure 11.4 The risk of late gastrointestinal injury in patients with supradiaphragmatic stage I or II Hodgkin's disease was increased by previous laparotomy. From Cosset *et al* (1988), with permission.

Figure 11.5 The frequency of arm oedema following post-mastectomy radiotherapy increased in relation to the number of lymph nodes previously excised from the axilla. Adapted from Bentzen *et al* (1989a), with permission.

An example of the influence of laparotomy on radiation-induced gastrointestinal damage is shown in Figure 11.4 (Cosset *et al*, 1988). A total of 345 patients with supradiaphragmatic stage I or II Hodgkin's disease were entered into two EORTC trials. All patients received 39–41 Gy in less than 35 days. Due to the workload in one of the participating institutions, other fractionation schedules than the standard 5 x 2 Gy per week were tried. No dose reduction was implemented to correct for the increase in biological effect from the larger dose fractions. There was a highly significant increase in late gastrointestinal injury with increasing dose per fraction ($p < 0.001$). Also patients having a previous laparotomy had a significantly increased risk of intestinal damage, whatever the dose per fraction ($p < 0.001$).

Figure 11.5. shows the incidence of arm oedema after post-mastectomy radiotherapy as a function of the number of lymph nodes excised from the axilla. Oedema was defined as more than a 2-cm difference in the circumferences of the ipsilateral (relative to the treatment field) and the contralateral arms. The total doses in the 12- and the 22-fraction groups were presumed to be equivalent according to the NSD formula (Chapter 12). In spite of this, large dose fractions considerably increased the incidence of late oedema of the arm. However, the extent of surgery, assessed by the number of lymph nodes excised from the axilla, was also highly significantly associated with increased arm oedema ($p = 0.01$ and $p < 0.001$ in the 12- and 22-fraction groups, respectively).

All of the above examples show how procedures that are not in themselves associated with significant morbidity may 'top up' the radiation-induced damage and make it clinically overt. This phenomenon may seriously confound the radiation dose–response relationship.

THE EFFECT OF TREATMENT VOLUME

Several factors contribute to the clinical volume effect and some of these are exploited in routine radiotherapy. Many organs, *e.g.* kidney, lung, liver, have a large physiological reserve capacity and this means that irradiation of a partial volume even to high doses may have minimal effect on organ function (Section 5.4). Also, patient tolerance may decrease with increasing irradiated volume. For example, confluent mucositis may be more tolerable to the patient if confined to a limited volume. Irradiated volume is thus an important factor in clinical radiotherapy, although more subtle effects of irradiated volume have been difficult to demonstrate convincingly in clinical studies. The available clinical data lack the statistical resolution needed for a critical test of some of the more elaborate mathematical models for volume effects. It is hoped that the technological advances in 3D treatment planning systems will provide more accurate dose–volume data that may in turn be used to establish the dependence of radiotherapy sequelae on dose and volume.

OTHER CO-FACTORS INVOLVED IN EXPRESSION OF RADIATION DAMAGE

Several patient characteristics and in particular various concomitant diseases have been proposed to cause a patient-to-patient variability in the response to radiotherapy (Bentzen and Overgaard, 1994; Baumann, 1995) but a critical review of the literature gives only weak support for the importance of many of these proposed risk factors. Table 11.7 summarizes some of these factors with a rough evaluation of whether the literature at the time of writing provides adequate support for the importance of these. This is definitely a field where more clinical research is needed.

Table 11.7 Some factors that have been proposed to affect normal-tissue reactions after radiotherapy

Factor	Remarks
Host-related	
Age	Growth-related or reduced tissue reserve capacity in the elderly
Haemoglobin level	High haemoglobin level may be associated with an increased risk of NT reactions* (based on a single study)
Smoking	Enhances a number of early and late reactions
Comorbidity	
Primary malignancy	Destruction of tumour tissue may interfere with NT reactions
Diabetes mellitus	Risk factor for retinopathy; doubtful influence for other NT reactions
Collagen vascular disease	Doubtful effect on NT reactions
Hypertension	Doubtful effect on NT reactions
Infection / immunosuppression	Infections may add to RT trauma, especially in immune-suppressed individuals
Genetic syndromes	Several of these are associated with clinical overreaction to radiotherapy**

* Normal-tissue reactions to radiotherapy.
** See Section 9.1.
Adapted from Bentzen and Overgaard (1994).

The study by Johansen *et al* (1988) showed how other medication may substantially affect the expression of radiation injury. The incidence of pharyngocutaneous fistulae after salvage laryngectomy in patients receiving primary radiotherapy increased with increasing total dose. In this situation, fistulae represent an important expression of normal-tissue injury after radiotherapy. The introduction of prophylactic metronidazole before surgery during the time of the study reduced the incidence of this complication from 60% to 13% ($p < 0.01$) in two groups of patients receiving the same radiotherapy schedule.

Age of the patient significantly influences the incidence of normal-tissue damage in a number of clinical situations. Figure 11.6 shows an example of this. Dose–response relationships for the proportion of patients with moderate and severe impairment of shoulder movement after post-mastectomy radiotherapy were documented according to patient age. In the clinical examination of these patients a relative measure of impaired shoulder movement was used and the ability to move the treated and the untreated shoulders was compared in individual patients. There was a significant tendency for the older patients to experience greater impairment.

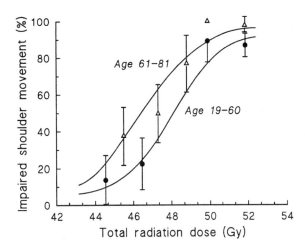

Figure 11.6 Dose–response relationships for moderate and severe impairment of shoulder movement after post-mastectomy radiotherapy in relation to the age of the patient. (△) patients 61–81 years of age; (●) patients 19–60 years of age. All patients received 12 dose fractions. The effect of age is significant ($p = 0.005$). From Bentzen *et al* (1989a), with permission.

> **Key points**
>
> 1. In any study of cancer treatment, the reporting of treatment-related morbidity using a well-defined scoring system is *essential*.
> 2. Previously published and generally used systems should be preferred to *ad hoc* systems.
> 3. Avoid throwing away valuable information. Graded responses should be reported using a clinically relevant grading system. The statistical methods should be chosen to exploit all of the available information.
> 4. Late radiation reactions occur over time-spans of several years. Actuarial statistical methods are needed for the statistically valid analysis of such data.
> 5. Patient characteristics, surgical and drug treatment and other interventions may influence the incidence and severity of radiotherapy sequelae.

BIBLIOGRAPHY

Baumann M (1995). Impact of endogenous and exogenous factors on radiation sequelae. In: *Late Sequelae in Oncology*. (Eds) Dunst J and Sauer R. Springer-Verlag; Berlin.

Bentzen SM, Overgaard M and Thames HD (1989a). Fractionation sensitivity of a clinical end-point: impaired shoulder movement after post-mastectomy radiotherapy. *Int J Radiat Oncol Biol Phys* 17:531–7.

Bentzen SM, Overgaard M, Thames HD *et al* (1989b). Early and late normal-tissue injury after postmastectomy radiotherapy alone or combined with chemotherapy. *Int J Radiat Biol* 56:711–15.

Bentzen SM, Thames HD, Travis EL *et al* (1989c). Direct estimation of latent time for radiation injury in late-responding normal tissues: gut, lung and spinal cord. *Int J Radiat Biol* 55:27–43.

Bentzen SM, Turesson I and Thames HD (1990). Fractionation sensitivity and latency of telangiectasia after postmastectomy radiotherapy. A graded response analysis. *Radiother Oncol* 18:95–106.

Bentzen SM, Vaeth M, Pedersen DE and Overgaard J (1995). Why actuarial estimates should be used in reporting late normal-tissue effects of cancer treatment ... NOW. *Int J Radiat Oncol Biol Phys* 32:1531–4.

Chasssagne D, Sismondi P, Horiot JC *et al* (1993). A glossary for reporting complications of treatment in gynecological cancers. *Radiother Oncol* 26:195–202.

Cosset JM, Henry-Amar M, Burgers JM *et al* (1988). Late radiation injuries of the gastrointestinal tract in the H2 and H5 EORTC Hodgkin's disease trials: emphasis on the role of exploratory laparotomy and fractionation. *Radiother Oncol* 13:61–8.

Cox JD, Stetz J and Pajak TF (1995). Toxicity criteria of the Radiation Therapy Oncology Group (RTOG) and the European Organization for Research and Treatment of Cancer (EORTC). *Int J Radiat Oncol Biol Phys* 31(5):1341–6.

Field SB and Upton AC (1985). Non-stochastic effects: compatibility with present ICRP recommendations. *Int J Radiat Biol* 48:81–94.

Holthusen H (1936). Erfahrungen über die Verträglichkeitsgrenze für Röntgenstrahlen und deren Nutzanwendung zur Verhütung von Schäden. *Strahlentherapie* 57:254–69.

Johansen LV, Overgaard J and Elbrønd O (1988). Pharyngo-cutaneous fistulae after laryngectomy. Influence of previous radiotherapy and prophylactic metronidazole. *Cancer* 61:673–8.

Machin D and Gardner MJ (1989). Calculating confidence intervals for survival time analyses. In: *Statistics with Confidence*, pp. 64–70. (Eds) Gardner MJ and Altman DG. British Medical Journal; London.

Overgaard M, Bentzen SM, Christensen JJ and Hjøllund Madsen E (1987). The value of the NSD formula in equation of acute and late radiation complications in normal tissue following 2 and 5 fractions per week in breast cancer patients treated with postmastectomy radiotherapy. *Radiother Oncol* 9:1–12.

Pavy J-J, Denekamp J, Letchert J *et al* (1995). Late effects toxicity scoring: the SOMA scale. *Radiother Oncol* 35:11–60.

Pedersen D, Bentzen SM and Overgaard J (1993). Reporting radiotherapeutic complications in patients with uterine cervical cancer. The importance of latency and classification system. *Radiother Oncol* 28:134–41.

Rubin P, Constine LS, Fajardo LF, Phillips TL and Wasserman TH (1995). Late effects of normal tissues (LENT) consensus conference. Special issue. *Int J Radiat Oncol Biol Phys* 31:1035–1364.

Sismondi P, Sinistrero G, Zola P et al (1989). Complications of uterine cervix carcinoma treatments: the problem of a uniform classification. *Radiother Oncol* 14:9–17.

Taylor JMG and Kim DK (1993). Statistical models for analysing time-to-occurrence data in radiobiology and radiation oncology. *Int J Radiat Biol* 64:627–40.

Taylor JMG, Withers HR, Vegesna V and Mason K (1987). Fitting the linear-quadratic model using time of occurrence as the end-point for quantal response multifraction experiments. *Int J Radiat Biol* 52:459–68.

Withers HR, Taylor JMG and Maciejewski B (1988). Treatment volume and tissue tolerance. *Int J Radiat Oncol Biol Phys* 14:751–9.

Bentzen SM and Overgaard J (1996). Clinical normal-tissue radiobiology. In: *Current Radiation Oncology*, Vol. 2. (Eds) Tobias JS and Thomas PRM. Arnold; London.

Dische S, Warburton MF, Jones D and Lartigau E (1989). The recording of morbidity related to radiotherapy. *Radiother Oncol* 16:103–8.

Miller AB, Hoogstraten B, Staquet M and Winkler A (1981). Reporting results of cancer treatment. *Cancer* 47:207.

Overgaard J and Bartelink H (1995). About tolerance and quality. An important notice to all radiation oncologists. *Radiother Oncol* 35:1–3.

Further Reading

Bentzen SM and Overgaard J (1994). Patient-to-patient variability in the expression of radiation-induced normal tissue injury. *Sem Rad Oncol* 4: 68–80.

12

Time–dose relationships in radiotherapy

Søren M. Bentzen and Jens Overgaard

12.1 From Strandqvist to Ellis

Early in the history of radiotherapy it became evident that the biological effect of a dose given as '*fraktioniert-protrahiert*' irradiation was less than the effect of the same total dose given as a single treatment. However, in the first half of this century treatments were given daily and there was no clear distinction between the effect of overall treatment time and the number or size of dose fractions. The monograph by Strandqvist (1944) presented the first attempt to establish a mathematical relationship between overall treatment time and response to radiotherapy. Strandqvist plotted recurrences and complications in patients treated for basal and squamous cell carcinomas of the skin and lip as a function of total dose and treatment time. He documented the complications of delayed wound healing and skin necrosis. In 91 patients treated between 1933 and 1937 there were 15 recurrences and 14 complications. Strandqvist's idea was to establish in a dose–time scattergram an exclusion line that would lie below most of the complications and above most of the recurrences. When dose and overall time were plotted on double-logarithmic co-ordinates the exclusion line was drawn as a straight line with a slope of 0.22 (Figure 12.1). Mathematically, the total dose (D) is related to overall treatment time (T) as:

$$D = k \times T^{0.22} \quad \text{(Eqn 12.1)}$$

where k is a constant. The *recovery exponent* of 0.22 corresponds to the slope of the exclusion line in double-logarithmic co-ordinates.

Strandqvist tested this relationship for other data sets as well and concluded that the formula described '*biologisch äkvivalente Gesamtdosen*' or, using modern terminology, isoeffective doses. Note that practically all treatments were single doses or fractionated courses lasting less than 15 days.

Strandqvist defined overall treatment time as the time between the first and the last fraction. With this convention, one problem with the use of logarithmic axes was the representation of single-dose treatments. Strandqvist used 0.35 days as the time for a single treatment (not shown in Figure 12.1) but was fully aware of the fact that this choice substantially influenced the steepness of the exclusion lines.

Cohen (1949) elaborated on Strandqvist's work and analysed three different sets of skin damage data using the type of representation shown in Figure 12.1. The end-points were weak or strong erythema and *skin tolerance*. Cohen found a recovery exponent for skin of 0.33, and concluded that the difference between this value and the 0.22 estimated for squamous cell carcinomas by Strandqvist was 'real and significant'. However, Cohen defined the time for a treatment to be numerically equal to the number of fractions. This meant that the single-dose treatments were plotted at time 1 day instead of the 0.35 days used by Strandqvist, and that two fractions on consecutive days were plotted at time 2 days instead of 1 day and so forth. This change in convention leads to an

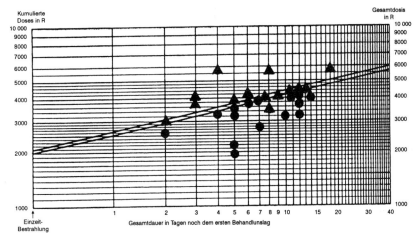

Figure 12.1 The classic data of Strandqvist (1944) on the radiotherapy of skin tumours. The ordinate shows the total radiation dose in roentgens (exposure units); the abscissa gives the overall treatment time in days after the first treatment (Einzeit-Bestrahlung = single dose). Both axes are logarithmic. The circles show recurrences and the triangles show skin complications. Adapted, with permission.

increase in the geometrical slope of the exclusion line, that is, the estimated recovery exponent becomes larger. If the same representation had been used for tumour and skin data, there would have been no significant difference between the two exponents. In other words, the alleged difference was an artefact (Liversage, 1971).

12.2 The Ellis NSD formula

In 1969 Frank Ellis published a paper with the title *'Dose, time and fractionation: a clinical hypothesis'*. Ellis was inspired by the difference in recovery exponent observed by Cohen, also by the experimental studies of Fowler on pig skin which showed that the number of fractions (N) was more important than the overall treatment time (T), at least when T was under 28 days. This led Ellis to the important realization that the 'time factor' was actually a composite effect of N and T. This is in accordance with our present knowledge, although current models emphasize the importance of dose per fraction rather than number of fractions. Ellis based his derivation of an isoeffect relationship on three assumptions, derived from his own earlier work:

(i) The healing of skin epithelium depends on the condition of the underlying connective tissue stroma.
(ii) Apart from bone and brain, connective tissues throughout the body are similar.
(iii) Within and around a malignant tumour, normal connective tissue elements make up the stroma.

'Therefore,' he wrote, 'apart from bone and brain, the tumour dose limited by the normal tissue tolerance dose, could be based on skin tolerance.' This is essentially a two-tissue-type model, distinguishing only between *normal tissues* (excepting brain and bone) and *tumours*. Furthermore, Ellis assumed from (ii) and (iii) above that the N exponent was the same for tumours and normal tissues and that the overall treatment time was of no importance for tumour control by radiotherapy. His argument for the latter was based on the idea that it was homeostatic control in normal tissues that gave rise to the time factor. His further reasoning is remarkable: 'Malignant cells are not susceptible to homeostatic control as are normal cells.... We know that some tumours are hormone sensitive to some extent. In so far as they are sensitive to hormones they are not behaving like malignant cells.'

So the recovery exponent from Strandqvist's work (0.22) was interpreted as the number-of-fractions exponent, and the difference between this value and the 0.33 observed for skin by Cohen was interpreted as a T exponent of 0.11 (remember that T was numerically equivalent to N in Cohen's work). Therefore Ellis suggested the following formula for the total dose, D, at normal-tissue tolerance:

$$D = NSD \times N^{0.22} \times T^{0.11} \qquad \text{(Eqn 12.2)}$$

where NSD, the *nominal standard dose*, is a constant. Later versions of the formula had a number-of-fractions exponent of 0.24 rather than 0.22 to adjust for schedules treating in 5 rather than 6 days per week. The NSD was expressed in units of rad-equivalent-therapy, *i.e.* in *rets*. A similar formula was proposed for tumours, with the same number-of-fractions exponent but omitting the time factor. An unfortunate side-effect of the general acceptance of the NSD formula was that many subsequent papers did not present dose fractionation details in terms of dose, number of fractions and overall treatment time, but only as the nominal standard dose in rets, thus preventing recalculation.

12.3 Variants of the NSD formula

A number of extensions and modifications of the NSD formula have been developed. The *cumulative radiation effect* (CRE) of Kirk *et al* (1971) was mathematically a simple rearrangement of the NSD formula, but more importantly it incorporated the notion of isoeffect at subtolerance doses. Ellis introduced the concept of *partial tolerance,* which has the advantage of being additive for two or more schedules. The disadvantage was that the partial effect value was calculated from clinical experience in each specific institution and therefore could not readily be compared. Orton and Ellis (1973) introduced the TDF (*time-dose-fractionation*) factors, mainly for computational convenience. It is important to realize that both of these formulae are derived from the basic NSD formula, and that therefore the criticisms of the NSD apply to these variations as well.

During the 1970s, the NSD formula and its variants came into widespread use and at the time of writing they continue to be used in many institutions throughout the world. As early as 1971, Liversage pointed to the dangers of a formula in which the therapeutic ratio is independent of the number of fractions so long as they are given in a constant overall time. The social and economic advantages of treating patients in fewer, larger, fractions thus became a temptation in many institutions. It is fair to note that Ellis actually regarded his contribution as an *hypothesis* which he encouraged others to test in the clinic.

12.4 Critique of the NSD formula

Liversage (1971) published a critical analysis of the NSD formula. He started by considering the general case of a power-law model:

$$D = k \times N^m \times T^\tau \qquad \text{(Eqn 12.3)}$$

where m and τ are constants that have to be determined from clinical data. Ellis' two fundamental assumptions were that m is the same for tumours and normal tissues and that τ is zero for tumours. Liversage's basic criticisms were:

(i) Isoeffect curves in the Strandqvist plots are curved, not linear as expected from Equation 12.3 (*see* Figure 13.1). Thus the mathematical form of the power-law model cannot be correct, no matter what the actual values of m and τ.
(ii) The difference in recovery exponents is, as indicated above, an artefact of the different time-scale conventions for single-fraction data.
(iii) m varies from one set of tumour data to another.
(iv) The value of τ varies from one data set to another. For example, Strandqvist's data yields $\tau = 0$.
(v) Two experimental animal studies, known to Liversage, gave substantially different values of τ.

These arguments led Liversage to the conclusion that the NSD formula could not be a valid description for all tumours and normal tissues; he also doubted its validity for squamous cell carcinoma and skin reactions. In hindsight, Liversage's criticism appears to be crucial, but it had a limited penetration at the time. Proponents of the model emphasized its apparent ability to *describe* clinical observations rather than claiming any deeper biological validity, a pragmatic attitude that made the model almost immune to biological criticism. The following 20 years have produced strong evidence from the laboratory and the clinic that the basic biology of the NSD formula is wrong for tumours and for early-responding and late-responding normal tissues, and that power-law models have a mathematical form that is inconsistent with basic radiobiological observations.

Table 12.1 Incidence of early and late reactions following an increase in fraction size. Cervical carcinomas treated with combined intracavitary and external-beam radiotherapy

Regimen*	No. of patients	Early reactions	Late sequelae
40/20/26 + IC +10/5/5	24	8 (33%)	8 (33%)
29/5/28 + IC + 6.7/1/1	24	8 (33%)	20 (83%)

* Dose in Gy/number of fractions/overall time in days.
IC = intracavitary radiotherapy.
From Singh (1978).

The shortcomings of the NSD formula may be listed as set out below.

(i) The NSD formula underestimates the incidence of late sequelae after large dose fractions. Table 12.1 shows the clinical experience of Singh (1978) with intracavitary plus external-beam radiotherapy for carcinoma of the uterine cervix. The intracavitary schedules were identical, delivering 40 Gy to point A of the Manchester system. The two external-beam schedules had an equivalent TDF value. Changing the number of fractions (and therefore the dose per fraction) had no effect on early reactions (33% in each case), but radiotherapy with fewer larger fractions gave a significant increase in late sequelae (33% → 83%, $p = 0.001$).

Figure 12.2 shows one example of evidence for dissociation between *early* and *late* tissue responses as a result of altered dose per fraction (Overgaard et al, 1987). It shows the relationship between acute skin reaction (erythema) and late subcutaneous fibrosis in 73 patients treated with 12 fractions, and 66 with 22 fractions. Each dot represents data for an individual patient. The total doses were equivalent on the basis of the NSD formula. The incidence of grade 3 erythema was similar (35% compared with 31%) but the incidence of marked subcutaneous fibrosis (grade ≥ 2) was much higher in the 12-fraction schedule (68% compared with 5%). Note that some patients developed late complications without a preceding early complication. Results of this type revealed shortcomings in the NSD approach and led to the more recent preference for the linear-quadratic model. These data represent a breakdown in the first two of Ellis' assumptions listed in Section 12.2. But once again it should be stressed that even if the exponents of the power-law model are allowed to vary from one tissue to another, this type of model has a mathematical form that is inconsistent with biological knowledge.

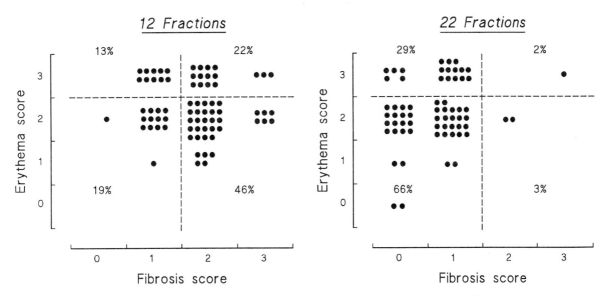

Figure 12.2 Relationship between acute skin reaction (erythema) and late subcutaneous fibrosis in breast cancer patients treated with an equivalent total dose according to the NSD formula but with very different fraction numbers. From M. Overgaard et al (1987), with permission.

Table 12.2 Lack of a time factor for late clinical end-points

End-point	Result	Reference
Telangiectasia	No recovery after a '3-week split'	Turesson and Thames (1989)
Late laryngeal oedema	$D_{rec} = 0.10 \pm 0.22$ Gy/day	Overgaard et al (1988)
Fistula secondary to salvage surgery	$D_{rec} = 0.09$ Gy/day	Overgaard et al (1988)
Rectosigmoid complications	$D_{rec} < 0.15$ Gy/day	Bentzen et al (1992)
Parotid gland function	$D_{rec} \approx 0$ Gy/day	Leslie and Dische (1991)
Bladder	Reduction of treatment time to 4 weeks gave no increase in bladder toxicity	Moonen (1994)

* D_{rec} = dose recovered per day, with 1 standard error where available.
From Bentzen and Overgaard (1996); for references, see the original.

(ii) There is no appreciable treatment-time factor for late sequelae. Several studies have looked for an effect of overall treatment time on the occurrence of late sequelae. Table 12.2 compiles the findings of some of these studies in terms of the dose recovered per day due to proliferation. None of these studies found a statistically significant effect of overall treatment time. One caveat exists after treatment with very high early toxicity. In this case, there is emerging evidence, at least for mucosal reactions in the gastrointestinal tract, that the level of late sequelae may be enhanced as a consequence of the early injury (see Section 11.2).

(iii) Prolonged overall treatment time gives poorer local tumour control in squamous cell carcinoma. One clinical example in support of this conclusion is shown in Table 12.3. It summarizes the results of three studies in head and neck cancer where split-course radiotherapy was instituted as a treatment policy for a period of time. The estimated loss of local tumour control was 7–10%. Other estimates of prolonged treatment are found in the literature but in most of these biases cannot be ruled out in the sense that patients with poor prognosis tended to have had prolonged treatment (Bentzen, 1993). The dose recovered per day in squamous cell carcinoma of the head and neck has typically been estimated in the range 0.5–0.7 Gy/day. More extensive data are shown in Figure 14.4.

(iv) The number-of-fractions exponent (m) is not constant, even for a specific end-point. Support for this conclusion comes mainly from radiobiological data on experimental animals. Figure 12.3 (Fowler, 1984) shows isoeffect curves for radiation damage

Table 12.3 The estimated fall in tumour control probability for 1 week extra overall treatment time in squamous cell carcinoma of the head and neck

Tumour site	Authors	Estimated loss per week (%)
Larynx	Overgaard et al (1988)	10
Oropharynx	Bentzen et al (1991)	7
Various head and neck cancers	Parsons et al (1980)	7

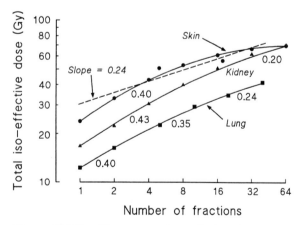

Figure 12.3 Isoeffect curves for radiation damage to mouse kidney, skin and lung. Local values of the number-of-fractions exponent (*m*) are shown underneath the curves. From Fowler (1984), with permission.

to mouse kidney, skin (*i.e.* desquamation) and lung (*i.e.* pneumonitis). The NSD formula predicts the relationship to be a straight line with slope 0.24 (the dashed line). In each case the data define a bending curve. Local values of the number-of-fractions exponent (m) are shown underneath the curves. The linear-quadratic model predicts that m increases with dose per fraction (and decreases with increasing fraction number) to a limiting value of 0.5.

The dependence of the isoeffective total dose on number of fractions has been found in a very large collection of normal-tissue data to correspond with the predictions of the LQ model (Thames and Hendry, 1987). Turesson and Notter (1975) have made a similar observation for telangiectasia after clinical post-operative radiotherapy. A correction was needed for the isoeffective dose estimated from the NSD formula. This again can be interpreted as an increasing value of m with increasing dose per fraction. Thus, a power-law model cannot describe the outcome of radiotherapy for late radiation reactions, no matter what value is assigned to m.

(v) A power-law model for the effect of overall treatment time has the wrong mathematical form. A power-law relationship of the form T^τ (with $\tau < 1$) predicts a *decreasing* dose compensation for proliferation as the duration of treatment increases. This is contrary to biological data for both early-responding tissues and tumours (Withers and Peters, 1980; Figure 13.8).

12.5 The cell population kinetics (CPK) model of Cohen

The CPK model of Cohen (1971) represented a new point of view, completely different from the empirical Strandqvist-NSD approach. Many of the ideas incorporated in the CPK model still play a role in more recent approaches to time–dose relationships. Cohen's biological starting-point was the target-cell hypothesis: that radiation reactions in human tissues depend on the depletion of some critical cell population below a certain threshold level. Cell survival was assumed to be described by the two-component model (Section 7.4), but Cohen subsequently substituted the linear-quadratic model. Cohen also included the kinetics of cellular renewal in his model. In its basic formulation the CPK model employed seven parameters and a brief list of these provides a feeling for the biological effects that Cohen attempted to model. The two-component model contained three parameters (D_0, D_1, and n). The other four parameters were the growth rate of the target cell population, the number of available cell cycles in the tissue (a factor limiting repopulation), the critical threshold for cell depletion, and a field-size correction for both tumours and normal tissues. The field-size effect was intended as a correction for the number of target cells in the tumour and as a representation of the (possibly) reduced normal-tissue tolerance for larger field sizes.

The model parameters were to be estimated from clinical dose–response data and were assumed to be specific for the different normal tissues or tumour types. Cohen took full advantage of the computers that became generally available in the 1960s. He and his collaborators wrote a number of computer programs for obtaining best-fit values of the parameters in the model and for predicting response to radiotherapy for given treatment and tissue characteristics.

Some aspects of Cohen's model are inconsistent with biological observations, for example that no delay occurs before the onset of regeneration in normal tissues. But more seriously, and Cohen himself was aware of this, very few clinical data sets can provide sufficient information to allow the parameters of the CPK model to be estimated. A considerable variation in treatment characteristics is required, besides sufficiently accurate clinical observations of response to radiotherapy. For ethical reasons it is doubtful whether data of this kind can be obtained from current radiotherapy practices.

Cohen's model provided arguments in favour of split-course radiotherapy, *i.e.* fractionation schedules including typically a 2- to 4-week treatment-free interval. The basic rationale was that proliferation during the treatment gap would improve normal-tissue tolerance and thereby allow the tumour dose to be escalated. At the Department of Oncology in Aarhus, a 3-week split was introduced and the total dose was raised by 10–12 Gy. The expectation from the CPK model was that tumour control would improve and that the incidence of early and late normal-tissue complications would decrease. The clinical observation was that tumour cell proliferation during the gap

104 Time–dose relationships in radiotherapy

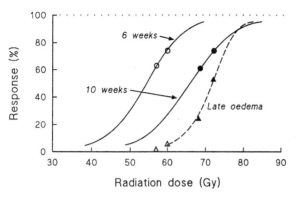

Figure 12.4 The influence of overall treatment time on the dose–response relationship for local control of squamous cell carcinoma of the larynx (circles) and late oedema (triangles). Open symbols show data for continuous-course radiotherapy (nominal duration 6 weeks); closed symbols show data for split-course radiotherapy (nominal duration 10 weeks). Proliferation during the additional 4 weeks shifted the dose–response curve for local control towards higher doses. In contrast, the data for late oedema fit a single dose–response curve independently of treatment time, implying that the split had no effect. Adapted from Overgaard et al (1988), with permission.

completely offset the gain in tumour control expected from the higher dose, and late sequelae increased (Figure 12.4).

In the early 1980s, the linear-quadratic model came into widespread use for calculating isoeffect relationships. This model was deliberately kept simple in order to describe quantitatively the clinically important effect of dose per fraction. The relative simplicity of this model matched the biological resolution of clinical and experimental animal studies. That is the subject of the next chapter.

Key points

1. The NSD formula and its derivatives may underestimate the incidence of late normal-tissue sequelae after radiotherapy.
2. Even if the exponents of the NSD formula are determined for each specific normal tissue and tumour type, the model incorrectly describes the effect of number of fractions or treatment duration.
3. Overall treatment time is important for tumour control after radiotherapy, at least for squamous cell carcinomas.
4. There is no appreciable proliferation in *late-responding* normal tissues during a standard 6- to 10-week course of radiotherapy.
5. The sensitivity to large fraction sizes is much higher for late-responding than for early-responding normal tissues.
6. It is possible to construct models of response to radiotherapy that contain many known biological processes but these inevitably become too complex to be defined by the available clinical data; hence the attraction of the linear-quadratic approach.

Bibliography

Bentzen SM (1993). Time-dose relationships for human tumors: estimation from non-randomized studies. In: *Current Topics in Clinical Radiobiology of Tumors*, pp. 11–26. (Ed) Beck-Bornholdt HP. Springer-Verlag; Berlin–Heidelberg.

Bentzen SM, Johansen LV, Overgaard J and Thames HD (1991). Clinical radiobiology of squamous cell carcinoma of the oropharynx. *Int J Radiat Oncol Biol Phys* 20:1197–1206.

Bentzen SM and Overgaard J (1996). Clinical normal-tissue radiobiology. In: *Current Radiation Oncology*, Vol. 2. (Eds) Tobias JS and Thomas PR. Arnold; London.

Cohen L (1949). Clinical radiation dosage, Pt.II. *Br J Radiol* 22:706–13.

Cohen L (1971). A cell population kinetic model for fractionated radiation therapy. I. Normal tissues. *Radiology* 101:419–27.

Ellis F (1969). Dose, time and fractionation: a clinical hypothesis. *Clin Radiol* 20:1–7.

Fowler JF (1984). Review: Total doses in fractionated radiotherapy – implications of new radiobiological data. *Int J Radiat Biol* 46:103–20.

Kirk J, Gray WM and Watson ER (1971). Cumulative radiation effect. Part I. Fractionated treatment regimes. *Clin Radiol* 22:145–55.

Liversage WE (1971). A critical look at the ret. *Br J Radiol* 44:91–100.

Orton CG and Ellis F (1973). A simplification in the use of the NSD concept in practical radiotherapy. *Br J Radiol* 46:529–37.

Overgaard M, Bentzen SM, Christensen JJ and Hjoellund Madsen E (1987). The value of the NSD formula in equation of acute and late radiation complications in normal tissue following 2 and 5 fractions per week in breast cancer patients treated with postmastectomy radiotherapy. *Radiother Oncol* 9:1–12.

Overgaard J, Hjelm-Hansen M, Johansen LV and Andersen AP (1988). Comparison of conventional and split-course radiotherapy as primary treatment in carcinoma of the larynx. *Acta Oncol* 27:147–52.

Parsons JT, Bova FJ and Million RR (1980). A re-evaluation of split-course technique for squamous cell carcinoma of the head and neck. *Int J Radiat Oncol Biol Phys* 6:1645–52.

Singh K (1978). Two regimes with the same TDF but differing morbidity used in the treatment of stage III carcinoma of the cervix. *Br J Radiol* 51:357–62.

Strandqvist M (1944). Studien über die kumulative Wirkung der Röntgenstrahlen bei Fraktionierung. *Acta Radiol* 55(Suppl.):1–300.

Turesson I and Notter G (1975). Skin reactions after different fractionation schedules giving the same cumulative radiation effect. *Acta Radiologica Therap* 14:475–84.

Further Reading

Bentzen SM and Thames HD (1995). A 100-year Nordic perspective on the dose-time problem in radiobiology. *Acta Oncol* 34: 1031–40.

Thames HD and Hendry JH (1987). *Fractionation in Radiotherapy*. Taylor & Francis. London.

Withers HR and Peters LJ (1980). Biological aspects of radiation therapy. In: *Textbook of Radiotherapy*. (Ed) Fletcher GH. Lea & Febiger; Philadelphia.

13

The linear-quadratic approach to fractionation and calculation of isoeffect relationships

Michael C. Joiner and Albert J. van der Kogel

13.1 Introduction

The linear-quadratic (LQ) cell survival model, introduced in Chapter 7, can also be used to describe the relationship between total isoeffective dose and the dose per fraction in fractionated radiotherapy. This is one of the most important recent developments in radiobiology applied to therapy. We describe here the theoretical background and the animal data that have led to the adoption of the LQ approach and we indicate how calculations can be made using this model.

13.2 Linear-quadratic vs. NSD

Experiments using animal models carried out over many years have shown that if isoeffective total dose (or tolerance dose) is measured as a function of decreasing dose per fraction or increasing number of fractions, then this relationship is steeper for late-responding than for early-responding normal tissues (Section 14.2). Figure 13.1 illustrates this for some studies on mice: skin is an early-responding tissue and kidney a late-responding tissue. The charts are isoeffect plots (the total radiation dose to give a fixed level of damage is plotted against dose per fraction and fraction number). Note that the curve for kidney is steeper than that for skin.

The solid lines in Figure 13.1 are calculated by an equation based on the LQ model:

$$\text{Total dose} = \frac{\text{constant}}{1 + d/(\alpha/\beta)} \quad \text{(Eqn 13.1)}$$

where d is the dose per fraction. See Section 13.4 for the derivation of this equation. The steepness and curvature of these lines are both determined by one parameter: the α/β ratio. For the skin data (Figure 13.1A), α/β is about 10. The units of α/β are grays, so the α/β ratio in this case is 10 Gy. For the kidney data α/β is about 2 Gy.

The LQ model fits these data very well. Also shown in Figure 13.1A is the fit of Ellis' Nominal Standard Dose model (Section 12.2) to the skin data. The equation is:

$$\text{Total dose} = \text{NSD} \cdot N^{0.24} \cdot T^{0.11}$$

where T is constant in this case. As indicated in the previous chapter, this gives a straight line in this type of plot. The NSD equation can be made to fit the data well from 4 to 32 fractions but is not as accurate as the LQ model for large doses per fraction or, more importantly, for doses per fraction of 1 – 2 Gy that are of current interest in hyperfractionation (Section 14.2). For late reactions, as illustrated by the kidney data (Figure 13.1B), the NSD formula again does not fit as well as the LQ formula and overestimates tolerance doses when the

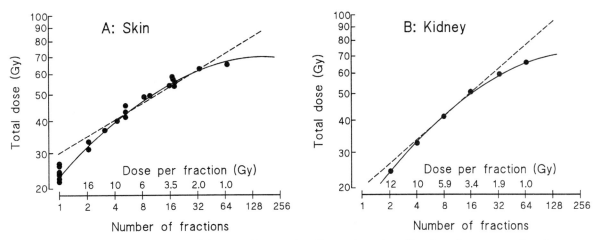

Figure 13.1 Relationship between *total dose* to achieve an isoeffect and *number of fractions*. **A**: acute reactions in mouse skin (Douglas and Fowler, 1976). **B**: late injury in mouse kidney (Stewart et al, 1984). Note that the relationship for kidney is steeper than that for skin. The broken lines are NSD formulae fitted to the central part of each data set. The solid lines show the LQ model, from which the guide to the dose per fraction has been calculated. Reproduced with permission.

dose per fraction is below 2 Gy. In addition, the N exponent needs to be raised from 0.24 to 0.35. A similar modification, but not necessarily by the same amount, must be made for all late-responding tissues if the NSD formulation is to be even approximately correct. The conclusion is that it is always better to use the linear-quadratic model, with correctly chosen α/β ratio, to describe isoeffect dose relationships, especially when considering doses per fraction below 2 Gy.

13.3 Cell-survival basis of the LQ model

What is the explanation for the difference between the fractionation response of early- and late-responding tissues? Figure 13.2 shows hypothetical single-dose (one-fraction) survival curves for the target cells in early- and late-responding tissues, drawn according to the LQ equation (see Figure 7.2B). E represents the reduction in cell survival equivalent to tissue tolerance. The total doses that would need to be given in two fractions are obtained by drawing a straight line from the origin through the survival curve at $E/2$ and measuring the intersection of this line with the dose axis. The total dose for three fractions is obtained in the same way by drawing a line through $E/3$ on the survival curve, and similarly for the other fraction numbers. Because the late-responding survival curve (Figure 13.2B) is more 'bendy' (it has a lower α/β ratio), there is a *larger* change in total dose with increasing number of fractions compared with the early-responding tissue where the survival curve bends less.

13.4 The LQ model in detail

The surviving fraction (SF_d) of target cells after a dose per fraction d is given in Chapter 7 as:

$$SF_d = \exp(-\alpha d - \beta d^2)$$

Radiobiological studies have shown that each successive fraction in a series is equally effective, so the effect (E) of n fractions can be expressed as:

$$E = -\log_e(SF_d)^n = -n\log_e(SF_d)$$
$$= n(\alpha d + \beta d^2) = \alpha D + \beta dD$$

where the total radiation dose $D = nd$. This equation may be rearranged into the following forms:

$$1/D = (\alpha/E) + (\beta/E)\,d \qquad \text{(Eqn 13.2)}$$
$$1/n = (\alpha/E)\,d + (\beta/E)\,d^2 \qquad \text{(Eqn 13.3)}$$
$$D = (E/\alpha)\,/\,[1 + d\,/\,(\alpha/\beta)] \qquad \text{(Eqn 13.4)}$$

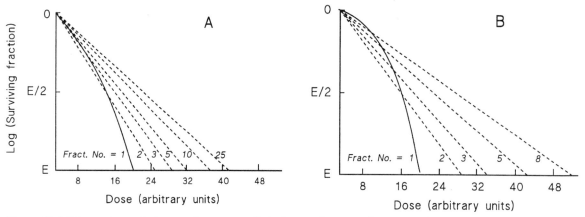

Figure 13.2 Schematic survival curves for target cells in **A**: acutely responding and **B**: late-responding normal tissues. The abscissa is radiation dose on an arbitrary scale. From Thames and Hendry (1987), with permission.

$$\frac{D_2}{D_1} = \frac{d_1 + (\alpha/\beta)}{d_2 + (\alpha/\beta)} \quad \text{(Eqn 13.5)}$$

Figure 13.3 shows the results of an experiment in which a measure of functional damage in the kidney is plotted against the total dose given in fractionated schedules of 1–64 fractions. To apply the LQ model to this example, first we measure off

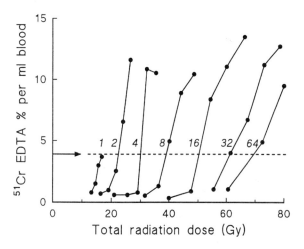

Figure 13.3 Dose–response curves for late damage to the mouse kidney with fractionated radiation exposure. Damage is indicated by EDTA clearance, curves determined for 1 to 64 dose fractions, illustrating the sparing effect of increased fractionation. From Stewart *et al* (1984), with permission.

from the graph the total doses at a fixed level of effect (arrowed) and then plot the reciprocal of these total doses against the corresponding dose per fraction. Equation 13.2 shows that this should give a straight line whose slope is β/E and whose intercept on the vertical axis is α/E. This plot is shown in Figure 13.4A. The points fit well to a straight line. This line cuts the X-axis at −3 Gy; it can be seen from Equation 13.2 that this is equal to $-\alpha/\beta$, thus providing a measure of the α/β ratio for these data. The relative contributions of α and β to the α/β ratio can be judged by comparing the reciprocal total dose intercept (α/E) and the slope of the line (β/E).

An alternative way of deriving parameter values from these data is to plot the reciprocal of the number of fractions against the dose per fraction as suggested by Equation 13.3. Figure 13.4B shows that this gives the shape of the target-cell survival curve with the y-axis proportional to $-\log(SF_d)$. [*Statistical note*: This method combined with non-linear least-squares curve fitting is preferred over the linear-regression method shown in Figure 13.4A for determining α/β, because the $1/n$ and dose-per-fraction axes are independent.]

Equations 13.4 and 13.5 are further ways of expressing the linear-quadratic model. Equation 13.4 shows the LQ model in the form used already to describe the relationship between total dose and dose per fraction (Figure 13.1). Equation 13.5 shows what change in total dose ($D_1 \rightarrow D_2$) is required when the dose per fraction is changed from $d_1 \rightarrow d_2$ in a fractionated schedule.

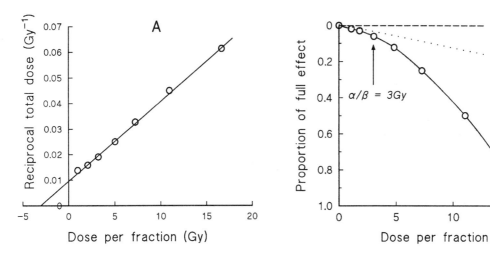

Figure 13.4 The data of Figure 13.3 after two different transformations. **A**: a reciprocal-dose plot according to Eqn 13.2. **B**: transformation according to Eqn 13.3 with the same data plotted as a proportion of full effect.

13.5 The value of α/β

Many detailed fractionation studies of the type shown in Figures 13.1 and 13.3 have been made in animals. Table 13.1 summarizes the α/β values obtained from these experiments. For acutely responding tissues which express their damage within a period of days to weeks after irradiation, the α/β ratio is in the range 7–20 Gy, while for late-responding tissues, which express their damage months to years after irradiation, α/β generally ranges from 0.5 to 6 Gy. It is important to recognise that *the α/β ratio is not constant and its value should be chosen carefully to match the specific tissue under consideration.*

The fractionation response of well-oxygenated tumours is thought to be similar to early-responding normal tissues, sometimes with an even higher α/β ratio. The values shown in Figure 13.5 were compiled by Williams *et al* (1985). Values calculated from data obtained in experiments under fully radiosensitized conditions (marked 'Miso' and 'Oxic' in the figure) are plotted directly, and values calculated from fractionation responses under hypoxic conditions (marked 'Clamp', 'Anoxic' and 'Hypoxic') are plotted after dividing by an assumed oxygen enhancement ratio (OER) of 2.7, because the α/β ratios for a tissue under anoxic and oxic conditions are in the same proportion as the OER. Error bars are estimates of the 95% confidence range on each value. Such experiments assayed the effect of radiation either *in situ* by regrowth delay or local tumour control, or by excising the tumour from the animal and measuring the survival of cells *in vitro* (Section 6.4).

Analysis of data from human tissues and tumours to determine α/β ratios has been made by Thames *et al* (1989) and the subsequently updated results are shown in Table 13.2. Estimates from these data are much less precise than the animal values because studies using a wide range of fraction numbers cannot be performed on human patients for obvious ethical reasons. However, the agreement between human and animal data is generally good and confirms that animal tissues are good models for comparing the effects of different fractionation schemes that might be used clinically.

Table 13.1 Values for the α/β ratio for a variety of early- and late-responding normal tissues in experimental animals

Early reactions	α/β	References	Late reactions	α/β	References
Skin			Spinal cord		
Desquamation	9.1 – 12.5	Douglas and Fowler (1976)	Cervical	1.8 – 2.7	van der Kogel (1979)
	8.6 – 10.6	Joiner et al (1983)	Cervical	1.6 – 1.9	White and Hornsey (1978)
	9 – 12	Moulder and Fischer (1976)	Cervical	1.5 – 2.0	Ang et al (1983)
Jejunum			Cervical	2.2 – 3.0	Thames et al (1988)
Clones	6.0 – 8.3	Withers et al (1976)	Lumbar	3.7 – 4.5	van der Kogel (1979)
	6.6 – 10.7	Thames et al (1981)	Lumbar	4.1 – 4.9	White and Hornsey (1978)
Colon				3.8 – 4.1	Leith et al (1981)
Clones	8 – 9	Tucker et al (1983)		2.3 – 2.9	Amols, Yuhas (quoted by
Weight loss	9 – 13	Terry and Denekamp (1984)			Leith et al, 1981)
Testis			Colon		
Clones	12 – 13	Thames and Withers (1980)	Weight loss	3.1 – 5.0	Terry and Denekamp (1984)
Mouse lethality			Kidney		
30 d	7 – 10	Kaplan and Brown (1952)	Rabbit	1.7 – 2.0	Caldwell (1975)
30 d	13 – 17	Mole (1957)	Pig	1.7 – 2.0	Hopewell and Wiernik (1977)
30 d	11 – 26	Paterson et al (1952)	Rats	0.5 – 3.8	van Rongen et al (1988)
Tumour bed			Mouse	1.0 – 3.5	Williams and Denekamp
45 d	5.6 – 6.8	Begg and Terry (1984)			(1984 a,b)
			Mouse	0.9 – 1.8	Stewart et al (1984 a)
			Mouse	1.4 – 4.3	Thames et al (1988)
			Lung		
			LD$_{50}$	4.4 – 6.3	Wara et al (1973)
			LD$_{50}$	2.8 – 4.8	Field et al (1976)
			LD$_{50}$	2.0 – 4.2	Travis et al (1983)
			Breathing rate	1.9 – 3.1	Parkins and Fowler (1985)
			Bladder		
			Frequency, capacity	5 – 10	Stewart et al (1984b)

α/β values are in grays.
From Fowler (1989); for references, see the original.

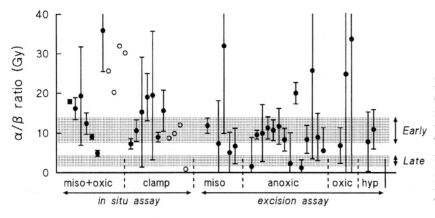

Figure 13.5 Values of α/β for experimental tumours, determined under a variety of conditions of oxygenation (see text). The stippled areas indicate the range of values for early- and late-responding normal tissues. From Williams et al (1985), with permission.

Table 13.2 α/β ratios for human normal tissues and tumours

Tissue/organ	End-point	α/β (Gy)	95% conf. lim. (Gy)	Reference
Early reactions				
Skin	Erythema	8.8	[6.9; 11.6]	Turesson and Thames, 1989
	Erythema	12.3	[1.8; 22.8]	Bentzen et al, 1988
	Desquamation	11.2	[8.5; 17.6]	Turesson and Thames, 1989
Oral mucosa	Mucositis	9.3	[5.8; 17.9]	Denham et al, 1995
	Mucositis	15	[-15; 45]	Rezvani et al, 1991
	Mucositis	~8	N/A	Chogule and Supe, 1993
Late reactions				
Skin/vasculature	Telangiectasia	2.8	[1.7; 3.8]	Turesson and Thames, 1989
	Telangiectasia	2.6	[2.2; 3.3]	Bentzen et al 1990
	Telangiectasia	2.8	[-0.1; 8.1]	Bentzen and Overgaard, 1991
Subcutis	Fibrosis	1.7	[0.6; 2.6]	Bentzen and Overgaard, 1991
Muscle/vasculature/ cartilage	Impaired shoulder movement	3.5	[0.7; 6.2]	Bentzen et al, 1989
Nerve	Brachial plexopathy	<3.5*	N/A	Olsen et al, 1990
	Brachial plexopathy	~2	N/A	Powell et al, 1990
	Optic neuropathy	1.6	[-7; 10]	Jiang et al, 1994
Spinal cord	Myelopathy	<3.3	N/A	Dische et al, 1981
Eye	Corneal injury	2.9	[-4; 10]	Jiang et al, 1994
Bowel	Stricture/perforation	3.9	±0.7	Deore et al, 1993
Lung	Pneumonitis	3.3	±1.5	van Dyk et al, 1989
	Fibrosis (radiological)	3.1	[-0.2; 8.5]	Dubray et al, 1995
Head and neck	Various late effects	3.5	±1.2	Rezvani et al, 1991
Supraglottic larynx	Various late effects	3.8	[0.8; 14]	Maciejewski et al, 1986
Oral cavity + oroph.	Various late effects	0.8	[-0.6; 2.5]	Maciejewski et al, 1990
Tumours				
Head and neck				
Larynx		14.5*	±4.9	Rezvani et al, 1993
Vocal cord		~13	wide	Robertson et al, 1993
Oropharynx		~16*	N/A	Horiot et al, 1992
Buccal mucosa		6.6	[2.9; infinity]	Maciejewski et al, 1989
Tonsil		7.2	[3.6; infinity]	Maciejewski et al, 1989
Nasopharynx		16	[-11; 43]	Lee et al, 1995
Skin		8.5*	[4.5; 11.3]	Trott et al, 1984
Melanoma		0.6	[-1.1; 2.5]	Bentzen et al, 1989
Liposarcoma		0.4	[-1.4; 5.4]	Thames and Suit 1986

* Reanalysis of original published data.
Compiled by Bentzen and Thames (unpublished). See also Thames *et al* (1990).

13.6 Derivation of LQ-based isoeffect formulae

BIOLOGICALLY EFFECTIVE DOSE (BED)

The linear-quadratic approach has led to various ways of calculating isoeffect relationships for radiotherapy, all based on similar underlying assumptions. Two principal formulations are the concept of extrapolated tolerance dose (*ETD*) introduced by Barendsen (1982) and total effect (*TE*) by Thames and Hendry (1987). Fowler (1989) has preferred the term Biologically Effective Dose (*BED*), which is mathematically identical to *ETD*. *BED* has the advantage that it can logically be calculated for levels of effect that are below normal-tissue tolerance, whereas *ETD* implies the full tolerance effect. Throughout this book we use the *BED* terminology.

These formulae seek to describe a range of fractionation schedules that are *isoeffective*. First, we must define a particular effect, or *end-point*. Although the validity of the LQ approach to fractionation depends principally on its ability to predict isoeffective schedules successfully, there is an implicit assumption that the isoeffect has a direct relationship with a certain level of cell inactivation (or final cell survival, $(SF_d)^n$). Generally, the fraction of surviving cells associated with an isoeffect is unknown and it is customary to work in terms of a level of tissue effect, which we denote as E. From Equation 13.4,

$$E/\alpha = D\,[1 + d/(\alpha/\beta)]$$
$$= \text{Biologically Effective Dose (BED)}$$

BED is a measure of the effect (E) of a course of fractionated or continuous irradiation; when divided by α it has the units of *dose* and is usually expressed in grays. Note that as the dose per fraction (d) is reduced towards zero, *BED* becomes $D = nd$ (i.e. the total radiation dose). *BED* thus has a simple conceptual significance: it is the theoretical total dose that would be required to produce the isoeffect E using an infinitely large number of infinitesimally small dose fractions. It is therefore also the total dose required for a *single* exposure at very low dose rate (Section 18.4). As will be illustrated below, values of *BED* from separate parts of a course of treatment may be added in order to calculate the overall *BED* value.

Note that the *TE* formulation is conceptually similar and is also used in the literature. In this case we divide E by β rather than α, to get $TE = E/\beta = D\,(\alpha/\beta + d)$. The units of *TE* are (gray)^2, which makes it less convenient than *BED*. But there is a simple conversion:

$$TE = (\alpha/\beta) \times BED.$$

EQUIVALENT DOSE IN 2-GY FRACTIONS (EQD_2)

One disadvantage of *BED* as a measure of treatment intensity is that it is numerically much greater than any prescribable radiation dose of fractionated radiotherapy. An alternative approach to isoeffect calculations is to calculate the equivalent dose in 2-Gy fractions. This can be done using equation 13.5, an approach advocated by Withers *et al* (1983):

$$EQD_2 = D\,\frac{d + (\alpha/\beta)}{2 + (\alpha/\beta)}$$

where EQD_2 is the dose in 2-Gy fractions that is biologically equivalent to a total dose D given with a fraction size of d gray. Values of EQD_2 may be added for separate parts of a treatment schedule, as is the case with *BED* values. They have the advantage that since 2 Gy is a commonly used dose per fraction, EQD_2 values will be recognized by radiotherapists as being of a familiar size. But there is the possible disadvantage that with the increasing use of dose-per-fraction values *below* 2 Gy, it may seem arbitrary to fix on this value as the reference point.

13.7 Tolerance calculations using the BED formula

GENERAL PROCEDURE: CHANGES IN FRACTION SIZE

We deal here with the situation in which fraction size is changed *without change in the overall duration of treatment*.

Formula: $BED = D[1 + d/(\alpha/\beta)]$
(D is the total dose in n fractions of size d)

First decide what is the dose-limiting normal tissue for this treatment.

1. Select a value for α/β for the specific tissue end-point in question; examples are given in Tables 13.1 and 13.2.
   Select the reference tolerance dose D_{ref}. This will depend on the dosage policy of the treatment centre.
   Select a fraction size for the reference treatment (d_{ref}).
2. Calculate for the reference treatment:
   $$BED_{ref} = D_{ref}[1 + d_{ref}/(\alpha/\beta)]$$
   This value of *BED* is constant for the chosen α/β and reference treatment.
   Assuming conditions of complete repair (*i.e.* high dose rate, one fraction per day), we can calculate as follows:

New fraction size, constant throughout treatment:

3. For new dose per fraction d, calculate total dose $D = BED_{ref}/[1 + d/(\alpha/\beta)]$.
   Alternatively, use equation 13.5, bypassing the calculation of BED values:

$$D = D_{ref}\frac{d_{ref} + (\alpha/\beta)}{d + (\alpha/\beta)} \quad \text{(Eqn 13.5)}$$

Change of fraction size during treatment:

4. For the first part of treatment, calculate the partial *BED* value (PE_1) from d_1 and D_1.

5. The partial tolerance remaining for second part of treatment is: $PE_2 = BED_{ref} - PE_1$.
6. For the new dose per fraction d_2, the remaining total dose is given by
   $$D_2 = PE_2 / [1 + d_2/(\alpha/\beta)].$$
   A similar procedure can be performed for more than two fraction sizes during treatment.
   A fundamental aspect of these calculations is that partial *BED* values (*PE*) are additive:
   $$BED = PE_1 + PE_2 + PE_3 + ...$$
   When fractions of radiotherapy are given so close in time that repair is incomplete, it is necessary to adjust the *BED* formulae as described in Section 13.9.

PRACTICAL CALCULATIONS – EXAMPLE 1

Background: Head and neck cancer. The planned treatment is 70 Gy in 35 fractions (abbreviated 70 Gy/35 fx). Due to a dosimetric error the first six fractions were given with 4 Gy/fx instead of 2 Gy/fx. The accumulated dose is thus 24 Gy in 6 fx. Treatment will be continued using 2 Gy/fx.

Question: How many fractions of 2 Gy should be given to maintain an equal probability of late fibrosis?

Assumption: α/β for late fibrosis = 3.5 Gy

Solution to Example 1:

1.	$BED = 70 \times (1 + 2/3.5) = 110$	calculate *BED*
2.	$PE_1 = 24 \times (1 + 4/3.5) = 51.4$	*PE* of first six fractions
3.	$PE_2 = BED - PE_1 = 58.6$	remaining *PE*
4.	$PE_2 = D_2 \times (1 + 2/3.5) = 58.6$	D_2 at 2 Gy per fraction
5.	$D_2 = 58.6/1.57 = 37.3$	remaining total dose
6.	At 2 Gy/fx: 37.3 / 2 = 18 or 19 fractions	remaining number of fractions

PRACTICAL CALCULATIONS – EXAMPLE 2

Background: Melanoma, elective radiotherapy of nodal area. The planned treatment is five fractions of 6 Gy (2 fx/week). After the first fraction it was discovered that by mistake a single dose of 12 Gy had been given. It was decided to complete the treatment with the same total number of five fractions.

Question: What fraction size should be used for the remaining four treatments? Is there a risk of radiation damage to the spinal cord?
Assumptions: α/β for late fibrosis = 3.5 Gy; α/β for damage to the spinal cord = 2 Gy.

114 The linear-quadratic approach to fractionation

Solution A to Example 2, Late fibrosis:

1. $BED = 30 \times (1 + 6/3.5) = 81.4$ — BED of planned treatment
2. $PE_1 = 12 \times (1 + 12/3.5) = 53.1$ — partial BED for first fraction
3. $PE_2 = 81.4 - 53.1 = 28.3$ — partial BED for remaining 4 fx
4. $PE_2 = D_2 \times [1 + D_2/(4 \times 3.5)]$
5. $28.3 = D_2 + D_2^2/14$ — filling in the known values
6. $0.0714 D_2^2 + D_2 - 28.3 = 0$ — quadratic equation:
$$D = \left\{-b + \sqrt{b^2 - 4ac}\right\}/2a$$

7. $D_2 = 14$ Gy; $d_2 = 3.5$ Gy

Solution B to Example 2, Spinal cord tolerance:

1. $BED_{ref} = 50 \times (1 + 2/2) = 100$ — reference BED for the cord; 2 Gy/fx
2. $BED_{plan} = 30 \times (1 + 6/2) = 120$ — planned treatment: 5×6 Gy
3. $BED_{plan} / BED_{ref} = 1.2$ — cord tolerance exceeded by 20%
4. $100 = D_{max} \times (1 + 6/2)$ — D_{max} (cord) for 6-Gy fractions
   maximum four fractions of 6 Gy
5. $D_{max} = 100/4 = 25$ Gy

Additional Exercise:

The actual treatment was 1×12 Gy $+ 4 \times 3.5$ Gy. Calculate how this dose compares with the spinal cord reference dose of 50 Gy in 2-Gy fractions.
[Note: it is much safer to use small doses per fraction for spinal cord tolerance; it is therefore better to reduce the dose for all fractions, rather than shielding after four large fractions of 6 Gy.]

13.8 Double trouble

There is a theoretical problem that may arise in radiotherapy and which illustrates some aspects of the use of isoeffect formulae: the problem of 'double trouble', as described by Withers (1992; see also Lee et al, 1995). We imagine a fractionated treatment in which the dose is prescribed to the centre of a deep-seated tumour. The best-achievable plan gives a hot spot, outside the target volume, of 110% of the planned dose. This region of normal tissue will receive 10% higher radiation dose than the tumour *and it will also receive a 10% higher dose per fraction*. The fraction number is of course the same as for the tumour. The higher dose per fraction means that the biological effect per gray will be increased (Eqn 13.4): double trouble. If the dose–response curve for the irradiated tissue is steep (as is the case with the spinal cord) this could lead to a significant extra risk of morbidity. Care must be taken to evaluate this risk in critical cases. Double trouble may arise in a number of other clinical situations, in particular within overdosed volumes at the border of abutted radiation fields, and in the undesirable practice of treating the separate fields of a multiple-field plan on alternate days.

A comforting thought is that the converse of double trouble also exists, and is more common. Regions of normal tissue that receive *less than* the prescribed tumour dose also have a lower dose per fraction and the biological effect of this will be less than would appear from the actual dose in grays.

13.9 Incomplete repair

The simple LQ model described by Equations 13.1 to 13.5 assumes that sufficient time is allowed between fractions for complete repair of sublethal damage to take place after each dose. This *full-repair* interval is at least 6 hours but in some cases (e.g. spinal cord) may be as long as 1 day. If the interfraction interval is reduced below this value, for example when multiple fractions per day are used, the overall damage from the whole treatment is increased because the repair (or more correctly, recovery) of damage due to one radiation dose

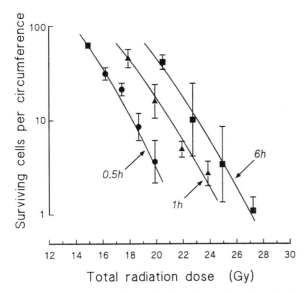

Figure 13.6 Effect of inter-fraction interval on intestinal radiation damage in mice. The total dose required in five fractions for a given level of effect is less for short intervals, illustrating incomplete repair between fractions. From Thames, Withers and Peters (1984), with permission.

calculations, an extra term is added to the basic BED formula:

$$BED = D\,[1 + d/(\alpha/\beta) + H_m \cdot d/(\alpha/\beta)]$$
(for fractionated radiotherapy)

Once again, d is the dose per fraction and D the total dose. Since repair from one day to the next is assumed to be complete, m is the number of fractions per day. Values of H_m are given in Table 13.3 for repair half-times up to 4 hours and for two or three fractions per day given with interfraction intervals down to 3 hours. Other values can be calculated using the formulae given in the Appendix. Table 13.4 shows values of $T_{1/2}$ for some normal tissues, but it has not been possible to derive more than a few estimates from human data. [Advanced note: In several cases, experiments have indicated than a proportion of the repair may occur rapidly and the balance occurs more slowly, leading to two repair half-time values, one for each of these different components. The BED equation above has to be reformulated in a more complex form to take account of these cases (see Guttenberger et al, 1992; Millar and Canney, 1993).]

Figure 13.7 demonstrates the fit of the incomplete repair LQ model to data for pneumonitis in mice following fractionated thoracic irradiation with intervals of 3 hours between doses (Thames et al, 1984). The end-point was mortality, expressed as the LD_{50}. On these reciprocal-dose plots, incomplete repair makes the data bow upwards away from the straight line (dashed) which shows the pure linear-quadratic relationship obtained when there is complete repair between successive doses, as will be the case with long time intervals between fractions. The repair half-time can be found by mathematically fitting the incomplete repair LQ model to data like those shown in Figures 13.6 and 13.7.

Another common situation in which incomplete repair occurs in clinical radiotherapy is during continuous irradiation. As described in Sections 18.1–3, irradiation must be given at a very low dose rate (below about 5 cGy/hour) for full repair to occur during irradiation. At the other extreme, a single irradiation at high dose rate may allow no significant repair to occur during exposure. As the dose rate is reduced below the range used in external-beam radiotherapy, the duration of irradiation becomes longer and the induction of damage is counteracted by repair, leading to an increase in the isoeffective dose. The corresponding BED formula for continuous irradiation incorporates the factor g to allow for incomplete repair:

may not be complete before the next fraction is given, and there is then interaction between residual unrepaired damage from one fraction and the damage from the next fraction. As an example of this phenomenon, Figure 13.6 shows data from mouse jejunum irradiated with five X-ray fractions in which the number of surviving crypts per gut circumference is plotted against total dose. Much less dose is needed to produce the same effects when the interfraction interval is reduced from 6 hours to 1 hour or 0.5 hour. This phenomenon is called incomplete repair.

The influence of *incomplete repair* is determined by the repair half-time $(T_{1/2})$ in the tissue. This is the time required between fractions, or during low dose rate treatment, for half the maximum possible repair to take place. Incomplete repair will tend to reduce the isoeffective dose and corrections have to be made for the consequent loss of tolerance. This can be accomplished by the use of the *incomplete repair model* as introduced by Thames (1985). In this model, the amount of unrepaired damage is expressed by a function H_m which depends upon the number of equally spaced fractions (m), the time interval between them and the repair half-time. For the purpose of tolerance

Table 13.3 Incomplete repair factors: fractionated irradiation (H_m factors)

Repair half-time (h)	Interval (h) for m = 2 fractions per day					Interval (h) for m = 3 fractions per day				
	3	4	5	6	8	3	4	5	6	8
0.5	0.0156	0.0039	0.0010	0.0002	0	0.0210	0.0052	0.0013	0.0003	0
0.75	0.0625	0.0248	0.0098	0.0039	0.0006	0.0859	0.0335	0.0132	0.0052	0.0008
1.0	0.1250	0.0625	0.0312	0.0156	0.0039	0.1771	0.0859	0.0423	0.0210	0.0052
1.25	0.1895	0.1088	0.0625	0.0359	0.0118	0.2766	0.1530	0.0859	0.0487	0.0159
1.5	0.2500	0.1575	0.0992	0.0625	0.0248	0.3750	0.2265	0.1388	0.0859	0.0335
2.0	0.3536	0.2500	0.1768	0.1250	0.0625	0.5547	0.3750	0.2565	0.1771	0.0859
2.5	0.4353	0.3299	0.2500	0.1895	0.1088	0.7067	0.5124	0.3750	0.2766	0.1530
3.0	0.5000	0.3969	0.3150	0.2500	0.1575	0.8333	0.6341	0.4861	0.3750	0.2265
4.0	0.5946	0.5000	0.4204	0.3536	0.2500	1.0285	0.8333	0.6784	0.5547	0.3750

From Thames and Hendry (1987), with permission.

Table 13.4 Half-times for recovery from radiation damage ($T_{\frac{1}{2}}$) in various normal tissues

Tissue	Species	Dose delivery*	$T_{\frac{1}{2}}$ (hours)	Source
Haemopoietic	Mouse	CLDR	0.3	Thames et al (1984)
Spermatogonia	Mouse	CLDR	0.3–0.4	Delic et al (1987)
Jejunum	Mouse	F	0.45	Thames et al (1984)
	Mouse	CLDR	0.2–0.7	Dale et al (1988)
Colon (acute injury)	Mouse	F	0.8	Thames et al (1984)
	Rat	F	1.5	Sassy et al (1988)
Lip mucosa	Mouse	F	0.8	Ang et al (1985)
	Mouse	CLDR	0.8	Scalliet et al (1987)
	Mouse	FLDR	0.6	Stüben et al (1991)
Tongue epithelium	Mouse	F	0.75	Dörr et al (1993)
Oral mucosa	Human	F	2.0–4.0	Bentzen et al (1996)
	Human	FLDR	0.3–0.5	Denham et al (1995)
Skin (acute injury)	Mouse	F	1.5	Rojas et al (1991)
	Mouse	CLDR	1.0	Joiner et al (unpublished)
	Pig	F	0.4 + 1.2**	van den Aardweg and Hopewell (1992)
	Pig	F	0.2 + 6.6**	Millar et al (1996)
	Human	F	0.35–1.2	Turesson and Thames (1989)
	Human	F	? + <1***	Nyman and Turesson (1995)
Skin (late injury)	Human	F	0.4 + 3.5**	Turesson and Thames (1989)
	Human	F	? + ?***	Nyman and Turesson (1995)
Lung	Mouse	F	0.4 + 4.0**	van Rongen et al (1993)
	Mouse	CLDR	0.85	Down et al (1986)
	Rat	FLDR	1.0	van Rongen (1989)
Spinal cord	Rat	F	0.7 + 3.8**	Ang et al (1992)
	Rat	CLDR	1.4	Scalliet et al (1989)
	Rat	CLDR	1.8	Pop et al (1997)
	Human	F	? + >5***	Dische and Saunders (1989)
Kidney	Mouse	F	1.3	Joiner et al (1993)
	Mouse	F	0.2 + 5.0	Millar et al (1994)
	Rat	F	1.6–2.1	van Rongen et al (1990)
Rectum (late injury)	Rat	CLDR	1.2	Kiszel et al (1985)
Heart	Rat	F	>3	Schultz-Hector et al (1992)

* F = acute dose fractions, FLDR = fractionated low dose rate, CLDR = continuous low dose rate.
** Two components of repair with different half-times.
*** Evidence for two components of repair with different half-times, but values uncertain.

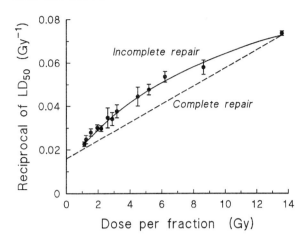

Figure 13.7 Reciprocal dose plot (compare Figure 13.4A) of data for pneumonitis in mice produced by fractionated irradiation; the points derive from experiments with different dose per fraction (and therefore different fraction numbers), always with 3 h between doses. The upward bend in the data illustrates lack of sparing due to incomplete repair. From Thames, Withers and Peters (1984), with permission.

$$BED = D[1 + d.g/(\alpha/\beta)]$$
(for continuous low dose rate radiotherapy)

where D is the total dose (= dose rate × time). The parameter d is retained, as in the equation for fractionated radiotherapy, in order to deal with *frac-tionated low dose rate* exposures. For a single continuous exposure $d = D$. This equation assumes that there is full recovery between the low dose rate exposures; if not, the H_m factor must be added. Table 13.5 gives values of the g-factor for exposure times between 1 hour and 4 days.

ADVANCED PRACTICAL CALCULATION – EXAMPLE 3 (INCOMPLETE REPAIR)

Background: Head and neck cancer, planned treatment of 70 Gy in 35 fractions of 2 Gy in 7 weeks. After 40-Gy in 4 weeks has been given, due to medical reasons the treatment is interrupted for 10 days.

Question: In order to maintain the same probability of local tumour control the intention is to give approximately the same total effective dose in the same overall treatment time as originally planned. In order not to increase the probability of late normal-tissue damage, fraction size should be reduced and an accelerated schedule with three fractions of 1.5-Gy per day is considered. What would be the minimum interval between the fractions on one day to allow a dose of 30 Gy to be given?

Assumptions: α/β for late damage = 3.0 Gy; $T_{1/2}$ for repair between fractions = 3 hours; fraction-size = 1.5 Gy

Solution:

1. $BED = 70 \times (1 + 2/3) = 116.7$ — BED of planned treatment
2. $PE_1 = 40 \times (1 + 2/3) = 66.7$ — partial BED for first 20 fractions
3. $PE_2 = BED - PE_1 = 50$ — partial BED for remaining schedule
4. $PE_2 = 30 \times (1 + d/3 + H_m.d/3)$ — BED calculation for incomplete repair
5. $50 = 45 + 15 H_m$ — filling in values, $d = 1.5$ Gy
6. $H_m = 5/15 = 0.333$
7. In Table 13.3, a value of $H_m = 0.333$ for three fractions per day corresponds to an interval slightly longer than 6 hours (an exact calculation with the formulae given in the Appendix gives 6.5 hours).

Table 13.5 Incomplete repair factors: continuous irradiation (*g* factors)

Repair half-time (h)	Exposure time (h)						Exposure time (days)						
	1	2	3	4	8	12	1	1.5	2	2.5	3	3.5	4
0.5	0.6622	0.4774	0.3671	0.2959	0.1641	0.1130	0.0583	0.0393	0.0296	0.0238	0.0198	0.0170	0.0149
0.75	0.7517	0.5888	0.4774	0.3983	0.2339	0.1641	0.0861	0.0583	0.0441	0.0354	0.0296	0.0254	0.0223
1	0.8040	0.6622	0.5571	0.4774	0.2959	0.2115	0.1130	0.0769	0.0583	0.0469	0.0393	0.0338	0.0296
1.25	0.8382	0.7137	0.6165	0.5394	0.3504	0.2555	0.1390	0.0952	0.0723	0.0583	0.0488	0.0420	0.0369
1.5	0.8622	0.7517	0.6622	0.5888	0.3983	0.2959	0.1641	0.1130	0.0861	0.0695	0.0583	0.0502	0.0441
2	0.8938	0.8040	0.7276	0.6622	0.4774	0.3671	0.2115	0.1475	0.1130	0.0916	0.0769	0.0663	0.0583
2.5	0.9136	0.8382	0.7720	0.7137	0.5394	0.4269	0.2555	0.1803	0.1390	0.1130	0.0952	0.0822	0.0723
3	0.9272	0.8622	0.8040	0.7517	0.5888	0.4774	0.2959	0.2115	0.1641	0.1339	0.1130	0.0977	0.0861
4	0.9447	0.8938	0.8471	0.8040	0.6622	0.5571	0.3671	0.2693	0.2115	0.1739	0.1475	0.1280	0.1130

From Thames and Hendry (1987), with permission.

13.10 The time factor

If the overall duration of fractionated radiotherapy is increased there will usually be greater repopulation of the irradiated tissues, both in the tumour and in normal tissues. So far we have not discussed the change in total dose necessary to compensate for changes in the overall duration of treatment. Overall time is included in the NSD model (Equation 12.2) but not in the basic LQ approach

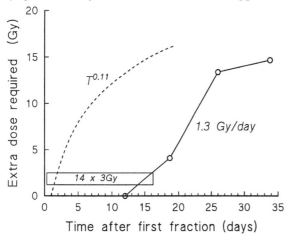

Figure 13.8 Extra dose required to counteract proliferation in mouse skin. Test doses of radiation were given at various intervals after a priming treatment with fractionated radiation. Proliferation begins about 12 days after the start of irradiation and is then equivalent to an extra dose of approximately 1.3 Gy/day. The broken line shows the prediction of the NSD equation. Adapted from Denekamp (1973), with permission.

described above. The reason is because the time factor in radiotherapy is now perceived to be more complex than has previously been supposed. For example, Figure 13.8 shows that the extra dose needed to counteract proliferation in mouse skin does not become significant until about 2 weeks after the start of daily fractionation. For this situation the time factor in the NSD formula (broken line; total dose $\propto T^{0.11}$) gives a false picture because it predicts a large amount of sparing if the overall time was increased from 1 to 12 days. These wrong time factors have not led to major clinical disasters because they predict much less effect on total dose compared with changing the number of fractions. Thus the $T^{0.11}$ factor in the NSD model predicts only an 8% increase in total dose for a doubling of overall time, for example from 3½ to 7 weeks.

Nevertheless, the $T^{0.11}$ factor is clearly misleading and should *not* be used (Section 12.4). The use of the LQ model in clinical practice with no time factor at all is probably the best strategy for late-reacting tissues because any extra dose needed to counteract proliferation does not become significant until beyond the overall time of treatment, even up to 6 weeks. This is illustrated in Figure 13.9 which diagrammatically shows the different effects of overall time in early- and late-responding tissues. Attempts are now being made to include time factors in the LQ model for early-responding normal tissues and tumours, but such factors depend in a complex way on the dose per fraction and interfraction interval as well as on the tissue type, and will need to take account of any delay in onset of proliferation which will probably depend in some way on these factors also.

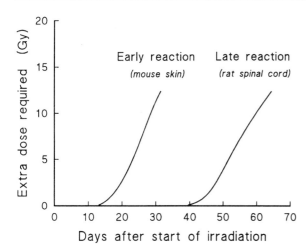

Figure 13.9 The extra dose required to counteract proliferation does not become significant until much later for late-responding normal tissues such as spinal cord, beyond the 6-week duration of conventional radiotherapy. From Fowler (1984), with permission.

13.11 How to respond to an unscheduled gap in treatment

It is unfortunately common for the schedule of radiotherapy treatments of patients to be interrupted and therefore to differ from that which was prescribed. A number of studies have shown that in Europe and the USA roughly one third of patients experience a gap of more than 6 days in treatment. Interruptions can result either from factors related to the condition of the patient (*e.g.* severe reactions or intercurrent disease) or from factors unconnected with the patient (*e.g.* machine failure or transport difficulties). They may range in duration from one missed fraction to some weeks of delay. Evidence that gaps in treatment may have a deleterious effect on tumour control rate has come from a number of clinical studies, some of which are summarized in Table 12.3. Further evidence for the importance of overall time is shown in Figure 14.4. The evidence is strongest for squamous cell tumours of the head and neck and uterine cervix. A frequent question in clinical departments is therefore how to adjust the remaining treatments when the patient returns after an unscheduled interruption.

There are several ways in which this can be done, and some are better in radiobiological terms than others. A common response is to carry on with daily-dose treatment to the prescribed total dose, now over a longer time than was planned. Local tumour control by such a treatment will be reduced as a result of greater tumour cell repopulation. An alternative would be to increase the dose per fraction for the remaining treatments in order to give the prescribed total dose within the intended overall time. This however involves the increased damage to late-responding normal tissues that will result from the higher dose per fraction (Section 14.2).

The management of treatment gaps has been considered in detail by a working party of the UK Royal College of Radiologists (Hendry *et al*, 1996). The main recommendation is to compensate for a gap by continuing to deliver the same number of fractions in the same overall time, by inserting the missed fractions (using the original dose per fraction) either during weekends and/or as extra fractions on previously planned treatment days. When multiple fractions are given in a day the interval between them should be at least 6 hours (Section 13.9). If the break in treatment is early on in the schedule or is short, then two fractions per day can be used. However, if there are not enough days left in the original schedule to catch up using two fractions on each of the remaining days, then the dose per day can be escalated using three fractions per day with reduced dose per fraction. Practical Calculation (Example 3) shows how to replan the remaining part of a treatment in this way, taking into account the problem of incomplete repair between fractions.

13.12 What degree of reliability can be placed on isoeffective dose calculations?

It is important to stress that the results of calculations like those demonstrated here must only be taken as a guide to clinical practice. The LQ approach to fractionation overcomes some of the deficiencies of the older NSD approach (Section 12.4) but it cannot be claimed to be universally correct. Indeed, it would be surprising if such simple equations satisfactorily described all of the possible effects of changing dose prescriptions in radiotherapy.

The validity of the equations is limited to more or less standard conditions. Deviations from the predictions of the incomplete-repair LQ model have become apparent under more extreme conditions, such as a reduced spinal cord tolerance in the CHART regime (three fractions per day continuous over 12 days). Most of the deviations that have so far been observed from the LQ model may have arisen from incorrect choice of the two basic parameters: α/β and $T_{1/2}$. As experience grows in the application of this method of calculation it may be expected that fewer deviations will be observed. *The exploration of new treatment schedules based on extrapolation from standard conditions using the current mathematical models must be done with caution.*

Key points

1. The linear-quadratic model satisfactorily describes the relationship between total isoeffective dose and dose per fraction over the range of dose per fraction from 1 Gy up to large single doses. In contrast, the NSD formulation can only be made to fit data over a limited range of dose per fraction.
2. The α/β ratio describes the *shape* of the fractionation response: a low α/β (0.5–6 Gy) is characteristic of late-responding normal tissues and indicates a rapid increase of total dose with decreasing dose per fraction and a survival curve for the putative target cells that is significantly curved.
3. A higher α/β ratio (7–20 Gy) is characteristic of early-responding normal tissues and tumours; it indicates a less rapid increase in total dose with decreasing dose per fraction and a less curved survival response for the target cells.
4. The BED formulae provide a simple and convenient way of calculating isoeffective radiotherapy schedules, based on the LQ model. Tolerance calculations require a reference tolerance dose and an estimate of the α/β ratio.
5. For short interfraction intervals, a correction may be necessary for incomplete repair. When using the BED formulae to calculate schedules with multiple fractions per day or continuous low dose rate, an estimate of the repair half-time is also needed.
6. The basic LQ model is appropriate for calculating the change in total dose for an altered dose per fraction, assuming the new and old treatments are given in the *same overall time*. For late reactions it is probably unnecessary to modify total dose in response to a change in overall time, but for early reactions (and tumour response) the effect of overall time is complex. The effect of this should be considered separately from the LQ calculation.

Bibliography

Barendsen GW (1982). Dose fractionation, dose rate, and isoeffect relationships for normal tissue responses. *Int J Radiat Oncol Biol Phys* 8:1981–97.

Dale RG (1985). The application of the linear quadratic dose effect equation to fractionated and protracted radiotherapy. *Br J Radiol* 58:515–28.

Denekamp J (1973). Changes in the rate of repopulation during multifraction irradiation of mouse skin. *Br J Radiol* 46:381.

Douglas BG and Fowler JF (1976). The effect of multiple small doses of X-rays on skin reactions in the mouse and a basic interpretation. *Radiat Res* 66:401–26.

Fowler JF (1984). What next in fractionated radiotherapy? *Br J Cancer* 49 (Suppl.VI) 285–300.

Guttenberger R, Thames HD and Ang KK (1992). Is the experience with CHART compatible with experimental data? A new model of repair kinetics and computer simulations. *Radiother Oncol* 25:280–6.

Hendry JH, Bentzen SM, Dale RG et al (1996). A modelled comparison of the effects of using different ways to compensate for missed treatment days in radiotherapy. *Clin Oncol* 8:297–307.

Lee SP, Leu MY, Smathers JB, McBride WH, Parker RG and Withers HR (1995). Biologically effective dose distribution based on the linear quadratic model and its clinical relevance. *Int J Radiat Oncol Biol Phys* 33:375–89.

Michael BD (1985). A simple graphical determination of α/β from reciprocal dose plots. *Int J Radiat Biol* 47:119–20.

Millar WT and Canney PA (1993). Derivation and application of equations describing the effects of fractionated protracted irradiation, based on multiple and incomplete repair processes. Part 1. Derivation of equations. *Int J Radiat Biol* 64:275–91.

Nilsson P, Thames HD and Joiner MC (1990). A generalized formulation of the 'incomplete-repair' model for cell survival and tissue response to fractionated low dose-rate irradiation. *Int J Radiat Biol* 57:127–42.

Thames HD (1985). An 'incomplete-repair' model for survival after fractionated and continuous irradiations. *Int J Radiat Biol* 47:319–39.

Thames HD, Bentzen SM, Turesson I et al (1989). Fractionation parameters for human tissues and tumors. *Int J Radiat Biol* 56:701–10.

Thames HD, Withers HR and Peters LJ (1984). Tissue repair capacity and repair kinetics deduced from multifractionated or continuous irradiation regimens with incomplete repair. *Br J Cancer* 49 (Suppl.VI) 263–9.

Thames HD, Withers HR, Peters LJ and Fletcher GH (1982). Changes in early and late radiation responses with altered dose fractionation: implications for dose survival relationships. *Int J Radiat Oncol Biol Phys* 8:219–26.

Turesson I and Thames HD (1989). Repair capacity and kinetics of human skin during fractionated radiotherapy: erythema, desquamation, and telangiectasia after 3 and 5 years' follow-up. *Radiother Oncol* 15:169–88.

Williams MV, Denekamp J and Fowler JF (1985). A review of α/β ratios for experimental tumors: implications for clinical studies of altered fractionation. *Int J Radiat Oncol Biol Phys* 11:87–96.

Withers HR (1992). Biologic basis of radiation therapy. In: *Principles and Practice of Radiation Oncology* (Eds) Perez CA and Brady LW. pp 64–69. Lippincott; Philadelphia.

Withers HR, Thames HD and Peters LJ (1983). A new isoeffect curve for change in dose per fraction. *Radiother and Oncol* 1:187–91.

FURTHER READING

Fowler JF (1989). The linear-quadratic formula and progress in fractionated radiotherapy. *Br J Radiol* 62:679–94.

Joiner MC (1989). The dependence of radiation response on the dose per fraction: in *The Scientific Basis for Modern Radiotherapy* (BIR Report 19), pp.20–6. (Ed) McNally NJ. British Institute of Radiology; London.

Thames HD, Bentzen SM, Turesson I, Overgaard J and van den Bogaert W (1990). Time-dose factors in radiotherapy: a review of the human data. *Radiother Oncol* 19:219–35.

Thames HD and Hendry JH (1987). *Fractionation in Radiotherapy*. Taylor & Francis; London.

Appendix: Summary of formulae

Basic Equations:

$$E = n(\alpha d + \beta d^2) = D(\alpha + \beta d)$$

d = dose per fraction
D = total dose
n = fraction number

$$SF = \exp(-E) = \exp[-(\alpha + \beta d)D]$$

$$BED = E/\alpha = D[1 + d/(\alpha/\beta)]$$

$$\frac{D}{D_{ref}} = \frac{d_{ref} + (\alpha/\beta)}{d + (\alpha/\beta)}$$

Low Dose Rate:

$$\mu = \log_e 2 / T_{1/2} \qquad T_{1/2} = \text{repair half-time}$$

$$g = 2[\mu t - 1 + \exp(-\mu t)]/(\mu t)^2 \qquad t = \text{exposure duration}$$

$$BED = D[1 + g \cdot d/(\alpha/\beta)]$$

Incomplete Repair Correction:

$$\Phi = \exp(-\mu \Delta T) \qquad \Delta T = \text{interval between fractions}$$

$$H_m = \left(\frac{2}{m}\right) \cdot \left(\frac{\Phi}{1-\Phi}\right) \cdot \left(m - \frac{1-\Phi^m}{1-\Phi}\right) \qquad m = \text{number of fractions per day}$$

$$BED = D[1 + (1+H_m) \cdot d/(\alpha/\beta)]$$

Incomplete Repair between Low Dose-rate Fractions:

$$\Phi = \exp(-\mu(t+\Delta T)) \qquad \begin{aligned} t &= \text{exposure duration per fraction} \\ \Delta T &= \text{interval between fractions} \end{aligned}$$

$$g = 2[\mu t - 1 + \exp(-\mu t)]/(\mu t)^2$$

$$H_m = \left(\frac{2}{m}\right) \cdot \left(\frac{\Phi}{1-\Phi}\right) \cdot \left(m - \frac{1-\Phi^m}{1-\Phi}\right) \qquad m = \text{number of fractions per day}$$

$$C = g + 2\frac{\cosh(\mu t) - 1}{(\mu t)^2} \cdot H_m$$

$$BED = D[1 + C \cdot d/(\alpha/\beta)]$$

(For a full derivation of these equations, see Nilsson *et al*, 1990)

14

Hyperfractionation and accelerated radiotherapy

Michael C. Joiner

14.1 Introduction

Major developments in radiotherapy fractionation have been taking place during the past few years. These have grown out of the developments in radiation biology, mainly within the areas covered by Chapters 3 and 13. The relationships between total dose and dose per fraction for late-responding tissues, acutely responding tissues and tumours provide the basic information required to optimize radiotherapy according to the dose per fraction and number of fractions. Much work still needs to be done to determine the time of onset and rate of repopulation in normal tissues and tumours during and after radiotherapy, but enough is now known about this proliferative response to support the view that a reduction in the overall duration of fractionated radiotherapy should be considered in a number of clinical situations. This chapter discusses changes in clinical practice that are being made to take account of these two issues.

14.2 Hyperfractionation

The dependence of isoeffective total dose on dose per fraction is addressed in Chapter 13 in terms of the linear-quadratic model. A key development in this area was the publication by Thames *et al* (1982) of a survey of isoeffect curves for various normal tissues, mainly in mice. Their summary is shown in Figure 14.1. Each of the investigations

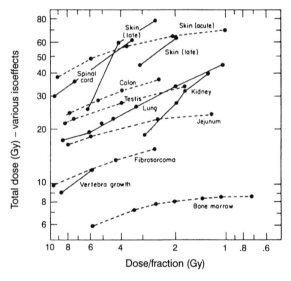

Figure 14.1 Relationship between total dose and dose per fraction for a variety of normal tissues in experimental animals. The results for late-responding tissues (full lines) are systematically steeper than those for early-responding tissues (broken lines). Chart from Hall (1988) quoting the data of Thames *et al* (1982), with permission.

contributing to this chart was a study of the response of a normal tissue to fractionated radiation treatment using a range of doses per fraction. In order to minimize the effects of repopulation, the survey was restricted to experiments in which the overall time was kept short by the use of multiple treatments per day, or 'where an effect of regeneration of target cells was shown to be unlikely'. This summary thus represents the influence of dose per fraction on response and excludes the influence of overall treatment time. It was possible in each study, and for each chosen dose per fraction, to determine the total radiation dose that produced some defined level of normal-tissue damage. These end-points of tolerance differed from one normal tissue or experimental study to another. Each line in Figure 14.1 is an isoeffect plot determined in this way. The dashed lines show isoeffect curves for acutely responding tissues and the full lines are for late-responding tissues. Note that the dose per fraction scale decreases from left to right; it is plotted in this way in order to produce plots that resemble the isoeffect curves of the 'NSD era' (Figures 12.1 and 12.3). Fraction *number* increases from left to right along the abscissa.

The results of this survey show that isoeffective total doses increase more rapidly with decreasing dose per fraction for late effects than for acute effects. A reminder from Chapter 13 is that, in consequence, late effects have low α/β ratios (0.5–6 Gy) and acute effects have high α/β ratios (7–20 Gy). The vertical axis can be regarded as the tissue tolerance dose and it can be seen that for low doses per fraction (right-hand end of the abscissa) late reactions tend to require a higher dose and are therefore spared. It was indicated in Section 13.5 that the available data on tumours suggest that in terms of fractionation response they may tend to behave like acutely responding normal tissues; thus, provided that late reactions are dose-limiting, small doses per fraction should give the best therapeutic index.

It is possible to summarize data of the type shown in Figure 14.1 by plotting the ratio of total dose for each dose per fraction to total dose for a reference treatment (usually 2 Gy per fraction). This is done in Figure 14.2A using calculations based on the LQ equation. The shaded areas enclose the range of these ratios, for α/β values of 1–4 Gy and 8–15 Gy, which apply to most late- and acute-responding tissues respectively. The change in total dose is greater for the lower α/β values, as is the potential for error if the wrong α/β value is used. α/β values should therefore be selected carefully and always conservatively when doing calculations involving changing dose per fraction.

An increase in dose per fraction relative to 2 Gy is termed *hyp*o*fractionation* and a decrease is *hyp*e*r*-*fractionation*. We can calculate a therapeutic gain factor (TGF) for a new dose per fraction from the ratio of the relative isoeffect doses for tumour and normal tissue. An example is shown in Figure

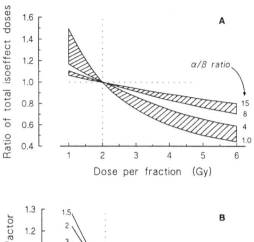

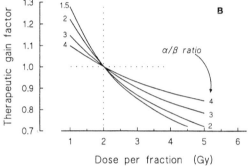

Figure 14.2 A: theoretical isoeffect curves based on the LQ model for various α/β ratios. The hatched areas enclose curves corresponding to early-responding and late-responding normal tissues. **B**: therapeutic gain factors for various α/β ratios of normal tissue, assuming an α/β ratio of 10 Gy for tumours. Redrawn from Withers *et al* (1989).

14.2B. Remember that we are assuming here that the new regimen is given in the same overall time as the 2-Gy regimen and that treatment is always limited by the late reactions. TGF curves are shown for a range of (late-responding) α/β values from 1.5 to 4, relative to 10 Gy for the tumours.

Hyp*e*rfractionation using a larger number of dose fractions below 2 Gy is predicted to give a therapeutic gain, and hyp*o*fractionation a therapeutic loss. Note, however, that this situation would be nullified, or even reversed, for specific tumours that have low α/β ratios (*e.g.* some melanomas and sarcomas). If an unacceptable increase in *acute* normal-tissue reactions prevented the total dose from being increased to the full tolerance of the late-responding tissues, the therapeutic gain for hyperfractionation would also be less than shown in Figure 14.2B.

Hyperfractionation has been tested in a multicentre randomized clinical trial with oropharyngeal cancer (EORTC No. 22791) and the results are shown in Figure 14.3. The right-hand panel (B) shows that a hyperfractionated treatment with 70 fractions of 1.15 Gy (two fractions per day with a 4- to 6-hour interval, total dose 80.5 Gy) produced a similar incidence of late tissue damage to a conventional schedule of 35 fractions of 2 Gy (70 Gy given in the same overall time of 7 weeks). However, the larger total dose in the hyperfractionated treatment produced an increase of about 19% in long-term local tumour control (Figure 14.3A). The results of this study are in good agreement with the linear-quadratic model and support hyperfractionation as a way of increasing the therapeutic benefit in radiotherapy.

14.3 The effect of tumour cell proliferation during radiotherapy

It is possible that tumour cell proliferation during treatment may reduce the effectiveness of radiotherapy and this is the rationale for accelerated radiotherapy. Studies of tumour cell proliferation had been made for many years (Chapter 3) but a key development was a further review by Withers *et al* (1988), this time of published clinical data on the radiation dose required to control head and neck tumours as a function of overall treatment time.

This review led to the diagram shown in Figure 14.4A. The data points show doses to achieve tumour control in 50% of cases (*i.e.* TCD_{50} values) plotted against overall treatment time, for squamous cell carcinomas of the head and neck. Since a variety of doses per fraction were used in the various original studies summarized in this plot, the LQ model was used with an α/β ratio of 25 Gy to convert from the actual doses per fraction used into equivalent doses with 2 Gy per fraction. The various studies summarized also achieved different tumour control rates and it was therefore necessary to interpolate or extrapolate to the 50% control level. This required an assumed value for the steepness of the dose–control relationship for the tumours; it was assumed that the dose to increase control from 40% to 60% was 2.9 Gy. Figure 14.4A clearly indicates that as overall time is increased a greater total radiation dose is required to control these tumours. The other significant observation is that in the analysis by Withers *et al* (1988) there seemed to be an initial flat portion to this relationship. This implied that for treatment times shorter than 3–4 weeks proliferation had little effect and that it takes time for accelerated repopulation (Section 17.4) to be switched on in human tumours. Withers *et al* also concluded that for treatment times longer than this, the effect of proliferation was equivalent to a loss of radiation dose of about 0.6 Gy per day.

This very significant report gave rise to considerable debate. A later analysis of these same original studies by Bentzen and Thames (1991) is shown in Figure 14.4B. This analysis made a different assumption about the steepness of dose–response curves for tumour control: the dose to increase

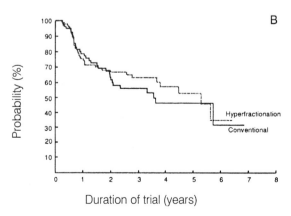

Figure 14.3 Results of the EORTC (22791) trial of clinical hyperfractionation. **A**: local tumour control (logrank, $p = 0.02$); **B**: patients free of late radiation effects, grade 2 or worse (logrank, $p = 0.72$). From Horiot *et al* (1992), with permission.

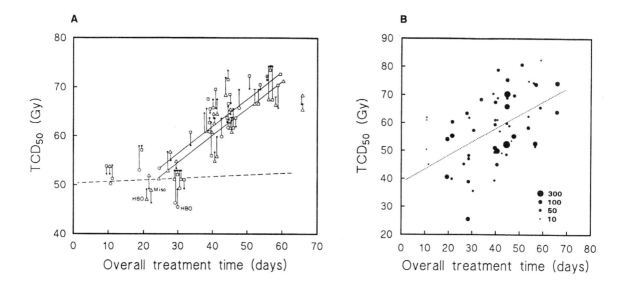

Figure 14.4 Tumour control dose (TCD$_{50}$) in head and neck cancer as a function of overall treatment time, normalized to a dose per fraction of 2 Gy. A large number of clinical studies have been summarized by (**A**) Withers *et al* (1988) and **B**: Bentzen and Thames (1991). In panel (**B**) each point indicates the result of a particular trial, the size of the symbol indicating the size of the trial. There is a trend for the curative radiation dose to increase with overall treatment time.

control from 40% to 60% was taken to be 10.5 rather than 2.9 Gy, thought to be more clinically realistic. In addition, the data points in Figure 14.4B have been drawn to indicate the size of the patient sample from which the estimate of TCD$_{50}$ had been made. Figure 14.4B suggests that the 'lag' period before commencement of repopulation may have been somewhat exaggerated by the plot shown in panel A, although this issue is still actively being discussed. The comparison between Figures 14.4A and B shows how difficult it is to make reliable retrospective analyses on clinical data.

The slope of the line through the data in Figure 14.4B is 0.48 Gy per day. From reasonable estimates of tumour cell radiosensitivity, this would correspond to a clonogen doubling time of less than 1 week, similar to the values of pre-treatment potential doubling time measured in human tumours (Section 25.5). The potential doubling time (T_{pot}, Section 3.2) is a cell kinetic parameter that indicates the rate at which cells are proliferating in an untreated tumour. Although there is much uncertainty about this, it has been suggested that *during* treatment the rate at which clonogenic cells within the tumour repopulate may also resemble the T_{pot} value. Thus *accelerated fractionation*, which uses a reduced overall treatment time below the conventional 6–7 weeks, should increase tumour cure rates by restricting the time available for tumour cell proliferation. From Figure 14.4B, for example, the dose in a 5-week schedule would be effectively larger than that in a 7-week schedule by a factor 0.48 × (7 − 5) × 7 = 6.7 Gy, or nearly 10% of a 70-Gy treatment.

Split-Course Radiotherapy

Intentional gaps in radiation therapy have sometimes been introduced in order to allow recovery of early-responding normal tissues. As will be clear from the foregoing paragraphs, the policy of split-course radiotherapy also faces the negative effects of tumour cell proliferation, which will tend to counteract the benefit to normal tissues of a gap in treatment. An example of the negative effect of a treatment gap is the study by Overgaard *et al* (1988) on the management of tumours of the larynx. The effect of a 4-week gap was investigated, increasing the overall treatment time from 6 to 10 weeks. The loss in tumour response that resulted from the introduction of the split is shown in Figure 12.4. The extra 4 weeks' increase in overall treatment time was worth about 10–12 Gy in total dose. Using reasonable estimates of tumour cell radiosensitivity, this would correspond to clonogen doubling times of about 4–5 days.

Modelling the Effect of Overall Time

Various calculations have been made of the efficacy of short schedules in radiotherapy. As an example, Figure 14.5 shows calculations made by J.F. Fowler comparing schedules with 5, 10 and 15 fractions per working week, *i.e.* 1, 2 and 3 fractions per day. Log(tumour cell kill) was determined using the linear-quadratic cell survival equation with $\alpha = 0.35$ Gy^{-1} and $\alpha/\beta = 10$ Gy (Section 13.4). As overall time increases (and thus the number of fractions) the dose per fraction has been continually adjusted to keep the same late effects as a reference schedule with 30 fractions of 2 Gy, using the LQ model with $\alpha/\beta = 3$ Gy. Tumour cell proliferation is assumed to occur after each fraction, with the clonogen doubling times shown against each curve. The star symbols show the maximum cell kill and hence the optimum overall time for each tumour cell doubling time and number of fractions per week. The horizontal dotted line shows the expected log cell kill needed to sterilize a 1-g tumour that is assumed to contain 10^9 cells. The following points should be noted:

1) For clonogen doubling times below 7 days, the optimum overall time is always less than 6 weeks and for such tumours accelerated radiotherapy should be better than conventional 6- to 7-week radiotherapy.
2) Whatever the value of the clonogen doubling time, it should always be better to use more than one treatment per day, to take advantage of the hyperfractionation effect (Figure 14.2) unless tumours have low α/β ratios.

It seems unlikely that accelerated radiotherapy will be appropriate for all patients in any given histopathological category. As indicated in Chapter 25, it will probably be necessary to employ a predictive test for tumour cell repopulation, on the basis of which to identify those patients who are likely to benefit. The future pattern of radiotherapy may well be to individualize treatments, using information such as the T_{pot} value.

Multiple Fractions per Day: How Close Together?

Hyperfractionation and accelerated radiotherapy both require multiple radiation treatments per day. In the case of hyperfractionation the reason for this is because if, as in the EORTC trial mentioned above, we were to reduce the fraction size and increase the fraction number to 70, this would require once-daily treatment over 14 weeks. This would be a bad option in view of the tumour cell

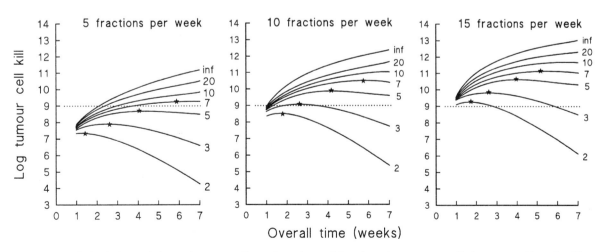

Figure 14.5 Calculations of the antitumour effect of accelerated fractionation over overall times up to 7 weeks. The ordinate shows the log cell kill for constant late normal-tissue reactions. Assumptions: $\alpha/\beta = 3$ Gy for late-responding normal tissues, 10 Gy for tumour; reference schedule is 30 fractions of 2 Gy. Lines are calculated for different assumed repopulation doubling times in days; the stars show the optimum schedule for each line. From Fowler (1990), with permission.

repopulation that this would allow. In the case of accelerated radiotherapy, shortening the overall time but still giving one fraction per day would require an increase in dose per fraction. This also is a bad option, as indicated in Section 14.2 and Figure 14.2B.

Multiple fractions per day is a non-standard approach to radiation therapy and most radiotherapy departments are unused to such treatments. There is an added radiobiological constraint in that the fractions given in one day should not be too close together because of incomplete repair (Section 13.9). As indicated in Section 6.7, repair of radiation damage occurs in almost all cell types (normal and malignant) and is a large phenomenon: much of the damage is normally repaired. We depend upon the repair of normal-tissue stem cells for their tolerance to radiation therapy. Repair occurs rapidly, with half-times in the region of 0.5–2 hours in tumours and many normal tissues. Nevertheless, it takes 6 half-times for the damage to decay to 1/64, *i.e.* 1–2% of its initial value. The problem is that repair in some late-responding normal tissues is slower than this (Table 13.4). In these cases, if we were to treat with multiple fractions spaced too close together, we would have a therapeutic disadvantage because the total radiation dose would have to be lowered in order to remain within tolerance. For this reason, current schedules involving multiple fractions per day are recommended to use a gap of at least 6 hours between fractions. It is possible that in some cases this might still be too short an interval, particularly if the radiation field includes the spinal cord, which has a significant proportion of its repair taking place slowly.

This requirement unfortunately creates logistic problems in radiotherapy departments. There appears to be a strong therapeutic case for using hyperfractionated or accelerated radiotherapy (or a combination of the two), but to put these into practice requires some extra expenditure of time and money by hospital authorities.

14.4 Clinical evaluation of accelerated radiotherapy

There are now a number of trials of accelerated radiotherapy taking place and this could prove to be one of the most significant advances in cancer treatment during the last 20 years. An example is EORTC trial No. 22851 (Horiot *et al*, 1997; see also Section 25.5) in which patients with head and neck cancer (excluding hypopharynx) were randomized to receive their treatment either conventionally in an overall time of 7–8 weeks (using 1.8–2 Gy per fraction each day, total 35–40 fractions, treatment on 5 days per week) or accelerated in 5 weeks (using 1.6 Gy per fraction three times per day with 4 hours' minimum interval, total 45 fractions, treatment on 5 days per week, overall time allocated to 8 days' radiotherapy, 12–14 days' gap, 17 days' radiotherapy). The report of this trial indicates that those patients receiving the accelerated treatment showed an approximately 12% increase in local tumour control ($p = 0.017$) although there was no increase in *survival* compared with patients receiving conventional treatment. However, both acute and late reactions were significantly increased in the accelerated arm; 'severe late toxicity' and 'severe connective-tissue damage' showed up to a 30% increase.

The strategy known as CHART is a much more radical approach to acceleration in radiotherapy. CHART means *Continuous Hyperfractionated Accelerated Radiotherapy* and it is now over 12 years since the first group of patients began receiving this treatment in the UK (Saunders *et al*, 1989; Saunders and Dische, 1989). The protocol is 36 fractions given over 12 consecutive days (including the weekend), using three fractions per day with an interval of 6 hours between the fractions within each day. Dose per fraction is 1.5 Gy to a total of 54 Gy. Total dose is therefore *reduced* compared with conventional therapy, in order to remain within the tolerance of acutely responding epithelial tissues.

Saunders *et al* (1996) have reported interim results of a Phase III clinical trial of CHART, comprising 563 patients with non-small-cell lung cancer and 918 patients with head and neck cancer. Patients were randomized between CHART (as described above) and conventional radiotherapy using 60 Gy for lung and 66 Gy for head and neck, in 2-Gy fractions given once per day on 5 days per week. For non-small-cell lung cancer (Figure 14.6), 30% of the CHART patients were alive at 2 years compared with 20% in the control group ($p = 0.006$). There was also a trend towards a greater gain with CHART in more advanced tumours, but this did not reach statistical significance. In the head and neck (Figure 14.7), there was a small (non-significant) improvement in disease-free interval when all cases were considered and a strong trend ($p = 0.065$) for improved benefit with

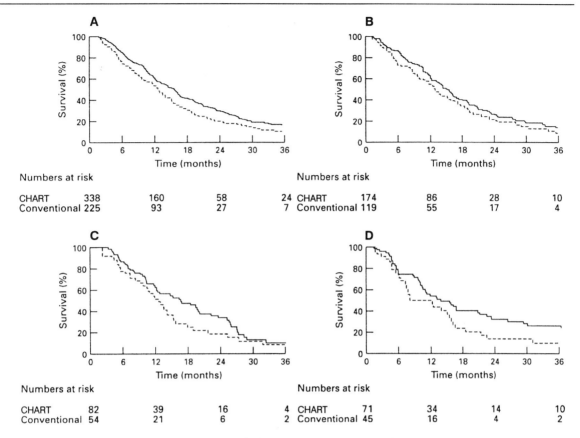

Figure 14.6 Results of a phase III randomized trial of CHART in non-small-cell lung cancer. Life-tables showing the overall survival of patients treated by CHART (——) and by conventional radiotherapy (----). **A**: all cases; **B**: T1 and T2; **C**: T3; **D**: T4. From Saunders *et al* (1996) with permission.

CHART for patients with more advanced disease, as in the lung cases. Early mucosal reactions appeared sooner and were more troublesome with CHART but they settled quickly and long-term morbidity with CHART was less than or equal to that in conventional therapy. Therefore, the conclusion overall was that CHART is at least as effective as conventional therapy with some evidence of a real gain especially in more advanced tumours.

CHART achieves this gain in tumour control despite a *reduction* in the total dose to 54 Gy. This means that the gain from using a shorter overall time (12 days compared with 6–7 weeks) leading to reduced tumour cell proliferation more than outweighs the reduction in cell kill due to the lower total dose. These results show clearly that tumour cell repopulation can occur during a conventional course of radiotherapy and be a cause of treatment failure. In the head and neck tumours (all cases), for which CHART and conventional radiotherapy are roughly equivalent, CHART uses 12 Gy less than conventional therapy in an overall time reduced by 33 days. This would correspond to 0.36 Gy per day lost through tumour cell proliferation, which is close to the value of 0.4 Gy per day implied by the clinical studies summarized in Figure 12.4 and the value of 0.48 Gy per day from the data in Figure 14.4B.

Although it could be that accelerated radiotherapy as in the CHART regime is an improved treatment for certain categories of cancer, the future may (as indicated above) lie with its use in subgroups of patients that are selected on the basis of a predictive assay for the rate of tumour cell repopulation (Section 25.4). Slowly growing tumours might be expected to do worse as a result of the reduced total dose compared with conventional radiotherapy and as a result accelerated radiotherapy might in such cases confer no advantage, or even a disadvantage.

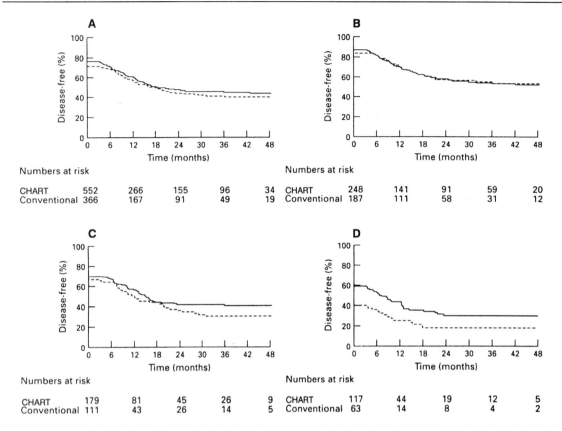

Figure 14.7 Results of a phase III randomized trial of CHART in head and neck cancer. Life-tables showing disease-free interval in patients treated by CHART (——) and by conventional radiotherapy (----). **A**: all cases; **B**: T1 and T2; **C**: T3; **D**: T4. From Saunders *et al* (1996) with permission.

Key points

1. *Hyperfractionation* is the use of a reduced dose per fraction over a conventional overall treatment time, employing multiple fractions per day. The therapeutic advantage is thought to derive from a more rapid increase in tolerance with decreasing dose per fraction for late-responding normal tissues than for tumours.
2. *Accelerated radiotherapy* is the use of a reduced overall treatment time with a conventional dose per fraction, achieved using multiple fractions per day. The aim is to reduce the protective effect of tumour cell repopulation during radiotherapy.
3. Multiple fractions per day should not be given closer than 6 hours apart.
4. There is encouraging clinical evidence that hyperfractionation and accelerated radiotherapy may both lead to improved treatments, especially if combined with appropriate patient selection.

BIBLIOGRAPHY

Bentzen SM and Thames HD (1991). Clinical evidence for tumor clonogen regeneration: interpretations of the data. *Radiother Oncol* 22:161–6.

Fowler JF (1990). How worthwhile are short schedules in radiotherapy?: A series of exploratory calculations. *Radiother Oncol* 18:165–81.

Hall EJ (1988). *Radiobiology for the Radiologist*. Lippincott; Philadelphia.

Horiot J-C, Bontemps P, van den Bogaert W *et al* (1997). Accelerated fractionation compared to conventional fractionation in the radiotherapy of advanced head and neck cancers: results of the EORTC 22851 randomized trial. *Radiother Oncol* 44 (in press).

Horiot J-C, Le Fur R, N'Guyen T *et al* (1992). Hyperfractionation *versus* conventional fractionation in oropharyngeal carcinoma: final analysis of a randomized trial of the EORTC cooperative group of radiotherapy. *Radiother Oncol* 25:231–41.

Overgaard J, Hjelm-Hansen M, Johansen LV and Andersen AP (1988). Comparison of conventional and split-course radiotherapy as primary treatment in carcinoma of the larynx. *Acta Oncol* 27:147–52

Saunders MI, Dische S, Barrett A *et al* (1996). Randomised multicentre trials of CHART vs. conventional radiotherapy in head and neck cancer and non-small-cell lung cancer: an interim report. *Brit J Cancer* 73:1455–62.

Saunders MI, Dische S, Hong A *et al* (1989). Continuous hyperfractionated accelerated radiotherapy in locally advanced carcinoma of the head and neck region. *Int J Radiat Oncol Biol Phys* 17:1287–93.

Thames HD, Withers HR, Peters LJ *et al* (1982). Changes in early and late radiation responses with altered dose fractionation: implications for dose-survival relationships. *Int J Radiat Oncol Biol Phys* 8:219–26.

Wilson GD, McNally NJ, Dische S *et al* (1988). Measurement of cell kinetics in human tumours *in vivo* using bromodeoxyuridine incorporation and flow cytometry. *Brit J Cancer* 58:423–31.

Withers HR, Taylor JMG and Maciejewski B (1988). The hazard of accelerated tumor clonogen repopulation during radiotherapy. *Acta Oncol* 27:131–46.

FURTHER READING

Begg AC, Hofland I, van Glabekke M *et al* (1992). Predictive value of potential doubling time for radiotherapy of head and neck tumor patients: results from the EORTC cooperative trial 22851. *Sem Radiat Oncol* 2:22–5.

Saunders MI and Dische S (1989). Continuous hyperfractionated accelerated radiotherapy in non-small-cell carcinoma of the bronchus. In: *The Scientific Basis of Modern Radiotherapy (BIR Report 19)*, pp. 47–51. (Ed) McNally NJ. British Institute of Radiology; London.

Thames HD, Peters LJ, Withers HR and Fletcher G. (1983). Accelerated fractionation vs hyperfractionation: rationales for several treatments per day. *Int J Radiat Oncol Biol Phys* 9:127–38.

Withers HR, Maciejewski B and Taylor JMG (1989). Biology of options in dose fractionation. In: *The Scientific Basis of Modern Radiotherapy*, pp. 27–36. (Ed) McNally NJ. British Institute of Radiology; London.

15

The oxygen effect

Michael R. Horsman and Jens Overgaard

15.1 The importance of oxygen

The response of cells to ionizing radiation is strongly dependent upon oxygen (Gray *et al.,* 1953; Wright and Howard-Flanders, 1957). This is illustrated in Figure 15.1 for mammalian cells irradiated in culture. Cell surviving fraction is shown as a function of radiation dose administered either under normal aerated conditions or under hypoxia, generally achieved by flowing nitrogen gas over the surface of the cell suspensions for a period of 30 minutes or more. The enhancement of radiation damage by oxygen is *dose-modifying, i.e.* the radiation dose that gives a particular level of survival is reduced by the same factor at all levels of survival. This allows us to calculate an *oxygen enhancement ratio* (OER):

$$\text{Oxygen enhancement ratio} = \frac{\text{Radiation dose in hypoxia}}{\text{Radiation dose in air}}$$

for the same level of biological effect. For most cells the OER for X-rays is around 3.0. However, some studies suggest that at radiation doses of 3 Gy or less the OER is actually reduced (Palcic and Skarsgard, 1984). This is an important finding because this is the dose range for clinical fractionation treatments.

It has been demonstrated from rapid-mix studies that the oxygen effect only occurs if oxygen is present either during irradiation or within a few milliseconds thereafter (Howard-Flanders and Moore, 1958; Michael *et al,* 1973). The dependence of the degree of sensitization on oxygen tension is shown in Figure 15.2. By definition, the OER under anoxic conditions is 1.0. As the oxygen level increases there is a steep increase in radiosensitivity (and thus in the OER). The greatest change occurs from 0 to about 20 mmHg; further increase in oxygen concentration, up to that seen in air (155 mmHg) or even to 100% oxygen (760 mmHg), produce a small though definite increase in radiosensitivity. Also shown in Figure 15.2 are the oxygen partial pressures typically found in arterial and

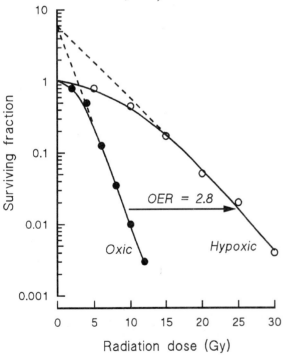

Figure 15.1 Survival curves for cultured mammalian cells exposed to X-rays under oxic or hypoxic conditions (diagrammatic), illustrating the radiation dose-modifying effect of oxygen. Note that the broken lines extrapolate back to the same point on the survival axis ($n = 5.5$).

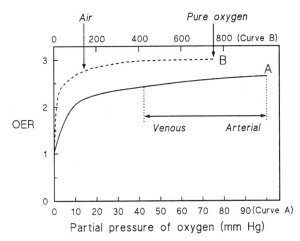

Figure 15.2 Variation of oxygen enhancement ratio (OER) with oxygen tension. The arrows indicate the range of physiological blood oxygen tensions on the lower scale (to convert mmHg to kPa, multiply by 0.133). Adapted from Denekamp (1989), with permission.

venous blood. Thus from a radiobiological standpoint most normal tissues can be considered to be well oxygenated, although it is now recognized that moderate hypoxia is a feature of some normal tissues such as cartilage and skin.

The mechanism responsible for the enhancement of radiation damage by oxygen is generally referred to as the oxygen-fixation hypothesis and is illustrated in Figure 15.3 (see also Section 8.1). When radiation is absorbed in a biological material, free radicals are produced. These radicals are highly reactive molecules and it is these which break chemical bonds, produce chemical changes, and initiate the chain of events that result in biological damage. They can be produced either directly in the target molecule (usually DNA) or indirectly in other cellular molecules and diffuse far enough to reach and damage critical targets. Most of the indirect effects occur by free radicals produced in water, since this makes up 70–80% of mammalian cells. It is the fate of the free radicals ultimately produced in the critical target, designated as $R^{\bullet}$ in Figure 15.3, that is important. If oxygen is present then it can react with $R^{\bullet}$ to produce $RO_2^{\bullet}$ which then undergoes further reaction ultimately to yield ROOH in the target molecule. Thus we have a change in the chemical composition of the target and the damage is chemically fixed. Subsequently this damage can be processed enzymatically and perhaps repaired (Section 8.5).

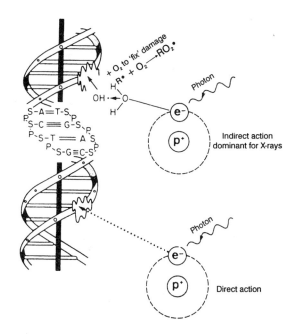

Figure 15.3 The oxygen fixation hypothesis. Free radicals produced in DNA either by a direct or indirect action of radiation can be repaired under hypoxia but fixed in the presence of oxygen. Adapted from Hall (1988), with permission.

In the absence of oxygen, or in the presence of reducing species, $R^{\bullet}$ can react with H, thus restoring its original form.

15.2 Tumour hypoxia

Oxygen plays an important role in the radiation response of tumours. The growth of solid tumours requires the induction of a blood supply, a process which is referred to as *angiogenesis*. This new blood supply is primitive in nature and it may be inadequate for meeting all the needs of the growing tumour. Nutrient-deprived and oxygen-deprived regions develop, yet the hypoxic cells existing in these areas may still be viable, at least for a time. The first clear indication that hypoxia could exist in tumours was made in 1955 by Thomlinson and Gray from their observations on histological sections of fresh specimens from human carcinoma of the bronchus (Figure 15.4). They observed viable tumour regions surrounded by vascular stroma

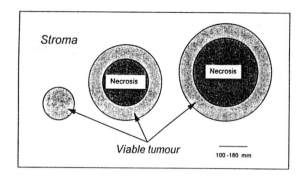

Figure 15.4 Idealized description of the development of microscopic regions of necrosis in tumours. Conclusions by Thomlinson and Gray from studies on histological sections of human bronchial carcinoma showing the development of necrosis beyond a limiting distance from the vascular stroma. Adapted from Hall (1988), with permission.

from which the tumour cells obtained their nutrient and oxygen requirements. As these regions expanded, areas of necrosis appeared at the centre. The thickness of the resulting shell of viable tissue (100–180 μm) was found to be similar to the calculated diffusion distance of oxygen in respiring tissues; it was thus suggested that as oxygen diffused from the stroma it was consumed by the cells and, while those cells beyond the diffusion distance were unable to survive, cells immediately bordering on to necrosis might be viable yet hypoxic.

Tannock (1968) made similar observations in mouse mammary tumours. The extent of necrosis in these tumours was much greater and each patent blood-vessel was surrounded by a cord of viable tumour cells outside which was necrosis. This 'corded' structure is also seen in other solid tumours and is illustrated in Figure 15.5. Cells at the edge of the cords are thought to be hypoxic and are often called *chronically hypoxic cells*. Tannock showed, however, that since the cell population of the cord is in a dynamic state of cell turnover these hypoxic cells will have a short lifespan, being continually replaced as other cells are displaced away from the blood-vessel and in turn become hypoxic. More recently it has been suggested that some tumour blood-vessels may periodically open and close, leading to *transient or acute hypoxia* (Figure 15.5). The mechanisms responsible for intermittent blood flow in tumours are not entirely clear. They might include the plugging of vessels by blood cells or by circulating tumour cells; collapse of vessels in regions of high tumour interstitial pressure; or spontaneous vasomotion in incorporated host arterioles affecting blood flow in downstream capillaries.

15.3 Hypoxia in experimental tumours

Since hypoxic cells are resistant to radiation, their presence in tumours is critical in determining the response of tumours to treatment with large doses of radiation. The presence of such cells in experimental tumours can easily be demonstrated, as shown in Figure 15.6. This shows the radiation survival response of KHT mouse sarcomas, irradiated *in situ* in air-breathing mice; or in nitrogen-asphyxiated mice (*i.e.* hypoxic); or as a single-cell suspension *in vitro* under fully oxic conditions. The studies in air-breathing mice were made both under normal and anaemic conditions. Cell survival was estimated immediately after irradiation using a lung colony assay (Section 6.4). The survival curves for tumours in air-breathing mice are biphasic. At low radiation doses the response is dominated by the aerobic cells and the curves are close to the oxic curve; at larger radiation doses the presence of hypoxic cells begins to influence the response and the survival curve eventually parallels the hypoxic curve. The proportion of hypoxic cells (the *hypoxic fraction*) can be calculated from the vertical separation between the hypoxic and air-breathing sur-

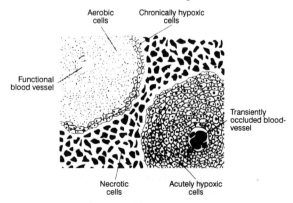

Figure 15.5 Schematic representation of diffusion-limited *chronic* hypoxia and perfusion-limited *acute* hypoxia within tumour cords. From Horsman and Overgaard (1992), with permission.

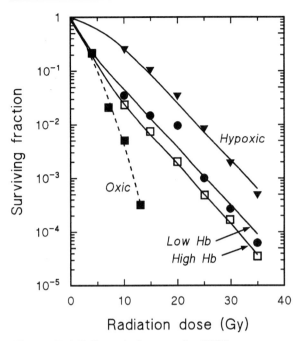

Figure 15.6 Cell survival curves for KHT mouse sarcomas irradiated under aerobic or hypoxic conditions. The hypoxic data were obtained by killing the mice shortly before irradiation. Two sets of data for tumours in air-breathing mice are shown, with high and low haemoglobin levels. The dashed curve shows the *in vitro* survival of oxic cells. From Hill *et al* (1971), with permission.

vival curves in the region where they are parallel. In this mouse sarcoma the hypoxic fraction was calculated to be 0.06 in mice with a high haemoglobin level ($\approx$ 16.5 g%) and 0.12 in anaemic mice (haemoglobin level $\approx$ 9.5 g%). These data thus illustrate not only the presence of hypoxic cells in these tumours but also the influence of oxygen transport.

Two other techniques are routinely used to estimate hypoxia in animal tumours. These are the so-called clamped tumour growth delay assay, which involves measuring the time taken for tumours to grow to a specific size after irradiation; and the clamped tumour control assay, in which the percentage of animals showing local tumour control at a certain time after treatment is recorded. For both techniques it is necessary to produce full radiation dose–response curves under normal and clamped conditions and the hypoxic fractions can then be calculated from the displacement of these dose–response curves. Using either of these assays, or the paired survival curve assay described above, it has been demonstrated that most experimental solid tumours contain hypoxic cells, with estimates of the hypoxic fractions ranging from below 1% to well over 50%.

Experimentally, a variety of procedures have been found to reduce or eliminate hypoxic cells in animal tumours. These include:

(i) Increasing oxygen availability by allowing animals to breathe oxygen or carbogen (*i.e.* 95% oxygen + 5% carbon dioxide) under normobaric or hyperbaric conditions; by introducing perfluorochemical emulsions into the vascular system to increase the oxygen-carrying capacity of the blood; by using drugs that modify the affinity of haemoglobin for oxygen; or by the use of agents that increase tumour blood perfusion.

(ii) Radiosensitizing the hypoxic cells either chemically using drugs such as misonidazole, etanidazole and nimorazole (Section 16.3), or by hyperthermia (Chapter 23).

(iii) Preferentially killing hypoxic cells by means of bioreductive drugs that are active under hypoxic conditions, such as mitomycin C, E09, tirapazamine and RSU-1069, or again using hyperthermia.

15.4 Hypoxia in human tumours

Although the presence of radiation-resistant hypoxic cells in animal tumours is well established and their impact on response to radiation can be reduced subtantially by the procedures listed above, the improvement of radiation response in human tumours as a result of these approaches has not been so successful. This could be because the phenomenon of reoxygenation (see below) is effective in some human tumours. Another possible explanation is that most of these treatments work well against chronically hypoxic cells, but apart from the radiosensitizing/bioreductive drugs, they have little or no influence on cells that are acutely hypoxic. More recent studies have now shown that there are agents that can specifically reduce acute hypoxia and these include nicotinamide, angiotensin II and flunarizine.

The failure to see significant clinical benefit from some manoeuvres designed to reduce the impact of hypoxia has led to questions about the presence and relevance of hypoxia in human tumours. Numerous attempts have been made to estimate the level of tumour hypoxia. Unfortunately, the direct procedures that have been used in animal tumours, such as the paired survival curve assay (Figure 15.6) or clamped tumour growth/control assays, are not applicable to the human situation. Clinical estimates of tumour hypoxia have been made indirectly by measurements of tumour vascularization, using such end-points as intercapillary distance, vascular density and the distance from tumour cells to the nearest blood vessel; tumour metabolic activity, measured with biochemical, HPLC, bioluminescent and NMR techniques; estimating the degree of DNA damage as with the comet assay; and the binding of specific radioactive or fluorescent-labelled compounds to hypoxic cells.

Direct measurements of hypoxia in human tumours have been made by determining oxygen partial pressure (pO_2) distributions using electrodes. Recent studies in a mouse tumour system have found a direct relationship between electrode estimates of tumour oxygenation and the actual percentage of hypoxic clonogenic cells. This is illustrated in Figure 15.7 in which the hypoxic fraction, determined using a clamped tumour-control assay, was altered in a C3H mouse mammary carcinoma by allowing the mice to breathe different gas mixtures. A strong correlation was found between hypoxic fraction and the percentage of measured pO_2 values that were equal to or less than 5 mmHg.

These oxygen electrodes have also been used to measure pO_2 distributions in human tumours, and in at least two sites (cervical cancers and tumours of the head and neck region) the pO_2 measurements have been related to local tumour response after radiation therapy. In general, good correlations between treatment outcome and pre-treatment pO_2 measurements were observed for both tumour types, with the less well oxygenated tumours showing the poorest results. This is clearly illustrated for head and neck tumours in Figure 15.8, in which those patients with tumours where the percentage of low pO_2 values measured was greater than the median value for all 35 tumours had a significantly poorer local regional tumour control.

Several other clinical trials have indirectly provided evidence that hypoxia exists in human tumours and can influence radiation response. Significant improvement in local tumour control has been seen, particularly in head and neck cancers, from hyperbaric oxygen, chemical radiation sensitizers and improved oxygen supply (Chapter 16). This is illustrated in Figure 15.9, in which local regional control of tumours is expressed as a function of pre-treatment haemoglobin concentration in male or female patients treated with radiotherapy for squamous cell carcinoma of the larynx and pharynx. Local tumour control was lower in those patients with reduced haemoglobin concentrations. Such a reduction in haemoglobin would make less oxygen available to the tumour and thus increase the level of tumour hypoxia.

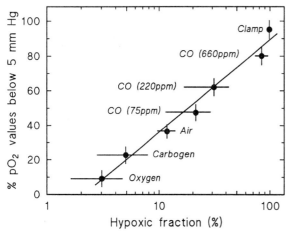

Figure 15.7 Relationship between pO_2 electrode measurements and the hypoxic fraction in a C3H mouse mammary carcinoma. Results were obtained from normal air-breathing mice, in clamped tumours, and in mice allowed to breathe oxygen, carbogen, or various concentrations of carbon monoxide (CO). From Horsman et al (1993), with permission.

15.5 Reoxygenation

The time-course of changes in the hypoxic fraction of a tumour before and after irradiation is illustrated in Figure 15.10. Tumours less than 1 mm in diameter have been found to be fully oxygenated (Stanley et al, 1977). Above this size they usually

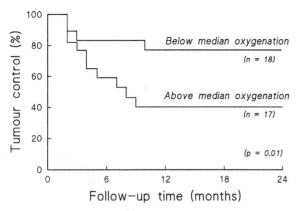

Figure 15.8 Local tumour control correlates with pre-treatment oxygen levels in tumours. Oxygen levels measured with an Eppendorf electrode in 34 lymph nodes and 1 primary advanced squamous cell carcinoma of the head and neck. Tumours were stratified by whether the fraction of pO$_2$ values less than 2.5 mmHg was above or below the median value for the whole group (*i.e.* 15%). The lines show Kaplan Meier estimates of actuarial tumour control probability up to 2 years after receiving conventional external radiotherapy (66–68 Gy in 33–34 fractions). From Nordsmark *et al* (1996), with permission.

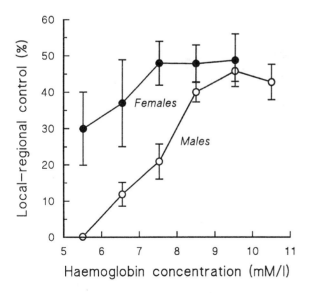

Figure 15.9 Local regional tumour control as a function of sex and pre-treatment haemoglobin value in 1112 patients treated with radiotherapy for squamous cell carcinoma of the larynx and pharynx. From Overgaard (1988), with permission.

become hypoxic. If tumours are irradiated with a large single dose of radiation, most of the radiosensitive aerobic cells in the tumour will be killed. The cells that survive will predominantly be hypoxic and therefore the hypoxic fraction immediately after irradiation will be close to 100%. Subsequently, the hypoxic fraction falls and approaches its starting value. This phenomenon is termed *reoxygenation*. The process of reoxygenation has been reported to occur in a variety of tumour systems, although the speed of reoxygenation varies widely, occurring within a few hours in some tumours and taking several days in others. Furthermore, the final level of hypoxia after reoxygenation can also be higher or lower than its value prior to irradiation.

The mechanisms underlying reoxygenation in tumours are not clearly understood. A number of possible processes are listed in Table 15.1. If reoxygenation occurs rapidly then it may be due either to recirculation of blood through vessels that were temporarily closed or to a decreased cellular respiration (which will increase the oxygen diffusion distance). Reoxygenation occurring at longer time intervals is probably the result of cell death leading to tumour shrinkage and a reduction in intercapillary distances, thus allowing oxygen to reach hypoxic cells.

Reoxygenation has important implications in clinical radiotherapy. Figure 15.11 illustrates the hypothetical situation in a tumour following fractionated radiation treatments. In this example, 98% of the tumour cells are considered well oxygenated and 2% are hypoxic. The responses of oxic and hypoxic cells to repeated large dose fractions are illustrated. If no reoxygenation occurs, then each dose of radiation would be expected to kill only a small number of the hypoxic cells and the resultant curve is therefore much shallower than that for oxic cells. At the end of treatment the tumour response will be dominated by the hypoxic cell population. However, if reoxygenation occurs between fractions then the radiation killing of initially hypoxic cells will be greater and the hypoxic cells then have less impact on response. It has not been possible to detect reoxygenation directly in human tumours, but its existence is supported by the fact that local control can be achieved in a variety of tumours given fractionated radiotherapy with 30 fractions of 2 Gy, consistent with the measured SF$_2$ values for oxic tumour cells (Section 17.5).

138 *The oxygen effect*

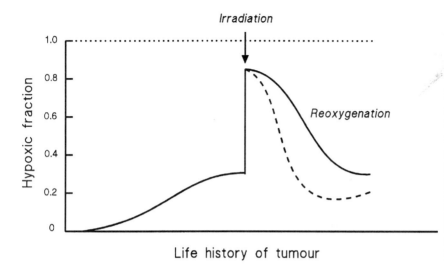

Figure 15.10 The time-course of changes in the hypoxic fraction during the life history of a tumour. Small lesions are well oxygenated but as the tumour grows the hypoxic fraction rises perhaps in excess of 10%. A large single dose of radiation kills oxic cells and raises the hypoxic fraction. The subsequent fall is termed reoxygenation.

15.6 Drug resistance and malignant progression

The presence of hypoxic cells in tumours not only has a significant impact on radiation therapy: there are also strong indications that these same cells may be responsible for tumour resistance to certain types of chemotherapy. Evidence from animal studies has shown that drugs like bleomycin, 5-fluorouracil, methotrexate and *cis*-platinum are less effective at killing tumour cells when they are hypoxic than when they are well oxygenated (Grau and Overgaard, 1992). Whether this is a consequence of hypoxia *per se* or because hypoxic cells are normally distant from blood-vessels, thus creating problems for drug delivery, or because such cells are typically non-cycling and exist in areas of low pH, both of which

Table 15.1 Mechanisms and time-scales of tumour reoxygenation

Recirculation through temporarily closed vessels	Minutes
Reduced respiration rate in damaged cells	Minutes to hours
Ischaemic death of cells without replacement	Hours
Mitotic death of irradiated cells	Hours
Cord shrinkage as dead cells are resorbed	Days

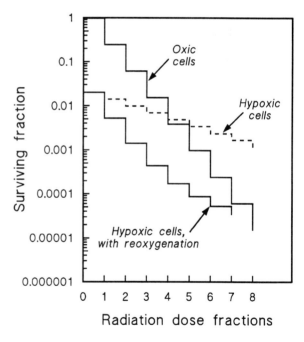

Figure 15.11 Calculated survival curves for a tumour containing 98% well-oxygenated cells and 2% hypoxic cells when given repeated fractions of radiotherapy. The upper two lines show the progressive depletion of oxic and hypoxic cells in the absence of reoxygenation. The lower line assumes that after each dose fraction reoxygenation restores the hypoxic fraction to its pre-treatment level.

can influence drug activity, has never been fully established.

More recent studies are now providing evidence that hypoxia may also contribute to malignant progression through its effects on signal transduction pathways and the regulation of transcription of various genes (Giaccia, 1996; Sutherland et al, 1996). Examples of the types of genes/proteins which have been shown to be upregulated by hypoxia are listed in Table 15.2. These include factors involved in cellular metabolism, as well as those that help cells deal with oxidative stress (induced by radiation, hydrogen peroxide and metals) and the detoxification of drugs and chemicals. More importantly, hypoxia has been shown to induce factors that are associated with apoptosis (e.g. p53) and angiogenesis (e.g. vascular endothelial growth factor, VEGF), both of which are likely to have profound influences on the growth and development of tumours.

Bibliography

Denekamp J (1989). Physiological hypoxia and its influence on radiotherapy. In: *The Biological Basis of Radiotherapy*, Second edition. (Eds) Steel GG, Adams GE and Horwich A. Elsevier Science; Amsterdam.

Giaccia A (1996). Hypoxic stress proteins: survival of the fittest. *Sem Radiat Oncol* 6:46–58.

Grau C and Overgaard J (1992). Effect of etoposide, carmustine, vincristine, 5-fluorouracil, or methotrexate on radiobiologically oxic and hypoxic cells in a C3H mouse mammary carcinoma *in situ*. *Cancer Chemother Pharmacol* 30:277–80.

Gray LH, Conger AD, Ebert M et al (1953). The concentration of oxygen dissolved in tissues at the time of irradiation as a factor in radiotherapy. *Br J Radiol* 26:638–48.

Hall EJ (1988). *Radiobiology for the Radiologist*. Lippincott; Philadelphia.

Table 15.2 Some examples of the genes/proteins that are upregulated by hypoxia

Glucose-regulated proteins	ORP80 (GRP78), ORP100 (GRP94)
Redox stress enzymes/molecules	ORP33 (hemeoxygenase I), ORP7 (metallothionein IIA)
Metabolic enzymes	ALDA, PGKI, PKM, PFKL, LDHA, others
Transcription factors and signalling molecules	cJun, cFos, Junb, AP-1, p53, NF-kB, pRB, HIF-1, SP-1, PKC
Growth factors/receptors/cytokines	EPO, VEGF (VPF), ET-1, EGFR, PDGF-B, IL-1α, GADD45, GADD153

Adapted from Sutherland *et al* (1996).

Key points

1. Hypoxic cells are much less sensitive to radiation than well-oxygenated cells.
2. Hypoxic cells probably occur in most animal and human tumours; they are believed to be an important cause of resistance to radiotherapy, especially using a small number of large dose fractions.
3. Hypoxia in tumours can be chronic or acute, for which the underlying mechanisms differ. Attempts to eliminate these radioresistant tumour cells require treatments that are effective against each of these types of hypoxia.
4. Reoxygenation has been shown to occur in animal tumours; some tumours reoxygenate rapidly, others more slowly. The evidence for reoxygenation in human tumours is less direct.
5. Hypoxic cells in tumours are also known to be resistant to certain chemotherapeutic agents. They may also play an important role in malignant progression.

Hill RP, Bush RS and Yeung P (1971). The effect of anaemia on the fraction of hypoxic cells in an experimental tumour. *Br J Radiol* 44:299–304.

Horsman MR, Khalil AA, Nordsmark M *et al* (1993). Relationship between radiobiological hypoxia and direct estimates of tumour oxygenation in a C3H mouse tumour model. *Radiother Oncol* 28:69–71.

Horsman MR and Overgaard J (1992). Overcoming tumour radiation resistance resulting from acute hypoxia. *Eur J Cancer* 28:717–8.

Howard-Flanders P and Moore D (1958). The time interval after pulsed irradiation within which injury in bacteria can be modified by dissolved oxygen. I. A search for an effect of oxygen 0.02 seconds after pulsed irradiation. *Radiat Res* 9:422–37.

Michael BD, Adams GE, Hewitt HB *et al* (1973). A post-effect of oxygen in irradiated bacteria: a submillisecond fast mixing study. *Radiat Res* 54:239–51.

Nordsmark M, Overgaard M and Overgaard J (1996). Pretreatment oxygenation status predicts radiation response in advanced squamous cell carcinoma of the head and neck. *Radiother Oncol* 41:31–40.

Overgaard J (1988). The influence of haemoglobin concentration on the response to radiotherapy. *Scand J Clin Lab Invest* 48 (Suppl 189):49–53.

Palcic B and Skarsgard LD (1984). Reduced oxygen enhancement ratio at low doses of ionizing radiation. *Radiat Res* 100:328–39.

Stanley JA, Shipley WU and Steel GG (1977). Influence of tumour size on hypoxic fraction and therapeutic sensitivity of Lewis lung tumour. *Br J Cancer* 36:105–13.

Sutherland RM, Ausserer WA, Murphy BJ and Laderoute KR (1996). Tumour hypoxia and heterogeneity: challenges and opportunities for the future. *Sem Radiat Oncol* 6:59–70.

Tannock IF (1968). The relation between cell proliferation and the vascular system in a transplanted mouse mammary tumour. *Br J Cancer* 22:258–73.

Thomlinson RH and Gray LH (1955). The histological structure of some human lung cancers and the possible implications for radiotherapy. *Br J Cancer* 9:539–49.

Wright EA and Howard-Flanders P (1957). The influence of oxygen on the radiosensitivity of mammalian tissues. *Acta Radiol* 48:26–32.

FURTHER READING

Brown JM (1979). Evidence for acutely hypoxic cells in mouse tumours and a possible mechanism of reoxygenation. *Br J Radiol* 52:650–6.

Chaplin DJ, Durand RE and Olive PL (1986). Acute hypoxia in tumors: implication for modifiers of radiation effects. *Int J Radiat Oncol Biol Phys* 12:1279–82.

Chapman JD (1984). The detection and measurement of hypoxic cells in solid tumors. *Cancer* 54:2441–9.

Dische S (1989). The clinical consequences of the oxygen effect. In: *The Biological Basis of Radiotherapy*, Second edition. (Eds) Steel GG, Adams GE and Horwich A. Elsevier Science; Amsterdam.

Graeber TG, Osmanian C, Jacks T *et al* (1996). Hypoxia-mediated selection of cells with diminished apoptotic potential in solid tumours. *Nature* 379:88–91.

Kallman RF and Rockwell S (1977). Effects of radiation on animal tumour models. In: *Cancer*, Vol 6. (Ed) Becker FF. Plenum Press; New York.

Moulder JE and Rockwell S (1984). Hypoxic fractions of solid tumors: experimental techniques, methods of analysis and a survey of existing data. *Int J Radiat Oncol Biol Phys* 10:695–712.

Overgaard J (1989). Sensitization of hypoxic tumour cells – clinical experience. *Int J Radiat Biol* 56:801–11.

Rasey JS and Evans ML (1991). Detecting hypoxia in tumours. In: *Tumour Blood Supply and Metabolic Microenvironment*. (Eds) Vaupel P and Jain RK. Gustav Fischer Verlag; Stuttgart.

Sutherland RM and Franko AJ (1980). On the nature of the radiobiologically hypoxic fraction in tumours. *Int J Radiat Oncol Biol Phys* 6:117–20.

Urtasun RC (1992). Tumor hypoxia, its clinical detection and relevance. In: *Radiation Research. A Twentieth Century Perspective*, Vol 2. (Eds) Dewey WC, Edington M, Fry RJ, Hall EJ and Whitmore GF. Academic Press; San Diego.

Vaupel P, Kallinowski F and Okunieff P (1989). Blood flow, oxygen and nutrient supply, and metabolic micro-environment of human tumors: a review. *Cancer Res* 49:6449–65.

16
Overcoming hypoxic cell radioresistance

Jens Overgaard and Michael R. Horsman

16.1 Introduction

There are two principal ways to overcome the radiobiological problem of tumour hypoxia: by increasing the delivery of oxygen (or oxygen-mimicking agents) to the cells, or by exploiting the special environmental conditions of hypoxic cells using agents that exercise their toxicity under those conditions. The first approach has been applied in radiotherapy by use of hyperbaric or normobaric oxygen, by attempts to increase blood flow or by the application of hypoxic cell radiosensitizers. Past experience has shown that even in tumours that have the same histology and degree of differentiation there may be substantial heterogeneity in hypoxia; thus, the identification of those tumours in which hypoxic radioresistance is a serious problem is currently a major goal of research. The second approach is illustrated by the use of bioreductive drugs or hyperthermia to destroy radioresistant hypoxic cells. This can be intensified by temporarily *increasing* the hypoxic state using agents such as hydralazine that *reduce* blood flow.

16.2 Hyperbaric oxygen

Following the identification of hypoxia as a potential source of tumour radioresistance, intensive efforts were made to overcome this problem especially by the use of hyperbaric oxygen therapy (Churchill-Davidson, 1968). Most trials were small and suffered from the use of unconventional fractionation schemes, but hyperbaric oxygen therapy did appear to be superior to radiotherapy given in air (Table 16.1), especially when a few large fractions were applied (Overgaard, 1989). Whether this gain also occurs with conventional fractionation has been the subject of considerable debate. However, the largest multicentre clinical trials of hyperbaric oxygen (those by the British Medical Research Council, MRC) showed both in uterine cervix and in advanced head and neck cancer a significant benefit in local tumour control and in subsequent survival. The same was not observed in bladder cancer, nor was this result confirmed by a number of smaller studies (Dische, 1985; Overgaard, 1989). The exploration of hyperbaric oxygen therapy was discontinued partly because of problems with patient compliance and partly due to the introduction of chemical radiosensitizers; in retrospect this may have been somewhat premature.

16.3 Hypoxic cell radiosensitizers

The concept of chemical radiosensitization of hypoxic cells was introduced by Adams (1977) and his co-workers. The largest group of sensitizing agents are the electron-affinic radiosensitizers, so-called because their efficiency of sensitization is directly related to their electron affinity. The rationale for their use is that they diffuse out of the

142 Overcoming hypoxic cell radioresistance

Table 16.1 Multicentre randomized trials with hyperbaric oxygen (HBO)

Site and reference	No. of patients	End-point	HBO	Air	
Head and neck carcinoma					
MRC (1st trial) (Henk et al, 1977)	294	Local/regional control (5 yrs)	53%	30%	($p < 0.01$)
MRC (2nd trial) (Henk, 1986)	106	Local/regional control (5 yrs)	60%	41%	($p < 0.05$)
Uterine cervix carcinoma					
MRC (Watson et al, 1978)	320	Local/regional control (5 yrs)	67%	47%	($p < 0.001$)
Bronchogenic carcinoma					
MRC (60 Gy/40 fx) (Cade and McEwen, 1978)	51	Survival (2 yrs)	15%	8%	(n.s.)
MRC (36 Gy/6 fx) (Cade and McEwen, 1978)	123	Survival (2 yrs)	25%	12%	($p < 0.05$)
Carcinoma of the bladder					
MRC (Cade et al, 1978)	241	Survival (5 yrs)	28%	30%	(n.s.)

n.s. = not significant.
See Overgaard (1989) for references.

tumour blood supply and are absorbed by, and thus sensitize, distant hypoxic cells. In principle, these drugs mimic the sensitizing effect of oxygen and therefore do not increase the radiation response of well-oxygenated cells, for instance in surrounding normal tissues.

The first electron-affinic compounds to show radiosensitization were the nitrobenzenes. They were followed by the nitrofurans and finally nitroimidazoles, the most potent of which was found to be the 2-nitroimidazole, *misonidazole*. Its *in vitro* activity is illustrated in Figure 16.1. Note that in these experiments misonidazole is radiation-dose-modifying: the survival curves have the same extrapolation number (*i.e.* 4). The radiation response of hypoxic cells can thus be enhanced substantially by irradiating the cells in the presence of misonidazole. In fact, at a drug concentration of 10 mM the radiosensitivity of hypoxic cells approaches that of aerated cells. The response of the aerated cells is unaffected.

Radiosensitizers such as misonidazole also enhance radiation damage in experimental tumours *in vivo*, as shown in Figure 16.2. The magnitude of the sensitizing effect is usually expressed by the sensitizer enhancement ratio (SER):

$$\text{Sensitizer enhancement ratio} = \frac{\text{Radiation dose } without \text{ sensitizer}}{\text{Radiation dose } with \text{ sensitizer}}$$

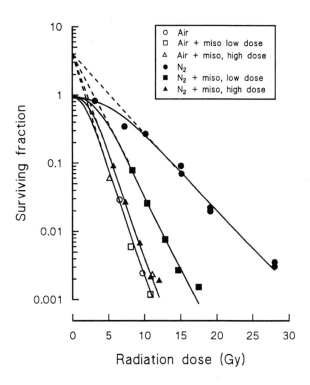

Figure 16.1 Survival curves for aerated and hypoxic Chinese hamster cells irradiated in the presence or absence of misonidazole. Low dose: 1 mM; high dose: 10 mM. From Adams (1977), with permission.

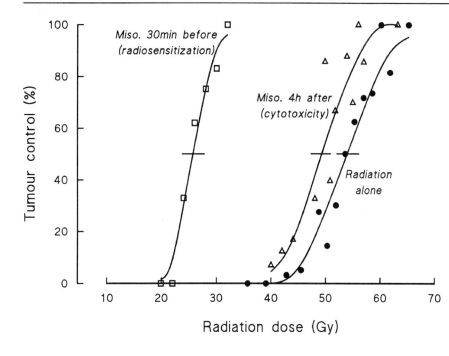

Figure 16.2 Local tumour control in C3H mouse mammary carcinomas measured 120 days after tumour irradiation. Mice were given misonidazole (1 g/kg, i.p.) either 30 minutes before or 4 hours after irradiation. The TCD_{50} dose was reduced from 54 Gy in control animals to 26 Gy in the misonidazole-pretreated mice, equivalent to an enhancement ratio (SER) of 2.1. Misonidazole given 4 hours *after* irradiation gave a TCD_{50} of 49 Gy, an SER of 1.1.

for the same biological effect. Large enhancement ratios (> 2.0) have been found in a variety of animal tumours following single-dose irradiation. Since the problem of tumour hypoxia increases with increasing fraction size (Figure 15.6) the benefits of radiosensitizers are clearest when tested by single-dose treatment. When misonidazole was combined with *fractionated* radiation, the SER values were smaller. This probably results from reoxygenation between radiation fractions reducing the therapeutic impact of hypoxia. Also shown in Figure 16.2 is the effect of giving misonidazole *after* irradiation, where a small but significant enhancement was seen. This obviously cannot be due to hypoxic cell radiosensitization; it is probably due to the well-demonstrated observation that mizonidazole is directly toxic to hypoxic cells, the level of cell killing increasing considerably with the *duration* of exposure to the sensitizer.

Following encouraging laboratory studies, misonidazole was introduced into clinical trials. As will be discussed in the next section, the results have been disappointing. This may largely be attributed to the fact that doses were limited to inadequate levels because of neurotoxicity. Figure 16.3 summarizes data in mice for the dependence of sensitization on misonidazole concentration, in comparison with *in vitro* results. The maximum tolerated dose of this drug that can be given with standard clinical fractionated radiotherapy is around 0.5 g/m², which results in a tumour concentration of only about 15 μg/g, and it is clear from the laboratory animal data that such a dose could only be expected to yield a small sensitizer enhancement ratio.

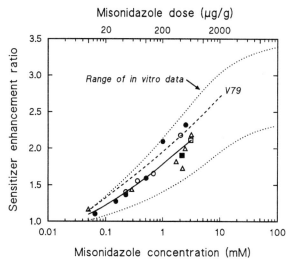

Figure 16.3 Sensitizer enhancement ratios determined *in vivo* using large single radiation doses as a function of misonidazole dose (upper scale). The symbols indicate different tumour types. The solid line shows the best fit to the *in vivo* results. The dotted lines enclose the range of *in vitro* data (lower scale); data on V79 cells indicated by the dashed line. From Brown (1989), with permission.

The difficulty of achieving sufficiently large clinical doses of misonidazole has led to a search for better radiosensitizing drugs (EORTC, 1991). Of the many compounds synthesized and tested, some of the most promising are etanidazole, pimonidazole and RSU-1069 (Overgaard 1994). Etanidazole was selected as being superior to misonidazole for two reasons. First, although it has a sensitizing efficiency equivalent to that of misonidazole, it does have a shorter half-life *in vivo*, which should reduce its toxicity. Second, it also has a reduced lipophilicity (a lower octanol/water partition coefficient) and thus is less readily taken up in neural tissue. This tends to reduce neurotoxicity, which was dose-limiting for misonidazole. Pimonidazole contains a side-chain with a weakly basic piperidine group. This compound is more electron-affinic than misonidazole and thus is more effective as a radiosensitizer; it is also uncharged at acid pH and this promotes its accumulation in ischaemic regions of tumours.

These new hypoxic cell sensitizers may be more potent than misonidazole but unfortunately they have a variety of side effects that may hamper their clinical use. Another strategy was to search for a less toxic drug that could be given in high doses. On this basis, nimorazole was evaluated (Overgaard *et al*, 1991). Although its sensitizing ability is less than theoretically can be achieved by misonidazole, the drug shows a flat dose–response curve so that at clinically relevant doses the SER is approximately 1.3. Furthermore, the drug can be given in association with a conventional radiation therapy schedule and is therefore amenable to clinical use.

RSU-1069 and its pro-drug RB-6145 (Adams *et al*, 1991) are dual-functional radiosensitizers because the presence of the nitro group confers radiosensitization while the aziridine group in the side-chain allows it to act also as cytotoxic alkylating agent. RSU-1069 is a more active radiosensitizer *in vivo* than any other nitroheterocyclic compound so far investigated and it also displays a particularly high cytotoxic activity towards hypoxic cells. Unfortunately RSU-1069 is also rather toxic to patients.

16.4 Hypoxic cell radio-sensitizers: clinical studies

Initial clinical studies using metronidazole in brain tumours generated enthusiasm for chemical radiosensitizers and these were followed by the widespread introduction of misonidazole. The late 1970s then saw a boom in clinical trials exploring the potential of misonidazole (Dische, 1985; Overgaard, 1989). Table 16.2 reviews the results of some of the larger multicentre studies. Overall, there was no clinical benefit and the general conclusion from the misonidazole era must be that this drug was unable to generate significant improvement in the results of radiotherapy. Dose-dependent peripheral neuropathy was a significant limiting factor.

Some trials did show improvement in certain subgroups of treated patients. This was especially prominent in the Danish head and neck cancer trial, DAHANCA 2 (Figure 16.4), which despite an overall insignificant effect of misonidazole showed a highly significant improvement in pharynx tumours.

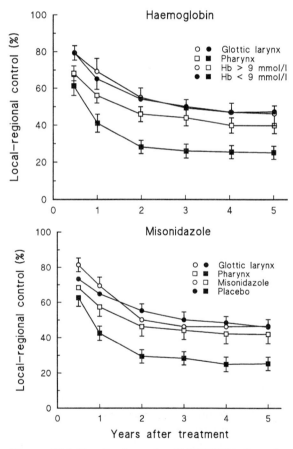

Figure 16.4 Results from the DAHANCA 2 study. Effect of misonidazole and haemoglobin concentration on loco regional tumour control in males with glottic larynx and pharynx carcinomas. From Overgaard *et al* (1989), with permission.

Table 16.2 Multicentre randomized trials with nitroimidazoles

Site and reference	No. of patients	Drug	End-point	RT+ Sens	RT alone	
Head and neck carcinoma						
DAHANCA 2 (Overgaard et al, 1989)	626	MISO	Control* (5 yrs)	41%	34%	$p < 0.05$
MRC (1984)	267	MISO	Control* (>2 yrs)	40%	36%	n.s.
EORTC (1986)	163	MISO	Control* (3 yrs)	52%	44%	n.s.
RTOG (Fazekas et al, 1987)	306	MISO	Control* (3 yrs)	19%	24%	n.s.
RTOG 79–04 (Fazekas et al, 1987)	42	MISO	Control* (2 yrs)	17%	10%	n.s.
DAHANCA 5 (Overgaard et al, 1992)	414	NIM	Control* (5 yrs)	49%	34%	$p < 0.002$
RTOG 85–27 (Lee et al, 1995)	500	ETA	Control* (2 yrs)	39%	38%	n.s.
European multicentre (Chassange et al, 1991)	374	ETA	Control* (2 yrs)	57%	58%	n.s.
Uterine cervix carcinoma						
Scandinavian study (Overgaard et al, 1989)	331	MISO	Control* (5 yrs)	50%	54%	n.s.
MRC (1984)	153	MISO	Control* (>2 yrs)	59%	58%	n.s.
RTOG (Leibel et al, 1987)	119	MISO	Control* (3 yrs)	53%	54%	n.s.
MRC (Dische et al, 1993)	183	PIM	Control* (4 yrs)	64%	80%	$p < 0.01$
Glioblastoma						
Scandinavian study (Hatlevoll et al, 1985)	244	MISO	Survival**	10	10	n.s.
MRC (1983)	384	MISO	Survival**	8	9	n.s.
EORTC (1983)	163	MISO	Survival**	11	12	n.s.
RTOG (Nelson et al, 1986)	318	MISO	Survival**	11	13	n.s.
Bronchogenic carcinoma						
RTOG (Simpson et al, 1987)	117	MISO	Survival**	7	7	n.s.
RTOG (Simpson et al, 1989)	268	MISO	Survival**	7	8	n.s.

n.s. not significant.
See Overgaard (1994) for references.
* Locoregional control (observation period)
** Median survival (months)
MISO = misonidazole; NIM = nimorazole; ETA = etanidazole; PIM = pimonidazole.

In contrast, the prognostically better glottic carcinomas showed no influence of misonidazole.

The overall impression of the misonidazole era is that problems related to hypoxia have not been overcome. The development of more efficient or less toxic hypoxic sensitizers has therefore continued and currently there are several promising options (Coleman, 1988; EORTC, 1991; Overgaard,

1994). Experience from the misonidazole trials has led to the selection of more homogeneous tumour populations in which hypoxia is more likely to be a problem. The pimonidazole trial in uterine cervix has been disappointing, and also the trials with etanidazole have been non-significant (Table 16.2). However, nimorazole, which is less toxic and can be given in sufficient doses (Overgaard, 1994), has in patients with supraglottic and pharyngeal carcinomas shown a highly significant improvement in locoregional tumour control and survival (DAHANCA 5; Figure 16.5; Overgaard et al, 1991).

16.5 Meta-analysis of controlled clinical trials of modified tumour hypoxia

The clinical role of hypoxia is one of the most thoroughly addressed issues in radiotherapy and has been under investigation for over two decades. Numerous clinical trials have been conducted, but most have been inconclusive. This may either be because no true difference exists among the treatment groups or because the trials have been too small to detect it. Meta-analysis seeks to distinguish between these alternatives: the results from *all* relevant trials are combined and analysed together.

A literature survey has identified 10 602 patients treated in 82 randomized clinical trials. The median number of patients per trial was 82 (range 14–620). The trials were directed at hyperbaric oxygen (29 trials), hypoxic cell radiosensitizers (52 trials), oxygen- or carbogen-breathing (3 trials) and blood transfusion (1 trial). The tumour sites were bladder (16 trials), uterine cervix (15 trials), CNS (13 trials), head and neck (25 trials), lung (10 trials) and mixed (1 trial). The trials were analysed with regard to local tumour control (57 trials), survival (68 trials), distant metastases (17 trials) and complications resulting from radiotherapy (23 trials). The overall results are given in Table 16.3. The most relevant end-point was considered to be local control, in view of the local nature of the radiation treatment, and this showed an improvement by 5% (51% vs. 46%; $p < 0.0001$). This improvement persisted when the trials were evaluated separately for radiosensitizer or hyperbaric oxygen treatment. When analysed according to site, especially head and neck cancer showed a significant improvement (7%, $p < 0.0001$). No other tumour sites demonstrated significant improvement.

Figure 16.6 shows the distribution of the statistical parameter Z in the 57 trials that gave local control data. Each point represents the results of a particular trial and the clear tendency for these to lie to the left of the zero value indicates the benefit of reducing the problem of tumour hypoxia. The overall survival probability also showed significant reduction ($p < 0.004$), once again dominated by head and neck tumours. No significant difference

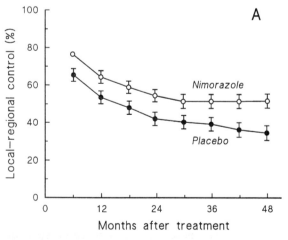 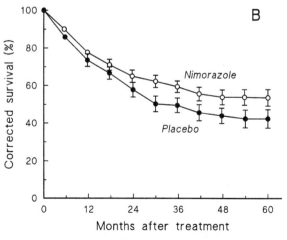

Figure 16.5 Results from the DAHANCA 5 study. Actuarial loco-regional tumour control (**A**) and corrected survival (**B**) in patients randomized to receive nimorazole (219 patients) or placebo (195 patients) in conjunction with radiotherapy for carcinoma of the pharynx and supraglottic larynx. From Overgaard et al (1991, 1992), with permission.

Table 16.3 Summary of meta-analysis of clinical modification of tumour hypoxia
A. Summary of all randomized trials

End-point	No. of Trials	No. of patients.	RT + modifier (%)	RT alone (%)	p-value	Odds ratio (95% CL)
Locoregional control	57	7921	51	46	<0.0001	1.23 (1.13–1.35)
Survival	68	9715	36	33	0.004	1.14 (1.05–1.24)
Distant metastases	17	4247	19	21	0.15	0.90 (0.77–1.04)
RT complications	22	3543	17	16	0.51	1.05 (0.88–1.26)

B. Locoregional tumour control as a function of type of modification

End-point	No. of Trials	No. of patients.	RT + modifier (%)	RT alone (%)	p-value	Odds ratio (95% CL)
HBO/oxygen*	19	2488	62	53	<0.0001	1.44 (1.23–1.69)
Hypoxic sensitizer*	38	5422	46	42	0.004	1.17 (1.05–1.30)
Transfusion	1	135	84	69	0.05	2.27 (1.00–5.20)

* Including 1 trial with HBO + miso (124 patients).

C. Locoregional control as a function of tumour type and location

End-point	No. of Trials	No. of patients.	RT + modifier (%)	RT alone (%)	p-value	Odds ratio (95% CL)
Head and neck	23	4106	47	40	<0.0001	1.30 (1.15–1.47)
Bladder	12	707	50	45	0.14	1.24 (0.93–1.67)
Uterine cervix	12	2292	66	62	0.06	1.18 (0.99–1.40)
Lung (NSCLC)	8	624	37	33	0.32	1.19 (0.85–1.65)
Oesophagus	2	192	30	26	0.49	1.25 (0.66–2.34)
All (group C) trials	57	7921	51	46	<0.0001	1.23 (1.13–1.35)*

* $p < 0.00001$.
RT = radiotherapy.

was observed in the incidence of distant metastases or radiotherapy complications. Of utmost importance was the relationship between improvement in local control and subsequent improvement in survival (Figure 16.7). The trials shown in this figure are all from epithelial carcinomas, but patients with head and neck tumours generally achieved the highest improvement in both local control and subsequent survival.

From the analysis, it appears that the hypoxic problem in radiotherapy may be marginal in most adenocarcinomas. Future efforts should therefore be focused on squamous cell carcinoma of the head and neck and to a lesser extent other squamous cell carcinomas and transitional carcinoma of the bladder, at least when radiotherapy is given in conventional treatment schedules. The variation in the results among the trials certainly points towards a considerable heterogeneity among tumours with the same localization and histology. Thus the need to predict the presence of hypoxia and especially the capacity for reoxygenation appears to be a key issue in order to optimize future clinical application. The observation by Nordsmark et al (1996) that polarographic pO_2 measurements were highly predictive for the outcome of radiotherapy in a prospective study of head and neck carcinoma indicates that a better selection of patients may be possible.

The significant improvement obtained by manipulation of the hypoxic status of squamous tumours of the head and neck indicates that the underlying biological rationale is probably sound, at least in this tumour site. It would be logical, therefore, to direct future clinical studies of the hypoxic problem at this tumour type and site.

16.6 Improving the oxygen supply to tumours

The relationship between haemoglobin concentration and local tumour control, which appears to be

148 Overcoming hypoxic cell radioresistance

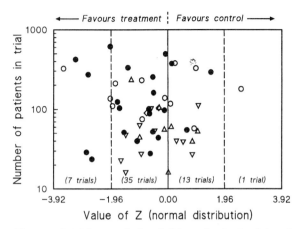

Figure 16.6 Meta-analysis of 57 randomized trials of radiotherapy combined with a treatment to modify tumour hypoxia, using the end-point of loco-regional control. (●) head and neck; (○) cervix; (▽) bladder; (△) lung; (◊) oesophagus. The value of Z represents the observed minus expected control rates divided by a measure of the uncertainty of this difference. Negative values denote results that favour the treatment arm, while positive values favour radiation alone. Results falling between 0 and ±1.96 indicate a tendency towards a difference and from ±1.96 to ±3.92 this difference becomes significant. Updated from Overgaard and Horsman (1996).

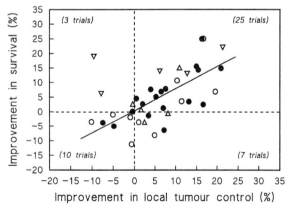

Figure 16.7 Results from 45 of the randomized trials shown in Figure 16.6 where both local control and survival data could be obtained. A positive improvement either in survival or local control indicates that the results of the hypoxic-modified arm were better than those of the controls. A strong correlation between the outcome of local control and subsequent survival is indicated ($r = 0.722$). The figure shows that patients with head and neck carcinoma seem to have the most impressive benefit. Updated from Overgaard and Horsman (1996).

most pronounced in larger tumours, raises the possibility of improving tumour control by manipulation of the haemoglobin concentration or, in a broader sense, the *oxygen unloading capacity* of the blood. Haemoglobin levels have been increased by blood transfusion prior to radiotherapy in a number of studies (Overgaard, 1989) and a small randomized study in the uterine cervix has shown that this led to a significant improvement of local tumour control probability (Bush, 1986). Transfusion is also part of the DAHANCA 5 trial (Overgaard *et al*, 1991). Although local tumour control may be associated with haemoglobin concentration, its association with the transfusion is complex. Nevertheless, the clinical results in patients with low haemoglobin who received both transfusion and hyperbaric oxygen are impressive (Overgaard, 1989).

Other approaches to the improvement of tumour oxygenation by manipulating the oxygen-unloading capacity of the blood are also being made. These include the use of artificial blood substances, such as perfluorocarbons, or manipulation of the oxyhaemoglobin dissociation curve (Coleman, 1988; Hirst, 1986; Horsman, 1993).

Smoking is clearly an important factor influencing tumour oxygenation, especially in patients with head and neck or lung cancer who are known to smoke during their treatment. Smoking may lead to the loss of more than 30% of the oxygen-unloading capacity of the blood and this could lead to a significantly reduced tumour oxygenation and thus cause reduced tumour control (Figure 16.8; Grau *et al* 1994; Overgaard and Horsman, 1996). Patients should strongly be encouraged to give up smoking, at least during radiotherapy.

16.7 Overcoming acute hypoxia in tumours

Although several of the procedures that have been used in patients to combat radiation resistance due to hypoxic cells have met with some success, the results are far from satisfactory. One explanation is that most of the procedures so far used clinically operate against diffusion-limited chronic hypoxia, and they have little or no influence on acute hypoxia (Section 15.2). Several years ago it was demonstrated that nicotinamide, a vitamin B_3 analogue,

Exploiting tumour hypoxia

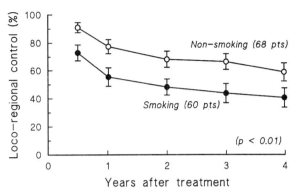

Figure 16.8 Influence of smoking during treatment on the outcome of radiotherapy in patients with advanced head and neck carcinoma. The local control probability was significantly poorer in patients who continued to smoke during radiotherapy, probably due to reduced oxygen delivery to the tumour. Results from a prospective study in patients treated with curative radiotherapy alone (Overgaard and Grau, unpublished).

could enhance radiation damage in a variety of murine tumour models (Figure 16.9; Horsman, 1991). Nicotinamide seems primarily to prevent the transient fluctuations in tumour blood flow that lead to the development of acute hypoxia (Horsman *et al*, 1990). It can be given in clinically effective doses and clinical studies are now under way.

It could be that the optimal approach to the hypoxic problem in human tumours is to combine treatments which independently attack chronic and acute hypoxia. Nicotinamide has been shown in experimental systems to improve tumour radiosensitization induced by hyperthermia, by the use of a perfluorochemical emulsion and by normobaric carbogen or oxygen breathing (Horsman, 1991; 1995). The potential of nicotinamide in combination with carbogen is now being explored in several phase I and II clinical trials, among others by the EORTC radiotherapy group. So far the studies have demonstrated that the strategy is feasible, but more data are needed to conclude on the ultimate clinical potential.

16.8 Exploiting tumour hypoxia

Although hypoxic cells are resistant to radiotherapy, they are *sensitive* to a number of other modalities, especially to bioreductive drugs (Adams *et al*, 1991) and to hyperthermia (Section 23.2). Both of these agents can directly sensitize hypoxic cells to radiation or specifically kill them. In the case of hyperthermia, cytotoxicity is not a consequence of hypoxia *per se*, but rather due to cellular metabolic changes leading to an increase in acidity, which is the result of prolonged oxygen deprivation (EORTC, 1991).

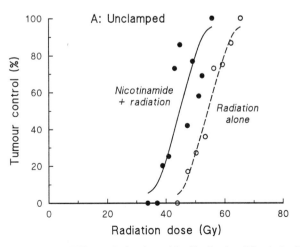
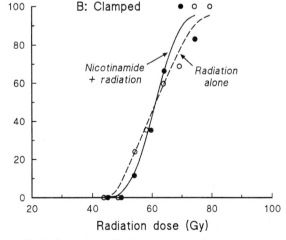

Figure 16.9 Effect of nicotinamide (1 g/kg i.p. 30 min before radiation) on the radiation response of a mouse mammary carcinoma. **A** shows that nicotinamide enhanced the radiation response of tumours treated under ambient conditions. When tumours were made fully hypoxic by clamping (**B**) no effect of nicotinamide was seen, indicating that the effect is due to hypoxic modification. Adapted from Horsman *et al* (1990), with permission.

Rather than seeking to reduce hypoxia in tumours, some studies suggest that it might be beneficial to *increase* the level of tumour hypoxia before subsequent exposure to agents that specifically kill hypoxic cells. There are a number of ways in which tumour hypoxia can be increased experimentally. One is to modify the oxygen-carrying capacity of the blood using either carbon monoxide breathing or haemoglobin-oxygen-affinity-modifying drugs like BW12C. Alternatively, tumour blood flow can be lowered using physiological modifiers like hydralazine or glucose, or by destroying tumour blood-vessels by photodynamic therapy or with drugs like flavone acetic acid. These and similar agents have been shown both to improve the radiation response of solid tumours and also to enhance (or act additionally with) the antitumour activity either of bioreductive drugs or hyperthermia (EORTC, 1991; Horsman, 1993).

Attention has recently been given to hypoxia-specific cytotoxins (Brown and Simm, 1996) such as quinones (*e.g.* mitomycin C, WO9), nitroaromatic compounds, and N-oxides (AQ4N, tirapazamide). Tirapazamide especially has attracted attention due to its strong differential between oxic and hypoxic toxicity. Ongoing phase I and II trials will evaluate its clinical tolerance and potential.

Key points

1. Hypoxic cell radioresistance is a significant cause of failure in the local control of tumours, in particular carcinomas of the head and neck.
2. Chemical radiosensitizers work well in experimental animal studies but less so in clinical trials, probably due to tumour heterogeniety and/or difficulty in achieving sufficient drug concentrations.
3. Pharmacological modification of tumour blood flow shows promise in experimental systems. Examples are nicotinamide, which appears to reduce the problem of acute (*i.e.* transient) hypoxia in tumours, and hydralazine, which *increases* tumour hypoxia and can enhance the activity of some bioreductive drugs that are toxic to hypoxic cells.
4. Hypoxic cytotoxins, taking advantage of tumour hypoxia, have a biological potential providing toxicity is acceptable.

Bibliography

Adams GE (1977). Hypoxic cell sensitizers for radiotherapy. In: *Cancer* (Vol 6.) Ed Becker FF. Plenum Press; New York.

Adams GE, Bremner J, Stratford IJ *et al* (1991). Nitroheterocyclic compounds as radiosensitizers and bioreductive drugs. *Radiother Oncol* 20 (Suppl 1):85–91.

Brown JM and Siim BG (1996). Hypoxia-specific cytotoxins. *Sem Rad Oncol* 6: 22–36.

Bush RS (1986). The significance of anemia in clinical radiation therapy. *Int J Radiat Oncol Biol Phys* 12:2047–50.

Churchill-Davidson I (1968). The oxygen effect in radiotherapy – historical review. *Front Radiat Ther Oncol* 1:1–15.

Coleman CN (1988). Hypoxia in tumours: a paradigm for the approach to biochemical and physiological heterogeneity. *J Natl Cancer Inst* 80:310–17.

Dische S (1988). Chemical sensitizers for hypoxic cells: a decade of experience in clinical radiotherapy. *Radiother Oncol* 3:97–115.

Grau C, Khalil AA, Nordsmark M, Horsman MR and Overgaard J (1994). The relationship between carbon monooxide breathing, tumour oxygenation and local tumour control in the C3H mammary carcinoma *in vivo*. *Br J Cancer* 69: 50–7.

Hirst DG (1986). Anemia: a problem or an opportunity in radiotherapy? *Int J Radiat Oncol Biol Phys* 12:2009–17.

Horsman MR (1991). Nicotinamide and the hypoxia problem. *Radiother Oncol* 22:79–80.

Horsman MR (1995). Nicotinamide and other benzamide analogs as agents for overcoming hypoxic cell radiation resistance in tumours. *Acta Oncol* 34: 571–87.

Horsman MR, Chaplin DJ and Overgaard J (1990). Combination of nicotinamide and hyperthermia to eliminate radioresistant chronically and acutely hypoxic tumor cells. *Cancer Res* 50:7430–6.

Kjellen E, Joiner MC, Collier JM *et al* (1991). A therapeutic benefit from combining normobaric carbogen or oxygen with nicotinamide in fractionated x-ray treatments. *Radiother Oncol* 22:81–91.

Nordsmark M, Overgaard M and Overgaard J (1996). Pretreatment oxygenation predicts radiation response in advanced squamous cell carcinoma of the head and neck. *Radiother Oncol* 41: 31–40.

Overgaard J (1994). Clinical evaluation of nitroimidazoles as modifiers of hypoxia in solid tumors. *Oncology Research* 6: 507–16.

Overgaard J and Horsman MR (1996). Modification of hypoxia induced radioresistance in tumours by the use of oxygen and sensitizers. *Sem Radiat Oncol* 6: 10–21.

Overgaard J, Sand Hansen H, Andersen AP *et al* (1989). Misonidazole combined with split-course radiotherapy in the treatment of invasive carcinoma of larynx and pharynx. Final report from the DAHANCA 2 study. *Int J Radiat Oncol Biol Phys* 16:1065–8.

Overgaard J, Sand Hansen H, Lindeløv B *et al* (1991). Nimorazole as a hypoxic radiosensitizer in the treatment of supraglottic larynx and pharynx carcinoma. First Report from the Danish Head and Neck Cancer Study (DAHANCA) Protocol 5–85. *Radiother Oncol* 20 (Suppl 1):143–50.

Overgaard J, Sand Hansen H, Overgaard M *et al* (1992). The Danish Head and Neck Cancer Study Group (DAHANCA) randomized trials with radiosensitizers in carcinoma of the larynx and pharynx. In: *Radiation Research. A Twentieth-Century Perspective*, Vol II, pp. 573–7. (Eds) Dewey WC, Eddington M, Fry RJM, Hall EJ and Whitmore GF. Academic Press; New York.

Further Reading

Brown JM (1989). Hypoxic cell radiosensitizers: where next? *Int J Radiat Oncol Biol Phys* 16:987–93.

EORTC Cooperative Group for Radiotherapy (1991). Consensus Meeting on Tumour Hypoxia. *Radiother Oncol* 20 (Suppl 1):1–159.

Horsman MR (1993). Hypoxia in tumours: its relevance, identification and modification. In: *Current Topics in Clinical Radiobiology of Tumours*, pp. 11–26. (Ed) Beck-Bornholdt HP. Springer-Verlag; Berlin.

Overgaard J (1989). Sensitization of hypoxic tumour cells – clinical experience. *Int J Rad Biol* 56:801–11.

Raleigh JA, (Ed) (1996). Hypoxia and its clinical relevance. *Sem Radiat Oncol* 6: issue 1.

17

The radiobiology of tumours

G. Gordon Steel

17.1 The clinical picture: tumour control probability

The level of success with radiotherapy for cancer varies considerably from one disease and tumour stage to another. It is common clinical experience that some tumours are highly curable, others not. A quantitative description of this picture is provided by a comparison of tumour control probability curves. The collection shown in Figure 17.1 was made in 1986 by J.M. Deacon. The curves are taken from the clinical literature; they are of variable reliability because the data do not come from randomized clinical trials, which would generally be unethical over such a wide range of dose levels. A variety of fractionation schedules were used, and also a range of field sizes, and the results have been brought together on a standardized dose scale.

Overall, it can be seen that doses required for 50% tumour control range widely, varying on the dose scale from 6 to 8 Gy for some lymphomas to over 20 Gy for some epithelial tumours. Gliomas are well known to be refractory and it is not possible to give a high enough radiation dose to approach 50% tumour control, nor even to see a clear dose–response relationship. The picture would be similar with osteosarcomas. The steepness of each dose–response curve reflects a number of variables, which include:

(i) An underlying Poisson relationship (Section 10.2). If the average number of clonogenic tumour cells that survive treatment is m, then the probability that no cells survive and that the tumour is cured is e^{-m}. Tumour control

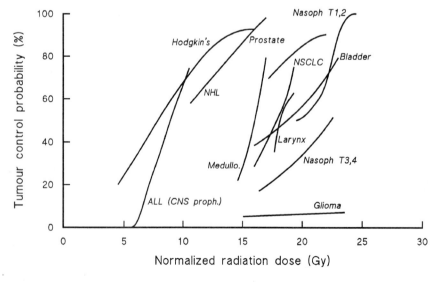

Figure 17.1 Dose–response curves for tumour control taken from a variety of clinical studies. From J.M. Deacon, unpublished.

probability therefore has a sigmoid dependence on m, with a finite maximum slope.
(ii) Variability in curability among tumours of the same type, arising from differences in cellular radiosensitivity, repopulation, hypoxia *etc.*, some of which may be genetic.
(iii) Inter-patient variation in the quality of radiation dose delivery.

17.2 Experimental tumour systems

Research into the radiation response of tumours has been performed on a wide variety of tumour cell systems. *In vivo* studies have mainly been carried out on tumours in experimental animals, usually in mice and rats. In addition to studies carried out on permanent *in vitro* tumour cell lines, *in vitro* techniques have often been used to investigate the radiosensitivity of cells from experimental tumours (so-called *in vivo* – *in vitro* experiments) or on cells taken directly from human tumours. The range of cell systems that have been used may be described under the following headings, arranged roughly in order of increasing closeness to cancer in man.

IN VITRO CELL LINES

Although it is possible to grow cells from normal tissues in cell culture systems the cultures usually fail after perhaps 20–30 passages. This is known as the *Hayflick limit*. It is clearly a disadvantage for most laboratory research that cultures should die out and it has been overcome in three main ways. First, some normal-cell cultures have been grown through the Hayflick limit and have become immortal. They retain some characteristic features of normal-tissue cells (contact-inhibition of growth in crowded cultures or failure to grow in immune-deficient mice etc.) and they are usually not known to be virally transformed. Widely used examples are CHO (Chinese hamster ovary) cells and V79 fibroblasts (also from hamster). Second, there are normal-tissue cells that have been virally transformed (for instance by Simian virus SV40). These usually lose contact-inhibition and retain viral sequences in their genome. The third approach is to use human or animal tumour cells. Tumour cells appear not to be subject to the Hayflick limit and can often be grown continuously without deterioration. A wide variety of cell lines have been used, including HeLa (derived from a human cervix carcinoma), L5178Y mouse lymphoma cells and the human tumour cell lines referred to in Section 17.5 below. All types of 'permanent' cell lines are subject to genetic drift and the various sublines in different laboratories may not be identical. See Freshney (1987) for a general text on tissue culture.

MULTICELLULAR SPHEROIDS

Some cell lines readily form aggregates in tissue culture that can grow up to a diameter of a fraction of a millimetre. They have been widely studied as *in vitro* models of tumours; they of course have no vascular system but depend on diffusion of oxygen and other nutrients through the surface of the 'spheroid'. Small spheroids are fully oxygenated but as they grow they develop a hypoxic core. In larger spheroids three or more concentric zones can sometimes be distinguished among which cell proliferation, oxygenation and drug and radiation sensitivity may vary (Durand and Sutherland, 1973).

TRANSPLANTED TUMOURS IN EXPERIMENTAL ANIMALS

A wide variety of tumours have been conditioned to repeated transplantation in laboratory animals. Some of these arose spontaneously; some were induced by chemicals, radiation or viruses. During repeated passage, usually subcutaneously, they grow faster and more uniformly and often lose differentiated characteristics. They are attractive as reproducible and convenient *in vivo* tumour systems but may in some respects have deviated away from the original primary tumour. Well-known examples are L1210 leukaemia, Lewis lung tumour, B16 melanoma, R1 rhabdomyosarcoma (see Kallman, 1987).

A matter of some importance is the immune status of the grafts. Tumours grown in non-inbred strains of animals or in an inbred colony that differs from that in which they arose will inevitably be subject to variable host rejection (Scott, 1991). If the tumours grow fast they may beat the developing rejection processes and the investigator may imagine that rejection is unimportant. But if the tumour is treated with radiation or cytotoxic drugs the influence of the immune response on residual tumour may become significant; more importantly, if the antitumour effects of two agents are compared and one is more immune-suppressive than the other, misleading results can be obtained. This is especially the case if tumour cure is used as the end-point, for immune responses are particularly effective against a small amount of tumour remaining after therapy.

Primary Animal Tumours

The word *autochthonous* is sometimes used for tumours that are studied in the host within which they arose. They may be *spontaneous*, or they may have been *induced* by radiation, carcinogenic chemicals, viruses or other means. A primary tumour would be expected to bear greater resemblance to human tumours than one that has been transplanted many times: its immune status should be more realistic and it will not have undergone the growth acceleration and other changes that occur on repeated transplantation. The main drawbacks are first that spontaneous tumours require large numbers of animals to be kept for a long period of time and are therefore expensive; and second, if potent carcinogens are used to increase the induction frequency the resulting tumours tend to be immunogenic. Mammary tumours arising spontaneously in C3H mice and their isotransplants have been used for some important radiobiological studies (Suit *et al*, 1965, 1992; Suit and Maeda, 1967).

Human Tumour Xenografts

A xenograft is a transplant from one species to another. In the cancer field this usually refers to a human tumour grown in a laboratory animal. If the recipient animal has a normal immune system then a xenograft should not grow, but there are two main ways in which growth has been achieved. First, animal strains have been developed that are congenitally immune-deficient. Best known are *nude mice*, which in addition to being hairless also lack a thymus. Many human tumours will grow under the skin of nude mice (Sparrow, 1980). More recently there have been nude rats, also SCID mice which suffer from the severe combined immunodeficiency syndrome and which are deficient in both B-cell and T-cell immunity. Second, it is possible to severely immune-suppress mice to the point where they will accept human tumour grafts. It is important to recognize that neither type of host completely fails to reject the human tumour cells: rejection processes are still present and these complicate the interpretation of *in situ* tumour therapeutic studies. Nevertheless, xenografts do maintain many of the biological and therapeutic properties of their source tumours, and they constitute a valuable type of experimental system (Steel *et al*, 1983).

Clinical Radiobiology

The ultimate experimental system is the cancer patient. Testing of new treatments and randomized clinical trials have an essential role in the development of new therapies. In addition, although the scope for experiment on patients is limited, there are some radiobiological experiments which, while not of direct benefit to the individual patient, can ethically be carried out with appropriate informed consent. Examples are measurements of the distribution of oxygen concentration in tumours, and the testing of chemical radiosensitizers by measuring the response of metastases within the same patient that are irradiated before or after drug administration (Ash, 1980). Results of studies in clinical radiobiology are described in a number of parts of this book, especially Chapters 10, 11 and 16.

17.3 End-points for measuring radiation effects on tumours

Three principal methods have been used to document and compare the effects of radiation and cytotoxic drugs on experimental tumours.

Tumour Growth Delay

This is especially useful for tumours growing subcutaneously, the size of which can be measured accurately with callipers. A group of tumours are selected to have closely similar size. They are divided into similar groups and given different

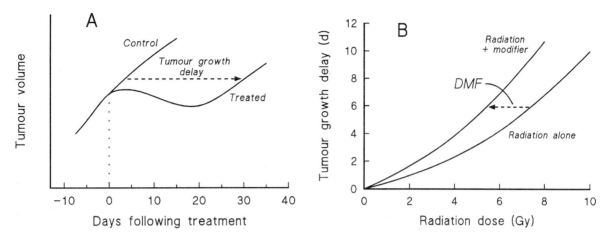

Figure 17.2 A: the measurement of tumour growth delay and **B**: the calculation of a dose-modifying factor (DMF).

treatments, one group being left untreated. Tumour volume is then followed at regular intervals. The usual pattern is that shown in Figure 17.2A: a period of regression followed by regrowth. In order to measure growth delay, we select an end-point (such as twice or four times the treatment size), determine the average time that treated and control tumours take to reach this size, and by subtraction find the delay.

Results obtained with different radiation doses can then be plotted as a *dose–response curve* (Figure 17.2B). If we wish to evaluate the effect on the tumour of some modification of treatment (adding a radiosensitizer, changing inspired oxygen level etc.) we would perform a larger experiment in which both modified and unmodified tumours are studied together. Thus we would obtain two dose–response curves and a good way of indicating the extent of modification is to calculate a *dose-modifying factor (DMF)*. This can be read off as the ratio of radiation doses that give the same effect: (without modification)/(with modification). If the modification is truly *dose-modifying*, then the DMF value should be the same at all levels of effect. It is then better to calculate the DMF (by computer) by asking what value would allow the two curves to be most nearly superimposed. A variety of terms similar to dose modifying factor have also been used and calculated in the same way: oxygen enhancement ratio (OER), sensitizer enhancement ratio (SER), thermal enhancement ratio (TER) etc.

Comparison of growth delay among tumours of different growth rate leads to uncertainties in interpretation. One approach (Kopper and Steel, 1975) is to calculate a *specific growth delay*:

$$\text{Specific growth delay} = \frac{TD_{treated} - TD_{control}}{TD_{control}}$$

where *TD* indicates the time within which treated or control tumours double in volume. Specific growth delay corresponds to number of volume doubling times saved by treatment. Since the doubling time for repopulation by clonogenic cells may not bear a constant relation to volume doubling time, this approach should be regarded as only roughly satisfactory. There is no ideal solution to the problem of comparing growth delay among tumours of different growth rate.

Tumour Control

As radiation dose is increased there will come a point where some tumours fail to regrow. We can then evaluate a modifying agent by determining its effect on this 'curative' radiation dose. The radiation dose that controls 50% of tumours is usually called the TCD_{50} and a DMF can be calculated by determining the ratio of TCD_{50} values without/with the modifier. It is important when using this assay to observe the tumour-bearing mice for a long enough time to detect almost all possible recurrences. Death of animals from metastases may frustrate this and the assay is therefore not appropriate for tumours that frequently metastasize. A further drawback is that the assay is greatly affected by any host immune reaction against the tumours (see previous section). The TCD$_{50}$ assay has been widely and successfully used by Suit and colleagues (Suit et al, 1965).

The use of this assay in evaluating the radiosensitization by misonidazole of first generation transplants of C3H mouse mammary tumours is illustrated in Figure 17.3 (Fowler et al, 1976). The objective was to ask: 'What duration of fractionated radiotherapy gives the best therapeutic response, in the presence or absence of the sensitizer?' Figure 17.3A shows examples of the dose–response curves for tumour control with and without the administration of the sensitizer 30 minutes before irradiation. Dose–response curves for the early skin reaction in the feet of mice were documented in parallel experiments. These allowed the radiation dose (with or without sensitizer and for any duration of treatment) that gave a fixed level of skin damage to be identified. The level of tumour control that corresponded to each of these doses (for instance, D) could then be read off from the corresponding tumour control curve. Figure 17.3B shows (in outline) the results: tumour control probability *for a fixed level of skin damage* as a function of the duration of fractionated treatment. Without misonidazole, short treatments were bad; there was an optimum treatment duration of around 8 days. Adding misonidazole greatly improved the results of short schedules but had little effect on the longer ones. The explanation may be one that is important for clinical radiotherapy: reoxygenation is incomplete for short treatment times and a radiosensitizer may counteract this deficiency; long treatment times may allow reoxygenation to occur but they may be suboptimal because of tumour cell repopulation.

Cell Survival

Cell survival studies have for many years underpinned the cellular basis of tumour response to treatment. In comparison with the other two endpoints described above, clonogenic assays have the advantage of removing the cells into a growth environment that is uniform and unaffected by treatment; there is therefore less opportunity for artefacts due to the effects of treatment on the host animal. Clonogenic cell survival is dealt with in Chapter 6.

17.4 Some conclusions from the in vivo radiobiology of experimental tumours

The large number of radiobiological studies done on animal tumours during the 1970s and 1980s pro-

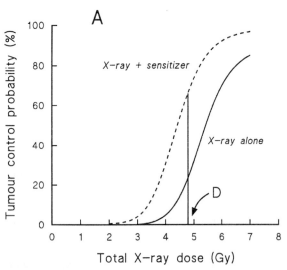
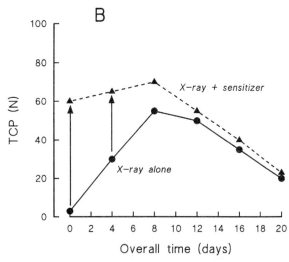

Figure 17.3 Tumour control in the evaluation of optimum fractionation with a radiosensitizer. **A**: examples of tumour control curves for a selected duration of treatment; **B**: tumour control probability, TCP(N), for a fixed level of early skin damage. The effect of the sensitizer (arrows) was greatest for short overall treatment times. A diagrammatic representation of the results of Fowler et al (1976), with permission.

vided the basis for many important concepts, including the following.

THE TUMOUR SIZE EFFECT

Large tumours are more difficult to cure than small tumours but this is mainly due to the number of clonogenic cells that have to be killed. A course of radiotherapy that can achieve a surviving fraction of say 10^{-9} could be curative in a tumour that has less than 10^9 clonogenic cells, but probably not in one that has 10^{12} (see Figure 2.1 and Section 6.6). The more intricate question is whether, in addition to this obvious effect, the clonogenic cells in large tumours are also *less sensitive* to therapy.

There is good evidence that this is the case. Stanley *et al* (1977) treated Lewis lung tumours with γ-radiation, BCNU or cyclophosphamide at a wide range of tumour sizes and produced the results shown in Figure 17.4. As tumour size increased, there was in each case a steep decline in cellular sensitivity (a rise in cell survival) over the size range from 1 to 100 mm³. This could have been due to the internal vascular supply becoming progressively poorer in the growing tumours, decreasing the access of chemotherapeutic agents and increasing the proportion of cells that were hypoxic and therefore resistant to radiation treatment.

ACCELERATED REPOPULATION

The notion that clonogenic cells which survive radiation treatment may quickly repopulate the tumour was suggested by the classic study of Hermens and Barendsen (1969). They irradiated rat R1 rhabdomyosarcoma transplants with a single 20-Gy dose of X-rays and measured both tumour volume and the fraction of surviving clonogenic cells over the following few weeks (Figure 17.5). Tumour volume showed a slight regression, followed by regrowth. Surviving fraction immediately after irradiation was around 10^{-2} and it remained at that level for 4 days. It then increased rapidly to reach unity at post-irradiation days 10–12. Although this appears to demonstrate rapid tumour repopulation, it should be noted that surviving fraction does not take into account the loss of countable but doomed cells from the tumour which presumably was increasing during this 12-day period. A rapid rise in surviving fraction could be due to the loss of non-clonogenic doomed cells.

More reliable, therefore, are the results of Stephens *et al* (1978), who in the Lewis lung tumour carefully measured not only the surviving fraction but also the number of viable cells per tumour and were thus able to make an estimate of the *total clonogenic cells per tumour* (Figure 17.6). Following single radiation doses in the range 15–35

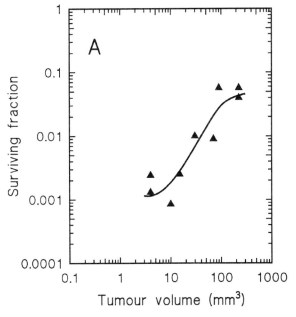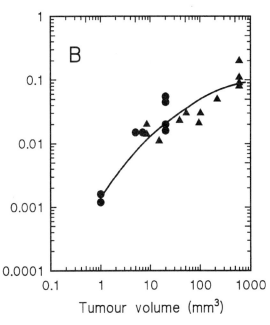

Figure 17.4 Variation with tumour size of cell survival in the Lewis lung tumour treated with **A** 75 mg/kg cyclophosphamide or **B** 10 Gy γ-rays. ▲ subcutaneous tumours; ● lung tumours. From Stanley *et al* (1977), with permission.

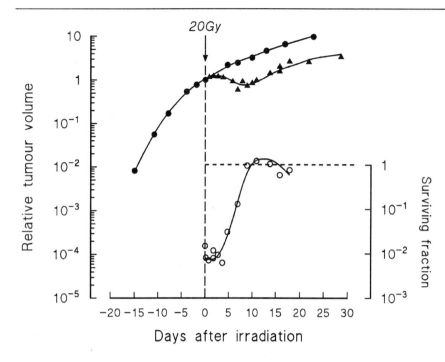

Figure 17.5 Response of the R1 tumour to 20 Gy X-rays showing: **A** the volume response and **B** the surviving fraction of clonogenic cells. From Hermens and Barendsen (1969), with permission.

Gy there was clear evidence for accelerated repopulation: the speed of repopulation was greater following the larger doses.

An important notion that is well illustrated by the data in Figure 17.5 is that tumour volume provides a poor reflection of clonogenic cell kill. A treatment that produces only slight and temporary regression may depress surviving fraction by as much as two decades (99% clonogenic cell kill). It is the slow rate of tumour regression that prevents this from being revealed in the volume change (Section 2.5).

THE TUMOUR BED EFFECT

Careful measurement of the growth rate of experimental tumours that recur following radiation treatment has often shown that it is slower than the growth rate of untreated tumours of the same size. This is called the *tumour bed effect*. A similar phenomenon is the reduced growth rate of tumours transplanted into previously irradiated subcutaneous sites. It is thought that both of these effects are due to radiation damage to stromal (including vascular) tissues. The reason why *accelerated repopulation* and *retarded regrowth* can occur together is because the former is seen in surviving clonogenic cells at a time when the tumour may be shrinking, while the latter is a property of the vol-

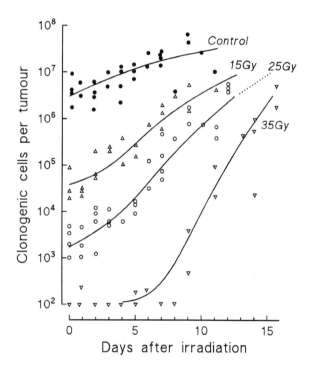

Figure 17.6 Estimated number of clonogenic cells per tumour in the Lewis lung tumour following various doses of γ-radiation: ∆ 15 Gy; ○ 25 Gy; ∇ 35 Gy. From Stephens et al (1978), with permission.

ume growth rate of irradiated tumours. See Begg and Terry (1985) and Kallman (1987: chapter 7).

Hypoxia and Reoxygenation

Most tumours contain clonogenic cells that are at various levels of oxygenation and this fact has a profound influence on tumour control by radiation and on ways of quantifying tumour response. These aspects are dealt with in Chapter 15.

17.5 The radiosensitivity of human tumour cells

The Initial Slope of the Cell Survival Curve

A key question in radiation biology applied to radiotherapy is: 'How radiosensitive are human tumour cells, and how does this relate to clinical radiocurability?' Prior to 1980 there was little information on this but interest was aroused by Fertil and Malaise (1981), who surveyed the published literature on *in vitro* human tumour cell survival curves and found evidence for a correlation with clinical response. This survey was repeated by Deacon *et al* (1984), who summarized data on 51 non-HeLa human tumour cell lines. These covered 17 different histopathological tumour types which (on the basis of data of the sort shown in Figure 17.1) were placed into five categories of local tumour radiocurability: A: lymphoma, myeloma, neuroblastoma; B: medulloblastoma, small-cell lung cancer; C: breast, bladder, cervical carcinoma; D: pancreatic, colorectal, squamous lung cancer; E: melanoma, osteosarcoma, glioblastoma, renal carcinoma. The placing of tumour types in this list is somewhat uncertain and the underlying clinical data do not allow this to be done unequivocally. However, the ranking A → E broadly reflects clinical experience. The *in vitro* cell survival data for each cell line were analysed to determine the surviving fraction at 2 Gy (which was termed SF_2), chosen as a measure of the *initial slope* of the cell survival curves. The results are shown in Figure 17.7. Within each category of clinical radiorespon-

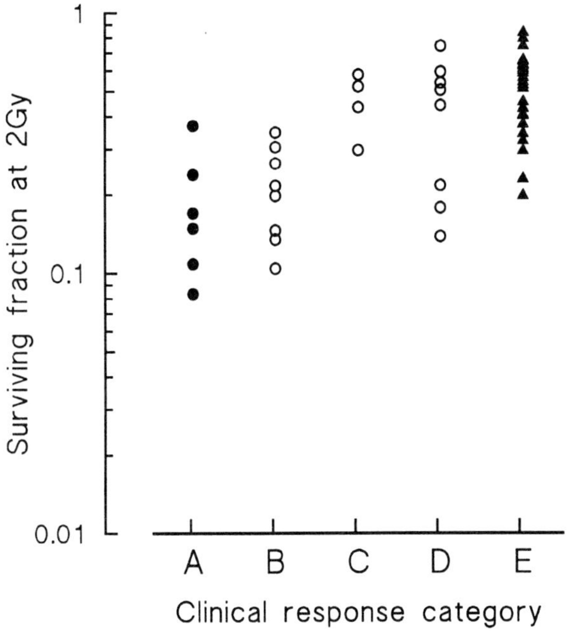

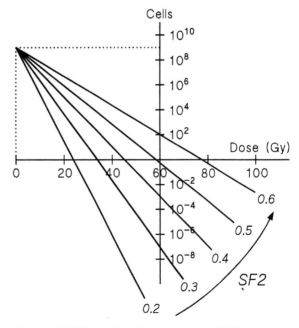

Figure 17.7 Surviving fraction at 2 Gy for 51 human tumour cell lines, arranged in five categories of clinical radioresponsiveness. From Deacon *et al* (1984), with permission.

Figure 17.8 Effect of multiple fractions of 2 Gy given to tumours whose SF_2 values range from 0.2 to 0.6. The full lines show the survival of an original 10^9 clonogenic cells during exposure to a sequence of 2-Gy doses.

siveness there was a considerable scatter (not surprising in view of the many different sources, cell lines and techniques used), but in confirmation of Fertil and Malaise's conclusions there was a significant trend in the data towards Group A having lower and Group E having higher SF_2 values. This important conclusion underlies the belief that the steepness of the initial slope is a significant factor in the clinical response of tumours to radiotherapy. A more recent analysis including additional data reached the conclusion that there was no significant difference between the survival data for groups C, D and E. The tumours in these groups, which include many of the solid tumours treated with radiation therapy, have an average SF_2 value of roughly 0.5: we would expect that half the oxic clonogenic cells would be killed with each 2-Gy radiation dose. The extended analysis showed that the radiosensitivity of groups A and B was significantly greater than for the three more resistant groups.

Initial Slope and Tumour Cure

Cell lines in Group A of Figure 17.7 have SF_2 values that average around 0.2 or less, while those in Group E cluster around 0.5. Is this difference large enough to be of clinical significance? It may be, for the reasons shown in Figure 17.8. Imagine treating a tumour whose SF_2 is 0.5 with a succession of 2-Gy radiation doses. If the surviving fraction per dose remains constant, then the survival from 30 doses (*i.e.* 60 Gy) will be $(0.5)^{30} \approx 10^{-9}$. For an SF_2 of 0.2 the overall survival would be below 10^{-20}. Looking horizontally in the figure, an SF_2 of 0.5 requires a total dose of 60 Gy to reduce an initial 10^9 clonogenic cells down to one cell, whereas for an SF_2 of 0.2 this would be 25 Gy. The difference of a factor of 2 in these total doses is comparable with the difference in isoeffective total doses for the clinical response of most human tumours. This is a very simplistic argument which assumes constant effect per fraction and ignores the effects of tumour hypoxia; what it indicates is that the steepness of the initial slope could be an important determinant of the success of multifraction irradiation.

Cell Survival Curves for Human Tumour Cells

The survival curves shown in Figure 17.9A illustrate the range of radiosensitivity commonly found among human tumour cell lines. The doses that

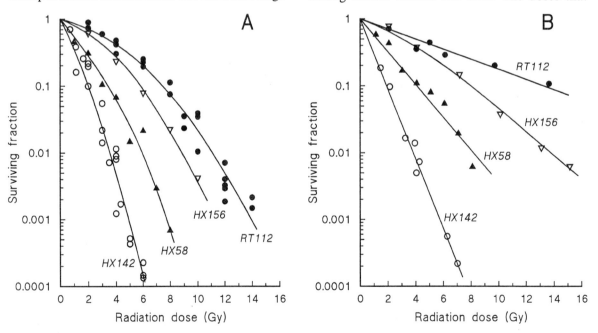

Figure 17.9 Cell survival curve for four representative human tumour cell lines irradiated (**A**) at high dose rate, 150 cGy/min; or (**B**) at low dose rate, 1.6 cGy/min. HX142, neuroblastoma; HX58, pancreas; HX156, cervix; RT112, bladder carcinoma. From Steel (1991), with permission.

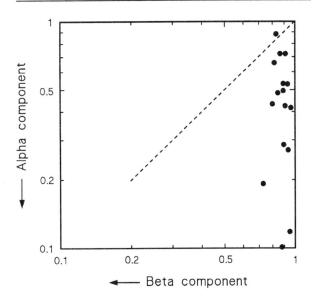

Figure 17.10 Contributions to the surviving fraction at 2 Gy from the linear and quadratic components of radiation cell killing, for 17 human tumour cell lines. From Steel and Peacock (1989), with permission.

correspond to a survival of 10^{-2} differ by approximately a factor of three. At low dose rate (panel B) the curves fan out and become straighter (Section 18.2); these curves tend to extrapolate the initial slopes of the high dose rate survival curves. The full lines in Figure 17.9A are linear-quadratic curves fitted to the data. With these and other data sets the fit is good: there is a clear initial slope (in contrast to the prediction of the multi-target equation, Figure 7.1) and the data are consistent with a continuously bending curve.

Since there may be a subcellular basis for regarding the separate terms of the linear-quadratic equation as mechanistically different (Section 8.1) it is interesting to examine their relative contributions to cell killing. We may separate them thus:

Linear or *alpha* component $\quad = \exp(-\alpha d)$
Quadratic or *beta* component $\quad = \exp(-\beta d^2)$

Having fitted the data with the linear-quadratic equation, as shown in Figure 17.9, we know the values for α and β, and for any chosen dose we can calculate these components. This has been done in Figure 17.10 for 17 human tumour cell lines (Steel and Peacock, 1989). The chosen dose is 2 Gy, a typical dose per fraction in clinical radiotherapy. It can be seen that the calculated points lie down the right-hand border of this diagram: the dispersion in radiosensitivity among these cell lines at 2 Gy is *entirely* due to differences in the steepness of the *linear* component of cell killing. The small amounts of cell kill due to the beta component (distance of the points from the right-hand boundary) do not seem to correlate with sensitivity.

The conclusion is one that derives from most of the models of radiation cell killing: cell killing at clinically realistic doses is dominated by the steepness of the linear component. The nature of this linear component is therefore a matter of considerable interest. One view is that it may be the result of DNA damage due to clusters of ionization events at the end of electron tracks, as illustrated in Figure 8.1.

How can we obtain an accurate measurement of the linear component of cell killing? By fitting the LQ model to a survival curve it is possible to derive estimates of α and β but these are often not very accurate. 'Trade-off' between α and β allows a range of combinations of the two parameters to fit the data almost equally well. A useful alternative is to use low dose rate irradiation. As indicated in Section 18.2, the cell survival curve at low dose rate seems to extrapolate the initial slope of the high dose rate curve. Furthermore, fitting data obtained at a variety of dose rates using the LPL model allows an estimate of the limiting slope at infinitely low dose rate (*e.g.* line B in Figure 18.2). This is the survival curve under conditions of full repair and is a useful indication of the linear component of cell killing (Steel, 1991).

Bibliography

Ash DV (1980). Growth delay studies in patients with multiple metastases. *Brit J Cancer* 41(Suppl IV):17–20.

Begg AC and Terry NHA (1985). Stromal radiosensitivity: influence of tumour type on the tumour bed effect assay. *Brit J Radiol* 58:93–6.

Deacon JM (1986). The Radiobiology of Human Neuroblastoma. *Thesis, University of London*.

Durand RE and Sutherland RM (1973). Dependence of the radiation response of an *in vitro* tumor model on cell cycle effects. *Cancer Res* 33:213–19.

Fertil B and Malaise EP (1981). Inherent radiosensitivity as a basic concept for human tumor radiotherapy. *Int J Radiat Oncol Biol Phys* 7:621–9.

Fowler JF, Sheldon PW, Denekamp J and Field SB (1976). Optimum fractionation of the C3H mouse mammary carcinoma using X-rays, misonidazole or neutrons. *Int J Radiat Oncol Biol Phys* 1:579–92.

Key points

1. Growth delay, tumour control and clonogenic cell survival are the principal experimental end-points for tumour response.
2. Human tumours differ in radiocurability and the initial slope of the cell survival curve for oxic tumour cells is one of the underlying factors.
3. In addition to the reduction in survival of clonogenic cells, tumour response is also influenced by stromal damage and factors that determine repopulation rate.
4. Cell survival curves for human tumour cells are well fitted by the linear-quadratic equation. Survival at radiation doses up to around 2 Gy is dominated by the linear term in this equation.

Freshney RI (1987). *Culture of Animal Cells*. Liss; New York.

Hermens AF and Barendsen GW (1969). Changes of cell proliferation characteristics in a rat rhabdomyosarcoma before and after X-irradiation. *Eur J Cancer* 5:173–89.

Kopper L and Steel GG (1975). The therapeutic response of three human tumour lines maintained in immune-suppressed mice. *Cancer Res* 35:2704–13.

Scott OCA (1991). Tumour transplantation and tumor immunity: a personal view. *Cancer Research* 51:757–63.

Sparrow S (1980). *Immunodeficient Animals for Cancer Research*. Macmillan; London.

Stanley JA, Shipley WU and Steel GG (1977). Influence of tumour size on a hypoxic fraction and therapeutic sensitivity of Lewis lung tumour. *Brit J Cancer* 36:105–13.

Steel GG (1989). Radiobiology of human tumour cells. In: *The Biological Basis of Radiotherapy*. (Eds) Steel GG, Adams GE and Horwich A. Elsevier; Amsterdam.

Steel GG, Courtenay VD and Peckham MJ (1983). The response to chemotherapy of a variety of human tumour xenografts. *Brit J Cancer* 47:1–13.

Steel GG and Peacock JH (1989). Why are some human tumours more radiosensitive than others? *Radiother Oncol* 15:63–72.

Stephens TC, Currie GA and Peacock JM (1978). Repopulation of gamma-irradiated Lewis lung carcinoma by malignant cells and host macrophage 1 precursors. *Brit J Cancer* 38:573–82.

Suit HD and Maeda M (1967). Hyperbaric oxygen and radiobiology of a C3H mouse mammary carcinoma. *J Nat Cancer Inst* 39:639–52.

Suit HD, Skates S, Taghian A *et al* (1992). Clinical implications of heterogeneity of tumor response to radiotherapy. *Radiother Oncol* 25:251–60.

Suit HD, Shalek RJ and Wette R (1965). Radiation response of C3H mouse mammary carcinoma evaluated in terms of radiation sensitivity. In: *Cellular Radiation Biology*. Williams & Wilkins; Baltimore.

Further Reading

Deacon J, Peckham MJ and Steel GG (1984). The radioresponsiveness of human tumours and the initial slope of the cell survival curve. *Radiother Oncol* 2:317–23.

Kallman RF (1987). *Rodent Tumor Models*. Pergamon; New York.

Steel GG (1991). Cellular sensitivity to low dose-rate irradiation focuses the problem of tumour radioresistance. *Radiother Oncol* 20:71–83.

18

The dose rate effect: brachytherapy

G. Gordon Steel

18.1 Mechanisms underlying the dose rate effect

The dose rates used for most radiobiological studies on cells and tissues tend to be in the range 1–5 Gy/min, as are dose rates used clinically for external-beam radiotherapy. Exposure times for a dose of say 2 Gy are therefore no more than a couple of minutes. Within this time, the initial *chemical* (*i.e.* free radical) processes that are generated by radiation can take place but such times are not long enough for the repair of DNA damage or for any other *biological* processes to occur (Section 1.3). As dose rate is lowered, the time taken to deliver a particular radiation dose increases; it then becomes possible for a number of biological processes to take place *during* irradiation and to modify the observed radiation response. These processes are best described by the *4 Rs of Radiobiology*: recovery (or repair), reassortment, repopulation, and reoxygenation (Section 6.9).

Figure 18.1 illustrates the operation of these processes in producing the dose rate effect. The range of dose rates over which each has an effect depends upon its speed. Repair is the fastest of these processes (half-time ~ 1 hour) and when the exposure duration is of the order of 1 hour considerable repair will take place. Calculations show that repair at this speed will modify radiation effects over the dose rate range from around 1 Gy/min down to ~ 0.1 cGy/min (see Figure 18.4). Even in the range of clinical external-beam dose rates, small effects on tolerance may arise from changes in dose rate. In contrast, repopulation is a much slower process. Doubling times for repopulation in human tumours or normal tissues cannot be less than 1 day; the range is probably very wide, from a few days to weeks (Sections 14.3, 25.4). Only when the exposure duration becomes a considerable fraction of a day will significant repopulation occur during a single radiation exposure. Repopulation, either in tumours or normal tissues, will therefore influence cellular response over a much lower range of dose rates, below say 2 cGy/min, depending upon the cell proliferation rate. Reassortment (*i.e.* cell cycle progression) will modify response over an intermediate range of dose rates, as will reoxygenation in tumours. The kinetics of reoxygenation are variable among tumour types and poorly understood. This could, nevertheless, be a significant factor in the effectiveness of brachytherapy.

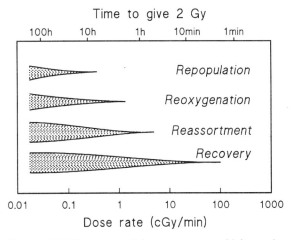

Figure 18.1 The range of dose rates over which repair, reassortment and repopulation modify radiosensitivity depends upon the speed of these processes. From Steel *et al* (1986), with permission.

18.2 Dose rate effect on cell survival

As radiation dose rate is lowered in the range 1 Gy/min down to 1 cGy/min, the radiosensitivity of cells decreases and the shouldered cell survival curves observed at high dose rate gradually become straighter. This is illustrated in Figure 18.2. At 150 cGy/min the survival curve has a marked curvature; at 1.6 cGy/min it is almost straight (on the semi-log plot) and seems to extrapolate the initial slope of the high dose rate curve. The amount of sparing associated with the dose rate reduction can be expressed by reading off the radiation doses that give a surviving fraction of say 0.01: these values are 7.7 Gy at 150 cGy/min and 12.8 Gy at 1.6 cGy/min. The ratio of these doses (*e.g.* 12.8/7.7 = 1.6) has been called the *dose–recovery factor (DRF)*. The data at all three dose rates in Figure 18.2 have been simultaneously fitted by the LPL model (Section 7.6), a model that is particularly useful for simulating the dose rate effect. This allows an estimate to be made of the half-time for cellular recovery (0.16 hour) and it also predicts survival under conditions of no repair (line A) or full repair (line B).

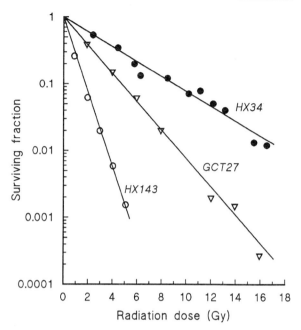

Figure 18.3 Cell survival curves for three human tumour cell lines irradiated at the low dose rate of 1.6 cGy/min. HX143, neuroblastoma; GCT27, germ cell tumour of the testis; HX34, melanoma. From Steel (1991), with permission.

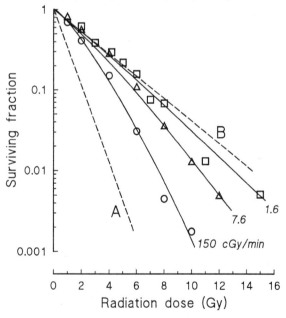

Figure 18.2 Cell survival curves for a human melanoma cell line irradiated at dose rates of 150, 7.6 or 1.6 cGy/min. The data are fitted by the LPL model, from which the lines A and B are derived (see text). From Steel *et al* (1987), with permission.

For four selected human tumour cell lines, Figure 18.3 shows the survival curves at these two dose rates (150 and 1.6 cGy/min). These four sets of data have been chosen to illustrate the dose rate effect within the range of radiosensitivities seen among human tumour cells (Steel *et al*, 1987). At high dose rate there is a range of approximately 3 in the radiation dose that gives a survival of 0.01. At low dose rate the curves fan out and become straight or nearly so: the range of $D_{0.01}$ values is now roughly 7. This illustrates an important characteristic of low dose rate irradiation: it discriminates better than high dose rate between cell lines of differing radiosensitivity.

18.3 Dose rate effect in normal tissues

Most normal tissues show considerable sparing as dose rate is reduced. An example is shown in Figure 18.4. The thorax of conscious mice was irra-

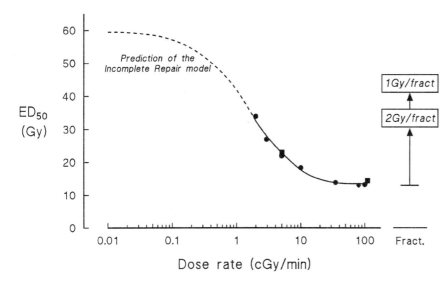

Figure 18.4 The dose rate effect for pneumonitis in mice. The full line fitted to the data was calculated on the basis of the incomplete repair model; the broken line shows its extrapolation to very low dose rates. The boxes on the right show the ED_{50} values for fractionated irradiation. From Down et al (1986), with permission.

diated with ^{60}Co γ-rays and damage to the lung was measured using a breathing-rate assay (Down et al, 1986). The radiation dose that produced acute pneumonitis in 50% of the mice (i.e. the ED_{50}) was 13.3 Gy at 100 cGy/min but it increased to 34.2 Gy at the lowest dose rate of 2 cGy/min (DRF = 2.6). Note that a similar degree of sparing could be achieved (in studies of other investigators) by fractionated high dose rate irradiation using 2 Gy per fraction, and even more sparing at 1 Gy per fraction. Note also that at 2 cGy/min the curve is still rising rapidly. It was not possible in these experiments to go down to dose rates below 2 cGy/min because of the difficulty of immobilizing the mice for long periods of time.

The data in Figure 18.4 have been fitted by the incomplete repair model (Thames, 1985; Sections 13.9, 18.4). This model simulates the effect of recovery on tissue sensitivity; it does not take account of cell proliferation during irradiation. The model fits the data well and it also allows extrapolation down to low dose rates. It predicts that dose-sparing due to recovery will continue to increase down to about 0.01 cGy/min at which the ED_{50} is 59 Gy and the recovery factor (i.e. DRF value) is 4.4. Proliferation of stem cells in the lung will lead to even greater sparing at very low dose rates.

The comparison between a single low dose rate exposure (2 cGy/min) and fractionated high dose rate irradiation (2 Gy/fraction) allows an important conclusion to be drawn. If the fractions are delivered once per day then the overall time to deliver an ED_{50} dose of 34 Gy is 17 days. The same effect is produced by a single low dose rate treatment in 28 hours. Continuous low dose rate exposure is thus the *most efficient way of allowing maximum tissue recovery in the shortest overall time*. It minimizes the effects of proliferation, which is an advantage in terms of damage to tumour cells but a disadvantage for the tolerance of early-responding normal tissues.

Figure 18.5 shows similar data from a number of other studies of the dose rate effect on normal tissues in mice. Considerable sparing of the intestine is observed, especially below 1 cGy/min, where proliferation has a marked effect. In contrast, there is little or no sparing of haemopoietic tissues. The dose rate effect on mouse lethality (LD_{50}, indicated by the full line without points) is comparatively slight, probably because this is predomi-

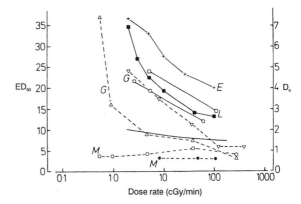

Figure 18.5 The dose rate effect in normal tissues of the mouse: L = lung, G = gut, E = epilation, M = bone marrow. Full lines refer to left-hand scale, dashed lines to right-hand scale. See Steel et al (1986) for sources.

nantly due to damage to bone marrow, no doubt with a component of damage to the intestine.

18.4 Isoeffect relationships between fractionated and continuous low dose rate irradiation

A variety of theoretical descriptions of the dose rate effect have been made but for clinical application the most widely used is the incomplete repair model of Thames (1985). The calculations of Dale (1985) make the same basic assumptions, although the formulation is slightly different. As indicated in Section 13.9, the basic equation of the incomplete repair model is:

$$E = \alpha D + \beta D^2 \cdot g \qquad \text{(Eqn 18.1)}$$

where E is the level of effect, α and β are parameters of the linear-quadratic equation, D is the total dose, and g is a function of time (both of the time between fractions and the duration of continuous exposure). Note that the time-dependent recovery factor modifies only the quadratic term in the LQ equation, a feature that is supported by experimental data (Wells and Bedford, 1983; Steel *et al*, 1987; Figure 18.2). Note also that repopulation is ignored in these calculations.

The *fractionated* case has been dealt with in Chapter 13. For the case of *continuous* exposure, the value of g depends upon the half-time for recovery ($T_{\frac{1}{2}}$) and the duration of exposure (t) according to the relation:

$$g = 2 [\mu t - 1 + \exp(-\mu t)] / (\mu t)^2 \qquad \text{(Eqn 18.2)}$$

where $\mu = 0.693 / T_{\frac{1}{2}}$

(Thames, 1985). This model allows isoeffect relationships to be calculated and as shown in Figure 18.4 it is successful in fitting experimental data over a range of dose rates. Further examples of calculated curves are shown in Figure 18.6. On the left is the fractionated case. The line in this chart corresponds to Equation 13.1, as shown in Figure 14.2. The interfraction intervals have here been assumed to be long enough to allow complete recovery between fractions, although the model can handle shorter intervals. The centre panel in Figure 18.6 shows isoeffect curves for a single continuous exposure at any dose rate, on the assumption that the half-time for recovery is 1.0, 1.5 or 2.0 hours. The three curves are slightly different and this illustrates the dependence of the isoeffect curve for continuous exposure on the speed of recovery: the curve shifts laterally to lower dose rates as the half-time is prolonged. Unfortunately the recovery half-time is seldom known in clinical situations, which limits the precision of calculations of this sort.

The curves in the panels of Figure 18.6 are *mutually isoeffective*. They are calculated for the same effect level and for the same values of α and β (the α/β ratio is 10 Gy), chosen to give an extrapolated dose (a BED value, Sections 13.6, 13.9) of 72 Gy at infinitely small dose per fraction or infinitely low dose rate. This example illustrates the equivalence that is predicted by the mathematical models between a particular continuous dose rate and a corresponding dose per fraction. For the parameters assumed here (as shown by the vertical arrows), a dose rate of around 1–2 cGy/min (roughly 1 Gy/hour) is equivalent to fractionated treatment with approximately 2 Gy per fraction, for both of which the isoeffective dose is 60 Gy.

A further important conclusion can be drawn from calculations of the type shown in Figure 18.6. In Chapters 13 and 14 (see Figure 14.2) we have seen how the use of large fraction sizes leads to a therapeutic *dis*advantage. The same is true for high continuous dose rate treatments. By drawing further horizontal lines between panels A and B in Figure 18.6 it can be seen that a dose rate of 5 cGy/min is equivalent to around 6–8 Gy per fraction and 10 cGy/min to over 10 Gy per fraction.

Figure 18.6c shows the results of model calculations for *fractionated low dose rate* irradiation. Once again using the incomplete repair model we have calculated isoeffect curves for treatment with 2–50 fractions, each given at the dose rate shown on the abscissa and with full recovery between fractions. Again, repopulation is ignored. This diagram indicates the basic feature of fractionated low dose-rate exposure: as we increase the number of fractions the dose rate effect is reduced (*i.e.* the curves become flatter), and as we lower the dose rate the effect of fractionation is reduced (as seen by the vertical spread between the curves). This results from a simple principle. As we protract irradiation it is *cellular recovery* that produces all these effects. There is a limit to how much recovery the cells can accomplish. If we allow recovery between

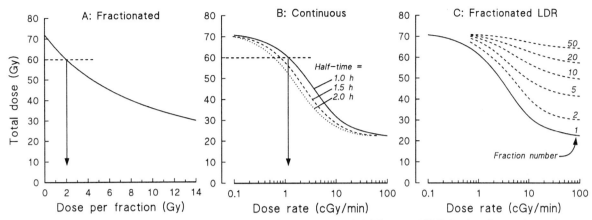

Figure 18.6 Isoeffect curves calculated on the incomplete repair model (Thames, 1985) for fractionated, continuous or fractionated low dose rate radiation exposure (the three panels are mutually isoeffective). Repopulation is ignored. The α/β ratio is 10 Gy and the extrapolated response dose (BED) is 72 Gy. Adapted from Steel (1991) and Steel et al (1989), with permission.

fractions then there is less to be recovered *during* each fraction, and *vice versa*.

An alternative approach to the description of the dose rate effect is the Lethal–Potentially Lethal (LPL) model of Curtis (1986). This is a mechanistic model which is described in Section 7.6. It has theoretical advantages for studies that seek to describe the cellular mechanisms of radiation cell killing but is less appropriate for clinical calculations than the empirical equations of Thames and Dale referred to above.

Effect of Cell Proliferation

The isoeffect relationships shown in Figure 18.6 assume no cell proliferation during treatment. They probably apply reasonably well to late-responding normal tissues and slowly repopulating tumours. In proliferating cell systems, either of tumours or early-responding normal tissues, cell multiplication during irradiation will counteract the killing effects of irradiation: the *isoeffective dose* for any treatment that extends over longer than a day or two will be higher than predicted by the regular incomplete repair model. The effect of proliferation during fractionated irradiation is dealt with in Sections 13.10 and 14.3. At the present time there is no consensus on how this should be accounted for in tolerance calculations. The following calculations are only approximate, designed to illustrate the relative roles of radiosensitivity changes and repopulation

as dose rate is modified. We will calculate using the biologically effective dose (BED, as described in Section 13.6). If we assume a constant proliferation rate, then for fractionated irradiation with full recovery between fractions:

$$\text{BED} = D[1 + d/(\alpha/\beta)] - K_r \cdot T \quad \text{(Eqn 18.3)}$$

where d is the dose per fraction and $D = n.d$ is the total dose. For a single continuous exposure:

$$\text{BED} = D[1 + g.D/(\alpha/\beta)] - K_r \cdot T \quad \text{(Eqn 18.4)}$$

where D is the total dose, and $f(t)$ is the time-dependent function given above. The effect of proliferation is described in both these equations by the term $K_r \cdot T$, where T is the overall duration of treatment. K_r has units of Gy.day^{-1} and can be regarded as the loss of biological effectiveness (*i.e.* BED) per day as a result of proliferation. Although repopulation may not begin at its full rate from the very start of treatment (Section 14.3), the assumption here is that it occurs at a constant rate during irradiation.

Unfortunately, the magnitude of K_r depends not only on the rate of repopulation in the tissue (doubling time T_r) but also on its radiosensitivity:

$$K_r = 0.693/(\alpha T_r)$$

where α is the coefficient of the linear term in the LQ equation. K_r will be greater for short doubling times, but it will also be greater for radioresistant tumours where α is small. Since the absolute values of α and T_r are not well known, these calculations may only be used as a guide to the relationships involved.

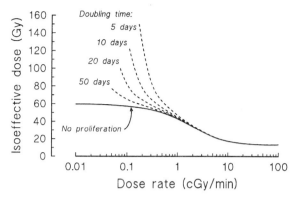

Figure 18.7 Illustrating the effect of cell proliferation as a function of dose rate. Isoeffect curves are shown for no proliferation or with the doubling times indicated. The calculations are based on a simple model of exponential growth, ignoring radiation effects on the rate of cell proliferation.

The effect of proliferation at very low dose rates is graphically illustrated in Figure 18.7. These calculations are made for a hypothetical cell population with an α/β ratio of 3.7 Gy and a repair half-time of 0.85 hour. Cell proliferation is assumed to occur with the doubling times shown in the figure and no account has been taken of radiation effects on the rate of proliferation (if this occurred it would reduce the effect of proliferation at the higher dose rates). For these parameter values there is no effect of proliferation at dose rates above 1 cGy/min but as the dose rate is lowered to 0.1 cGy/min dramatic effects are predicted. The implication for brachytherapy is that above 1 cGy/min repopulation effects can be ignored, but below this dose rate they will be substantial, in many tumours and in early-responding normal tissues.

Figure 18.8 shows the results of illustrative calculations of the therapeutic advantage of altered fractionation schedules, using the above equations. Following Dale (1989), we have assumed that 30 fractions of 2 Gy (total dose 60 Gy) is at the limit of *late* normal-tissue tolerance. The α/β ratios are taken to be 3 Gy for late normal-tissue damage and 10 Gy for tumour response (Fowler, 1989). The BED values for tumour and normal tissues are:

$$BED_{late} = 60(1 + 2/3) = 100 \text{ Gy}$$
$$BED_{tumour} = 60(1 + 2/10) - (0.6 \times 39) = 48.6 \text{ Gy}$$

We are assuming no proliferation in the late-reacting normal tissues but that repopulation in the tumour is associated with a K_r-value of 0.6 Gy/day (overall time 39 days including weekends).

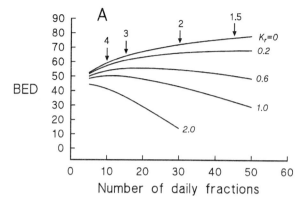

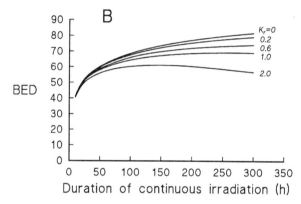

Figure 18.8 Tumour response for a fixed level of late normal-tissue damage as a function of (**A**) fraction number or (**B**) exposure time in a single continuous dose. These curves are recalculated and redrawn after Dale (1989), as described in the text. The arrows in panel **A** indicate the dose per fraction (Gy) at various points along the curve for no tumour cell proliferation ($K_r = 0$).

For modified fractionation, we vary the fraction number (for simplicity assumed to include weekends) and first calculate by Equation 18.1 the corresponding tolerance doses per fraction ($\alpha/\beta = 3$ Gy). Assuming that these are also the daily doses to the tumour, we then calculate BED_{tumour} values. This gives the tumour BED values that correspond to a fixed level of normal-tissue damage ($BED_{late} = 100$ Gy). In Figure 18.8A it can be seen that for no repopulation ($K_r = 0$) the curve rises continuously. This is the well-known gain from using large fraction numbers and small fraction sizes. For 30 fractions, BED_{tumour} is 72 Gy. As K_r is increased, repopulation counteracts this therapeutic gain and for $K_r = 0.6$ Gy/day the curve is almost flat: repopulation is just balancing the gain due to recovery between fractions. For more rapid repopulation and larger K_r values, there is a *disadvantage* in using large numbers of daily fractions. K_r values in excess of 2.0 Gy/day are unlikely in human tumours, though they may occur for tumours in experimental animals. Dale (1989) has also performed similar calculations for multiple fractions per day.

The effects of continuous irradiation can be calculated in a similar way. For any dose rate we calculate using Equation 18.4 the treatment time that is equivalent to a BED_{late} of 100 Gy without repopulation ($\alpha/\beta = 3$ Gy). The half-time for recovery has been taken as 1.5 hours. Then once again with $\alpha/\beta = 10$ Gy we obtain BED_{tumour} values (Figure 18.8B). Within the range of parameters used, lowering the dose rate almost always gives a therapeutic advantage, especially where the effect of proliferation is small. Proliferation has less effect in panel B than panel A of Figure 18.8, reflecting the distinctive advantage of continuous treatment: overall time is minimized, and so also is the detrimental effect of tumour cell repopulation.

The Inverse Dose Rate Effect

Although in situations affecting clinical practice it is a general rule that cellular sensitivity decreases with decreasing dose rate, exceptions to this rule have been noted. Mitchell and Bedford in early studies of cell killing in mammalian cell lines occasionally found a slight inversion which they attributed to a lower dose rate allowing cells to progress though the cell cycle into more sensitive phases, thus suffering greater damage. In studies of oncogenic transformation by high-LET radiations, the existence of an inverse dose rate effect is well recognized (Miller *et al*, 1993).

18.5 Radiobiological aspects of brachytherapy

The principal reasons for choosing interstitial or intracavitary radiotherapy in preference to external-beam treatment relate to dose delivery and dose distribution rather than to radiobiology. Irradiation from an implanted source *within* a tumour carries a distinct geometrical advantage for sparing the surrounding normal tissues that will inevitably tend to receive a lower radiation dose. They will also be exposed to a lower *dose rate*, which gives the additional advantage of 'negative double trouble' (Section 13.8).

Variation in Cell Killing around an Implanted Radioactive Source

The non-uniformity of radiation field around an implanted source has important radiobiological consequences. Close to the source the dose rate is high and the amount of cell killing will be close to that indicated by the acute-radiation survival curve. As we move away from the source, two changes take place: cells will be less sensitive at the lower dose rates, and within a given period of

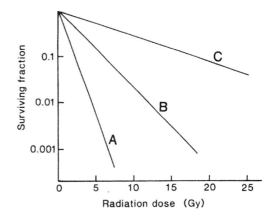

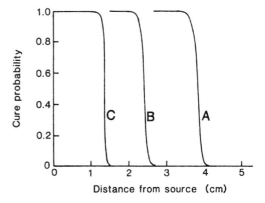

Figure 18.9 The likelihood of cure varies steeply with distance from a point radiation source. The radius at which failure occurs depends upon the steepness of the survival curve at low dose rate. From Steel *et al* (1989), with permission.

implantation the accumulated dose will also be less. These two factors lead to a very rapid change in cell killing with distance from the source. Within tissues (tumour or normal) that are close to the source the level of cell killing will be so high that cells of any radiosensitivity will be killed. Further out, the effects will be so low that even the most radiosensitive cells will survive. Between these extremes there is a critical zone in which differential cell killing will occur. As shown by Steel *et al* (1989), for cells of any given level of radiosensitivity there will be cliff-like change from high to low local cure probability, taking place over a radial distance of a few millimetres (Figure 18.9). The distance of the cliff from the source is determined by the radiosensitivity of the cells at low dose rate, nearer for radioresistant cells and further away for radiosensitive cells (Steel, 1991). There would

appear to be a strong case for introducing a predictive test of tumour cell radiosensitivity (Section 25.3) into brachytherapy practice. If the dose distribution or the total dose administered could be adjusted on this basis a therapeutic gain might be achieved.

Figure 18.10 contrasts this situation with external-beam radiotherapy where the aim is usually to deliver uniform radiation dose across the tumour. Only in a narrow zone around an implanted source (where the surviving fraction changes from say 10^{-20} to 10^{-6}) will radiobiological considerations be of interest or importance in relation to tumour control. The same principle will apply to normal-tissue damage: serious damage to normal structures depends on making sure that they are outside the corresponding 'cliff'.

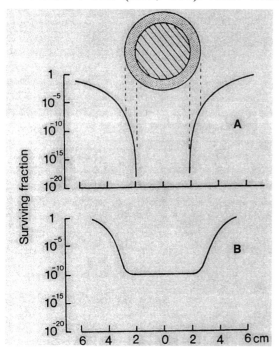

Figure 18.10A: Variation of cell kill around a point source of radiation. The source gives 0.87 Gy/min at 2 cm (*i.e.* 75 Gy in 6 d); there are 10^9 cells per cm^3, for which $\alpha = 0.35$ Gy^{-1}, $\beta = 0.035$ Gy^{-2}, half-time for recovery is 1 hour (Steel *et al*, 1989). The hatched area indicates the volume within which the surviving fraction is below 10^{-20}. The stippled area indicates the volume where survival is between 10^{-20} and 10^{-6}, which is the critical region for tumour control. For comparison, panel **B** shows the type of profile that would be aimed for with external-beam radiotherapy.

IS THERE A RADIOBIOLOGICAL ADVANTAGE IN LOW DOSE RATE RADIOTHERAPY?

The question of whether low dose rate irradiation itself carries a therapeutic advantage is an interesting one. There is a considerable volume of literature on the dose rate effect both in tumours and in normal tissues on the basis of which it would be difficult to claim that under all circumstances low dose rate treatment would have the best therapeutic index. As shown in Figure 17.9, cells that are the least sensitive to radiation and have the largest shoulder on the cell survival curve will show the greatest degree of dose-sparing. These are not necessarily cell lines of low α/β ratio, for Peacock *et al* (1992) have shown that a range of human tumour cell lines, including those shown in Figure 17.9, have similar α/β ratios: radioresistant tumour cells tend to have both a lower α *and* a lower β than more sensitive cells. In a particular therapeutic situation we could do a calculation comparing the relative dose reduction factors (DRF) between tumour and critical normal tissues. This would tell us whether the normal tissues might be spared more or less than the tumour cells if we were to lower the dose rate. But this does not answer the therapeutic question because to treat with one large high dose rate fraction is not normally a clinical option. In fact we know from clinical experience in the early history of radiotherapy that large high dose rate treatments are bad and that changing to a single low dose rate exposure would be better. This is similar to comparing hyp<u>o</u>fractionation with the use of conventional or reduced dose per fraction (Figure 14.2). The appropriate clinical question is whether a single continuous low dose rate treatment is better than using a conventional fractionation schedule.

As is illustrated in Figure 18.6, there is, on the basis of the incomplete repair model, an equivalence between dose per fraction in fractionated radiotherapy and dose rate for a single continuous exposure. Roughly speaking, for a given level of cell killing the total dose required at a continuous dose rate of 1 Gy/hour is the same as that required by fractionated treatment with 2 Gy/fraction. This equivalence depends upon the half-time for recovery but it is relatively independent of the α/β ratio (Fowler, 1989). In radiobiological terms, these two treatments should be equally effective. Lowering the (fractionated) dose per fraction will spare late-responding normal tissues whose α/β ratio is low, as also will lowering the dose rate (continuous) below 1 Gy/hour.

That high dose rate intracavitary therapy is widely thought to be acceptable may result from two factors: (1) the lower *volume* of normal tissue irradiated to a dose that discriminates between tissue sensitivities (*i.e.* the annular region referred to in relation to Figure 18.10); and (2) the practical and therapeutic benefits of short treatment times.

The clearest advantage for low dose rate irradiation is that for a given level of cell killing, and without hazarding late-responding normal tissues, this treatment will be complete within the *shortest overall time* (Section 18.4). Tumour cell repopulation will therefore be minimized. This could confer a therapeutic advantage for the treatment of rapidly repopulating tumours.

A potential *disadvantage* of low dose rate irradiation is that because of the short overall treatment time there may be inadequate time available for the reoxygenation of hypoxic tumour cells and therefore greater radioresistance due to hypoxia.

Pulsed Brachytherapy

The availability of computer-controlled high dose rate afterloading systems provides the opportunity to deliver interstitial or intracavitary radiotherapy in a series of pulses. The gaps between pulses allow greater freedom for the patient and increased safety for nursing staff. In principle, any move away from continuous exposure towards treatment with gaps carries a radiobiological *disadvantage*. This is equivalent to fractionation with a larger dose per fraction and the theoretical and

---

Key points

1. A larger radiation dose is usually required to produce a particular biological effect at low dose rate ($\sim$ 1 Gy/hour) compared with high dose rate ($\sim$ 1 Gy/min).
2. Low dose rate irradiation discriminates better than high dose rate between effects on cells of differing radiosensitivity.
3. The dose rate effect observed between 1 Gy/min and 1 cGy/min is predominantly due to *repair* during irradiation; the effect seen below 1 cGy/min is mainly due to *cell proliferation* during irradiation.
4. Continuous low dose rate exposure is the most efficient way of achieving full recovery from radiation damage in the shortest overall time. It minimizes the effects of proliferation, which is an advantage in terms of damage to tumour cells but a disadvantage for the tolerance of early-responding normal tissues.
5. High continuous dose rates (*i.e.* above say 2 Gy/hour) are biologically equivalent to the use of a large dose per fraction in fractionated radiotherapy; greater late damage to late-responding normal tissues for a given level of tumour response is therefore to be expected. This disadvantage may be offset in brachytherapy by the lower *volume* of normal tissue irradiated to high dose.
6. Cell killing varies steeply around an implanted radiation source. Close to the source, all cells will be killed. The *radius* at which tumour control will fail depends critically on the radiosensitivity of tumour cells at low dose rate.

experimental evidence that this will lead to a relative increase in late normal-tissue reactions is strong (Section 14.2). The magnitude of this effect has been considered by Brenner and Hall (1991), who concluded that for gaps between pulses of up to 60 minutes the radiobiological deficit may be acceptable.

A further warning about the use of high dose rate afterloading systems relates to overall time. High dose rate afterloading systems create the temptation to shorten the overall time and, as indicated above, this could lead to increased early reactions to radiotherapy and greater tumour radioresistance due to hypoxia.

Bibliography

Brenner DJ and Hall EJ (1991). Fractionated high dose rate *versus* low dose rate regimens for intracavitary brachytherapy of the cervix. *Brit J Radiol* 64:133–41.

Curtis SB (1986). Lethal and potentially lethal lesions induced by radiation – a unified repair model. *Radiat Res* 106:252–70.

Dale RG (1985). The application of the linear-quadratic dose-effect equation to fractionated and protracted radiotherapy. *Brit J Radiol* 58:515–28.

Dale RG (1989). Time-dependent tumour repopulation factors in linear-quadratic equations – implications for treatment. *Radiother Oncol* 15:371–82.

Down JD, Easton DF and Steel GG (1986). Repair in the mouse lung during low dose-rate irradiation. *Radiother Oncol* 6:29–42.

Miller RC, Randers-Pehrson G, Hieber L *et al* (1993). The inverse dose rate effect for oncogenic transformation by charged particles is dependent on linear energy transfer. *Radiat Res* 133:360–4.

Peacock JH, Eady JJ, Edwards SM *et al* (1992). The intrinsic a/b ratio for human tumour cells: is it a constant? *Int J Radiat Biol* 61:479–87.

Steel GG, Down JD, Peacock JH and Stephens TC (1986). Dose-rate effects and the repair of radiation damage. *Radiother Oncol* 5:321–31.

Steel GG, Kelland LR and Peacock JH (1989). The radiobiological basis for low dose-rate radiotherapy. In: *Brachytherapy 2*, pp.15–25. Proceedings of the 5th International Selectron Users' Meeting 1988. (Ed) Mould RF. Nucletron International BV; Leersum, The Netherlands.

Thames HD (1985). An 'incomplete-repair' model for survival after fractionated and continuous irradiation. *Int J Radiat Biol* 47:319–39.

Wells RL and Bedford JS (1983). Dose rate effects in mammalian cells. IV: Repairable and nonrepairable damage in noncycling C3H $10T_{1/2}$ cells. *Radiat Res* 94:105–34.

Further Reading

Fowler JF (1989). Dose rate effects in normal tissues. In: *Brachytherapy 2*, pp. 26–40. Proceedings of the 5th International Selectron Users' Meeting 1988. (Ed) Mould RF. Nucletron International BV; Leersum, The Netherlands.

Steel GG (1991). Cellular sensitivity to low dose rate irradiation focuses the problem of tumour radioresistance. (The ESTRO Breur Lecture). *Radiother Oncol* 20:71–83.

Steel GG, Deacon JM, Duchesne GM *et al* (1987). The dose-rate effect in human tumour cells. *Radiother Oncol* 9:299–310.

19

Particle beams in radiotherapy

Michael C. Joiner

19.1 Introduction

Radiotherapy is usually given with ^{60}Co γ-rays, or high-energy X-rays produced by linear accelerators at energies of 4–25 MV. These are uncharged electromagnetic radiations, physically similar in nature to radio waves or visible light except that the photons ('packets' of energy) are energetic enough to ionize molecules in tissues that they penetrate. It is this ionization that results in the biological effects seen in radiotherapy. Although there is *some* energy-dependence, the biological effect per unit dose of these radiations is similar. Electron beams are quantum-mechanically similar to X-rays and produce similar biological effects. Two other classes of radiations for use in radiotherapy are often referred to as:

- *Light particles* – e.g. protons, neutrons and α-particles.
- *Heavy particles* – e.g. fully stripped carbon, neon, silicon or argon ions.

These light and heavy particles may have a greater effect per unit dose compared with the conventional radiations. The *charged* particles have, in addition, very different depth–dose absorption profiles compared with the uncharged particles (*i.e.* neutrons) or conventional electromagnetic radiations, and this enables more precise dose distributions to be achieved in radiotherapy. This chapter focuses on these newer types of radiation for use in cancer therapy.

19.2 Microdosimetry

It is possible to build up a picture of the submicroscopic pattern of ionizations within a cell nucleus using special techniques for measuring ionization in very small volumes, together with computer simulations: this is the field of *microdosimetry*. Figure 19.1 shows examples of microdosimetric calculations of ionization tracks from γ-rays or α-particles passing through a cell nucleus (Goodhead, 1988; 1989). At the scale of the cell nucleus, the γ-rays deposit much of their energy as single isolated ionizations or excitations and much

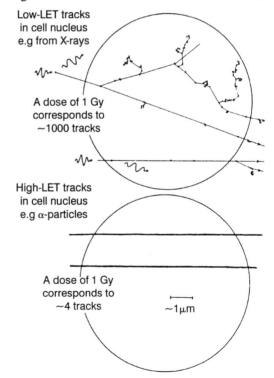

Figure 19.1 The structure of particle tracks for low-LET radiation (above) and α-particles (below). The circles indicate the typical size of mammalian cell nuclei. Note the tortuous tracks of low-energy secondary electrons, greatly magnified in this illustration. From Goodhead (1988), with permission.

of the resulting DNA damage is efficiently repaired by enzymes within the nucleus (Section 8.1). About 1000 of these sparse tracks are produced per gray of absorbed radiation dose. The α-particles produce fewer tracks but the intense ionization within each track leads to more severe damage where the track intersects vital structures such as DNA. The resulting DNA damage may involve several adjacent base pairs and will be much more difficult or even impossible to repair; this is probably the reason why these radiations produce steeper cell survival curves and allow less cellular recovery than X-rays. At the low doses of α-particle irradiation that are encountered in protection from environmental exposure only some cells will be traversed by a particle and many cells will be unexposed.

Linear energy transfer (LET) is the term used to describe the density of ionization in particle tracks. LET is the average energy (in keV) given up by a charged particle traversing a distance of 1 μm. In Figure 19.1, the γ-rays have an LET of about 0.3 keV μm^{-1} and are described as low-LET radiation. The α-particles have an LET of about 100 keV μm^{-1} and are an example of high-LET radiation.

Why are neutrons described as high-LET radiation when they are uncharged? Neutrons do not interact with the orbital electrons in the tissues through which they pass and they do not directly produce ionization. They do, however, interact with atomic nuclei from which they eject slow, densely ionizing protons. It is this secondary production of knock-on protons that confers high LET.

19.3 Biological effects depend upon LET

As LET increases, radiation produces more cell killing per gray. Figure 19.2 shows the survival of human T1g cells plotted against dose for eight different radiations, with LET varying from 2 keV μm^{-1} (250-kVp X-rays) to 165 keV μm^{-1} (2.5-MeV α-particles). The radiations become more efficient per gray at killing cells as the LET increases. As LET increases, the survival curves also become straighter with less shoulder, which indicates either a higher ratio of lethal to potentially lethal lesions (in lesion-interaction models) or that high-LET radiation damage is less likely to be repaired correctly (in repair saturation models; *see* Chapter 7). For particles of identical atomic composition, LET generally increases with decreasing particle energy. However, notice that 2.5-MeV α-particles are *less* efficient compared with 4.0-MeV α-particles even though they have a

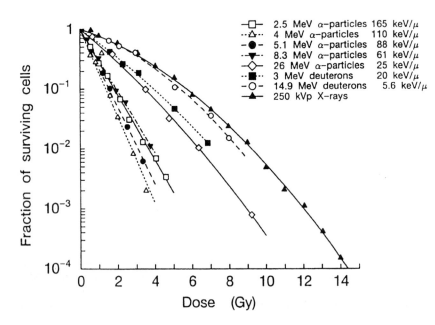

Figure 19.2 Survival of human kidney cells exposed *in vitro* to radiations of different LET. From Barendsen (1968), with permission.

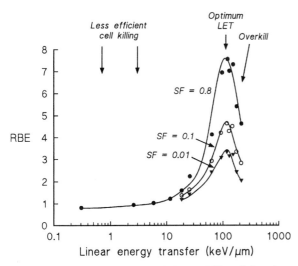

Figure 19.3 Dependence of RBE on LET and the phenomenon of overkill by very high LET radiations. From Barendsen (1968), with permission.

higher LET; this is due to the phenomenon of overkill indicated in Figure 19.3.

The *relative biological effectiveness (RBE)* of a high-LET radiation is defined as:

$$RBE = \frac{\text{dose of reference radiation}}{\text{dose of high-LET radiation}}$$

to give the same biological effect. The *reference low-LET radiation* is usually 250-kVp X-rays. Figure 19.3 shows RBE values for the T1g cells featured in Figure 19.2. Curves have been calculated at cell survival levels of 0.8, 0.1 and 0.01, illustrating the fact that RBE is not constant but *depends on the level of biological damage* and hence on the dose level. RBE rises to a maximum at an LET of about 100 keV μm^{-1}, then falls for higher values of LET due to *overkill*. For cells to be killed, energy must be deposited in a number of critical sites in the cell (Section 7.3). Sparsely ionizing, low-LET radiation is inefficient because more than one particle may have to pass through the cell to kill it. Densely ionizing, very high LET radiation is also inefficient because it deposits more energy than necessary in critical sites. These cells are *overkilled* and per gray there is then less likelihood that *other* cells will be killed, leading to a reduced biological effect. Radiation of optimal LET deposits just enough energy per cell to inactivate the critical targets. This optimum LET is usu-

ally around 100 keV μm^{-1} but it does vary between different cell types and depends on the *spectrum* of LET values in the radiation beam as well as the mean LET.

As LET increases, the oxygen enhancement ratio (OER, Section 15.1) decreases. The measurements shown as an example in Figure 19.4 were also made with cultured T1g cells of human origin (Barendsen, 1968). The sharp reduction in OER occurs over the same range of LET as the sharp increase in RBE (Figure 19.3).

19.4 Relative biological effectiveness (RBE) depends on dose

As indicated in Figure 19.3, the RBE is higher if measured at lower radiation doses, corresponding to higher levels of cell survival. Figure 19.5 shows in more detail the RBE for 4.0-MeV α-particles plotted against dose of 250-kVp X-rays, for the T1g human cells irradiated *in vitro*. The data points were derived from Figure 19.2 by reading off from the α-particle survival curve the dose required to achieve the same cell survival as obtained for each X-ray dose tested. The RBE for the 4.0-MeV α-particles increases with decreasing dose because

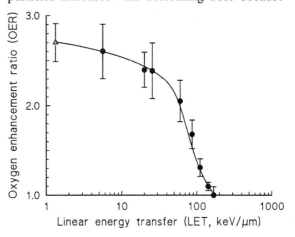

Figure 19.4 The oxygen enhancement ratio (OER) decreases with increasing LET. Closed circles refer to monoenergetic α-particles and deuterons and the open triangle to 250-kVp X-rays. From Barendsen (1968), with permission.

the low-LET X-ray survival response is more curved and has a bigger shoulder compared with the high-LET survival response. If linear-quadratic equations are used to model both the low- and the high-LET responses, RBE can also be predicted mathematically from the α/β ratios and the ratio $\alpha_{\text{high-LET}} / \alpha_{\text{low-LET}}$. This prediction is shown by the solid line.

RBE can also be measured *in vivo*. In normal tissues this may be done by comparing the relationships between damage and dose both for high- and low-LET radiations. This may be done for any end-point of damage, including tissue breakdown or loss of tissue function. Figure 19.6A shows the results of an experiment to study the loss of renal function in mice after external-beam radiotherapy. This was done by measuring the increased retention of ^{51}Cr-radiolabelled EDTA in the plasma 1 hour after injection; normally functioning kidneys completely clear this substance from the body within this time. For neutrons (produced by bombarding beryllium with 4-MeV deuterons, designated d(4)-Be), fractionation makes almost no difference to the tolerance dose but for X-rays a much higher total dose is required to produce renal damage when the treatment is split into two, five or ten fractions. This difference in the fractionation response for high- and low-LET radiations *in vivo* reflects the shape of the survival curves for the target cells in the tissue: almost straight for neutrons, and downwards-bending for X-rays (Figure 19.2). In this situation, RBE is calculated from the ratio of X-ray to neutron total doses required to produce the same biological effect *in the same number of fractions*. This is plotted against X-ray dose per fraction in Figure 19.6B. It can be seen that the *in vivo* RBE increases with decreasing dose per fraction in exactly the same way as for cells *in vitro* shown in Figure 19.5.

19.5 Response of different tissues to high-LET radiation

The response to neutrons shown in Figure 19.6A suggests that for a fixed level of biological effect there should be much less change in total dose with fractionation for high-LET radiation compared with low-LET radiation. Figure 19.7 summarizes isoeffect curves relating total dose to dose per fraction for early-responding (dashed lines) and late-responding (full lines) tissues exposed to fractionated neutron irradiation (Withers *et al*, 1982). This figure should be compared with Figure 14.1, which shows these relationships for a similar range of tissues exposed to fractionated X- or γ-rays. The following conclusions can be drawn from this comparison:

(i) There is much less change in isoeffective total dose with fractionation for neutrons, either in early- or late-responding tissues. This reflects the straighter survival curves for the target cells for high-LET radiation.

(ii) For photons, the total dose increases more steeply with decreasing dose per fraction for late-responding than early-responding tissues, reflecting the smaller α/β ratios for late tissues (Section 13.5). RBE therefore rises rapidly with decreasing dose per fraction for late-responding tissues and more gradually for early-responding tissues.

(iii) Comparing the same tissues exposed to both photons and neutrons, RBE values for late tissue responses are *not* intrinsically higher than for early responses, but because of their faster increase as dose is reduced the RBE values for late tissue response *tend* to be higher than for early tissue response at low

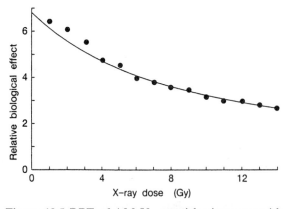

Figure 19.5 RBE of 4-MeV α-particles increases with decreasing dose for cell lines irradiated *in vitro*. RBE values were calculated from the cell survival data shown in Figure 19.2. The full line is calculated as described in the text.

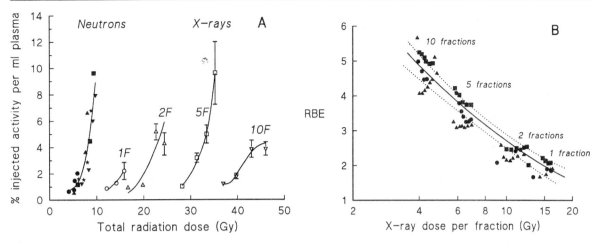

Figure 19.6 The RBE for kidney damage increases with decreasing dose per fraction. RBE values are derived from graphs similar to panel **A**, which shows dose–effect curves for ^{51}Cr-EDTA clearance following irradiation with 1, 2, 3, 5 and 10 fractions of neutrons or 1, 2, 5 and 10 fractions of X-rays. The RBE values in panel **B** were obtained with various renal damage end-points: isotope clearance (circles), reduction in haematocrit (squares) increase in urine output (triangles). From Joiner and Johns (1987), with permission.

doses per fraction, especially at or below 2 Gy per (X-ray) fraction.

To emphasize this last point, Figure 19.8 demonstrates the rise in neutron RBE (compared with X-rays) with decreasing dose per fraction in skin (an early-responding tissue) and kidney (a late-responding tissue). In this example, the RBE for d(16)-Be neutrons in kidney was greater than in skin at an X-ray dose per fraction of 2 Gy, but *lower* for a more modern, more highly penetrating, p(62)-Be neutron therapy beam. Therefore compared with conventional photon therapy, late renal damage would be increased relative to acute reactions (and perhaps relative to tumour response) by treating with a low-energy neutron beam, but late renal injury would actually be spared on the high-energy machine. It is very important to understand that these relationships are specific to these tissues and these neutron beams. Similar relationships between other early-responding and late-responding tissues may not follow the same pattern and so must be evaluated individually in each case and for each treatment site to determine whether neutrons would deliver a therapeutic gain. It is not true that late reactions are *always* worse after neutron therapy for the same level of acute injury, but they may be in some cases.

19.6 The biological basis for high-LET radiotherapy

We have seen (Figure 19.4) that the differential radiosensitivity between poorly oxygenated (more resistant) and well-oxygenated (more sensitive) cells is reduced by using high-LET radiations. Therefore, tumour sites in which hypoxia is a problem in radiotherapy (some head and neck tumours, for example) might benefit from high-LET radiotherapy in the same way as from chemical hypoxic-cell sensitizers (Section 16.3).

The effect of low-LET radiation on cells is strongly influenced by their position in the cell cycle, with cells in S-phase being more radioresistant than cells in G2 or mitosis (Section 6.8). Cells in stationary (i.e. plateau) phase also tend to be more radioresistant than cells in active proliferation. Both of these factors act to increase the effect of fractionated radiotherapy on more rapidly cycling cells compared with those cycling slowly or not at all, because the rapidly cycling cells which survive the first few fractions are statistically more likely to be caught later in a sensitive phase and so be killed by a subsequent dose. This differential radiosensitivity due to cell cycle position is considerably reduced with high-

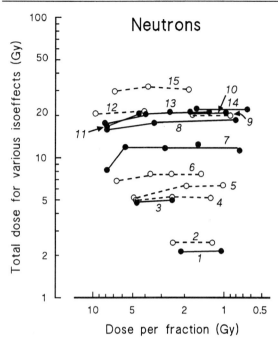

Figure 19.7 Summary of published data on isoeffect curves for neutrons as a function of dose per fraction in various tissues of mice and rats. Broken lines indicate data on early-responding tissues; full lines are for late-responding tissues. Compare with Figure 14.1. Key: 1, thyroid function; 2, haemopoietic colonies; 3, vertebral growth; 4, spermatogenic colonies; 5, fibrosarcomas; 6, jejunum colonies; 7, lung LD_{50} ; 8, lumbar nerve root function; 9, 12 skin desquamation; 10, skin contraction; 11, skin late changes; 13, spinal cord; 14, oral mucosa necrosis; 15, skin necrosis. From Withers *et al* (1982), with permission.

ficial in some slowly growing, X-ray resistant tumours.

A third biological rationale for high-LET therapy is based on the observation that the *range* of radiation response of different cell types is reduced with high-LET radiation compared with X-rays. This is shown in Figure 19.9, which summarizes the *in vitro* response of 20 human cell lines to photon and neutron irradiation (Britten *et al*, 1992). This reduced range of response affects the benefit expected, which is the balance between tumour and normal-tissue responses. Thus, if tumour cells are already more radiosensitive to X-rays than the critical normal-cell population, high-LET radiation should not be used since this would reduce an already favourable differential. Possible examples are seminomas, lymphomas and Hodgkin's disease. However, if the tumour cells are more resis-

Table 19.1 Hypothetical clinical trials which show the importance of patient selection in determining the value of high-LET radiotherapy

A	Total number	Successful response to		*p* value
		Neutrons	Photons	
Subgroup	40	14	6	0.03
Remainder	160	40	40	–
Total	200	54	46	0.3

B	Total number	Successful response to		*p* value
		Neutrons	Photons	
Subgroup	40	14	6	0.03
Remainder	160	36	44	0.3
Total	200	50	50	–

From Bewley (1989).

LET radiation (Chapman, 1980) and is a reason why we might expect high-LET radiotherapy to be bene-

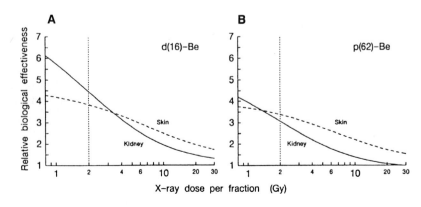

Figure 19.8 Comparison of RBE values for mouse skin and kidney exposed to two different neutron beams. From Joiner (1988), with permission.

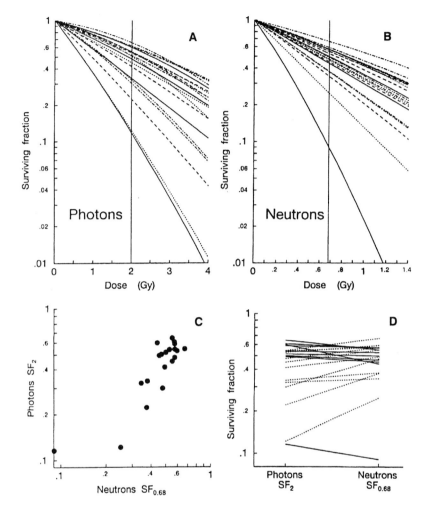

Figure 19.9 Response of 20 human tumour cell lines to (**A**) 4-MVp photons, or (**B**) p(62.5) – Be neutrons. The vertical lines show the photon (2-Gy) and neutron (0.68-Gy) doses that give the same *median* cell survival; the average RBE is therefore 2/0.68 = 2.94. Panel **C** shows that the range of cell survival at the reference neutron dose of 0.68 Gy is less than the range of photon SF_2 values. In 9/20 of the cell lines neutrons gave lower cell survival than photons at these doses (**D**).

tant to X-rays than the critical normal cells, high-LET radiation might reduce this difference in radiosensitivity and thus would effectively 'sensitize' the tumour cell population relative to a fixed level of normal-tissue damage. High-LET radiation would be advantageous in this case.

These radiobiological arguments lead us to expect that high-LET radiotherapy might be of benefit to some cancer patients but not to others. Clinical trials of neutrons that have so far been performed have generally failed to detect such an advantage in the whole cancer patient populations that have been studied. If high-LET therapy is going to be of clinical use, it should therefore only be given to patients who are likely to respond poorly to conventional X-ray therapy, on the basis of the results of assays for tumour oxygenation, cell kinetics and radiosensitivity. The principles of these predictive assays are described in Chapter 25. Table 19.1 demonstrates hypothetically the importance of patient selection for neutron therapy, or indeed for any new therapy that is only expected to benefit a subgroup of patients. Suppose that 200 patients enter a trial of high-LET *vs.* photon therapy with 100 patients in each arm. Of these, 80% respond equally well to the two treatments (and we suppose that they have a 50% cure rate), but the other 20% do better with the high-LET therapy, and this small subgroup responds with a 70% cure rate after high-LET and 30% cure after photons. The results for the total group fail to achieve significance; if the subgroup of 40 patients had been selected out they would have demonstrated a margin in favour of

Table 19.2 Summary of clinical indications for fast neutron therapy

1. Salivary gland tumours (locally extended)
2. Prostatic adenocarcinoma (locally extended)
3. Soft-tissue sarcoma (slowly growing, inoperable)
4. Paranasal sinuses (adenocarcinoma, adenoid cystic ca.)
5. Melanoma and rectal carcinoma (palliative treatment)

From Wambersie et al (1994).

neutrons. It would be even more confusing if the majority (160) of the patients actually did *worse* with high-LET therapy: suppose 45% were cured with high-LET but 55% were cured with photons. Now the whole trial (Table 19.1B) reveals nothing at all but would still conceal an important subgroup. Rational patient selection is thus an important principle in relation to any modality that may improve treatment in only a minority of cases.

In spite of the potential difficulties demonstrated by Table 19.1, some clinical indications for fast neutron therapy have emerged. These have been summarized by Wambersie et al (1994) as listed in Table 19.2. Neutrons may be of some benefit in treating X-ray-resistant tumours, slowly growing tumours and some advanced cancers that perhaps contain a high proportion of hypoxic cells. Wambersie has estimated that up to 10–15% of patients currently receiving radiotherapy would benefit from neutron therapy using modern high-energy isocentric machines, *if those patients could be identified reliably*.

19.7 The physical basis for charged-particle therapy

With conventional X-ray therapy, absorbed dose increases very rapidly within the short distance in which electronic equilibrium ('build-up') occurs, and then decreases exponentially with increasing penetration. Figure 19.10A shows central-axis depth doses from ^{60}Co γ-rays and from X-rays generated by a 6-MV linear accelerator (Fowler, 1981). Neutrons are also uncharged and their depth–dose characteristics are similar. Modern high-energy neutron therapy beams have a penetration that is comparable to 6-MV X-rays. The only rationale for neutron therapy is therefore radiobiological, as discussed earlier.

In contrast, ion beams (*i.e.* incident beams of *charged* particles) *increase* their rate of energy deposition as they slow down with increasing penetration, finally stopping and releasing an intense burst of ionization called the Bragg peak. As an example, curve *1* in Figure 19.10B shows the depth–dose distribution of a primary beam of 160-MeV protons. The broad peak is obtained by superimposing on curve *1* four other beams of different intensities and ranges (curves *2, 3, 4, 5*), achieved by passing the primary beam through a rotating wheel with sectors of different thickness of plastic sheet. This spread-out peak (*Sum*) can be adjusted to cover the tumour volume and therefore increase the ratio of tumour-to-normal-tissue dose compared with conventional photon therapy (Raju, 1980).

Figure 19.11 shows some possible treatment plans with heavy-ion beams of helium and carbon nuclei, using carcinoma of the pancreas as an example. The improvement given by the He ions over 18-MV X-rays is as dramatic as the comparison between 18-MV and 250-kVp X-rays. The mean doses to the spinal cord and kidney are almost zero for He ions, 50% for 18-MV X-rays and 70% for 250-kVp X-rays. Uniformity over the tumour is 2–3%, 5% and 15% respectively.

Carbon ions give a similar dose distribution to He ions but in addition they have a higher LET and RBE in the Bragg peak, which in suitable tumours (see above) might confer a radiobiological advantage. The LET of a charged particle is proportional to the square of its charge divided by the square of its velocity. Therefore in the Bragg peak, where the particles are slowing down rapidly, heavy ions such as carbon, neon and silicon have very high LET, with the potential for a greatly increased biological effect. To illustrate this, Figure 19.12 shows depth–dose curves for beams of heavy ions accelerated to two different energies giving maximum penetrations in tissue of about 14 or 24 cm. In each case the solid line represents the pattern of dose produced by a ridge filter designed to spread out the Bragg peak to cover imaginary tumours of 4 or 10 cm respectively. This is a similar 'peak-spreading' technique to that described in Figure 19.10B. However, the dotted line shows the distribution of *biologically effective dose*, which is physical dose multiplied by RBE. The RBE values are those for Chinese hamster cells corresponding to an X-ray dose of about 2 Gy (Hall, 1988). This demonstrates that for heavy ions (*not* high-energy protons or

The physical basis for charged-particle therapy 181

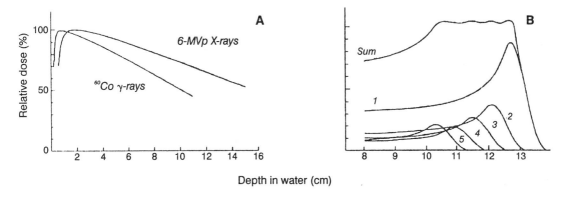

Figure 19.10 The different depth–dose characteristics of (**A**) photons and (**B**) proton beams of different intensities and ranges, achieved by passing a primary beam (*1*) through plastic absorbers.

helium ions) the physical advantage of better dose distribution in the spread-out Bragg peak can be further enhanced by the radiobiological advantage from the higher LET.

Figure 19.13 conveniently summarizes the relative physical and radiobiological properties of different radiations and charged particles (Fowler, 1981). Protons have superb depth–dose distributions and have radiobiological properties similar to orthovoltage X-rays: it is highly probable that light-ion beams of protons and perhaps helium will play a key role in better radiotherapy during the next 20 years. Neutrons have no dose distribution advantage over megavoltage X-rays but may be useful because of their high LET. The heavy ions give better dose distributions than X-rays and also a higher

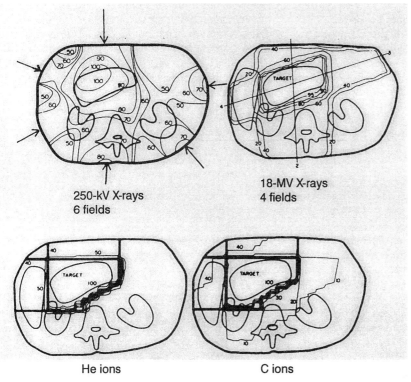

Figure 19.11 Comparison of treatment plans for the radiotherapy of a case of pancreas carcinoma using charged particle beams or photons. From Bewley (1989), with permission.

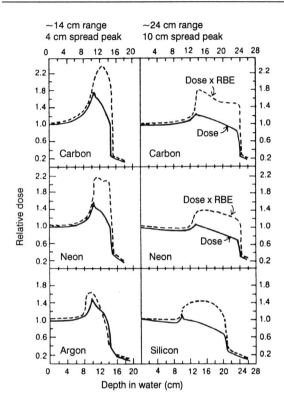

Figure 19.12 The biological effect of charged particle beams is increased further in the Bragg peak. Depth–dose curves are shown for three types of ion beam, each with a 4-cm or 10-cm spread peak. Full lines show the dose distribution; broken lines show the *biologically effective dose* (i.e. dose × RBE). From Blakely (1982), with permission.

LET, depending on the particle. Argon ions have a high LET but in practice they break up so readily that only limited penetration is useful. Carbon, neon and silicon ions seem to be the most promising of the heavy ions at the present time and if heavy-ion therapy has a future it will probably be with these particles (Castro, 1995).

19.8 Summary of therapeutic conclusions

Neutrons are the commonest type of particle beam so far used for radiotherapy but the clinical results obtained are controversial. To some extent this reflects the poor dose distributions and beam delivery systems of the early experimental neutron facilities. It is therefore essential that any further work with clinical neutron radiotherapy should use facilities with isocentric beam delivery and multileaf or multirod collimation. The benefits that have been claimed for neutrons are indicated in Table 19.2. Much of the clinical work with heavy ions has taken place at the Lawrence Berkeley Laboratory in California. Some benefit in some tumours has been claimed although the high cost of this treatment means that it will be restricted to a few large centres each serving a large population (Wambersie *et al*, 1994). Proton beams are probably the most attractive radiation at the present time. They are not excessively expensive and have greatly improved dose distributions compared with photons.

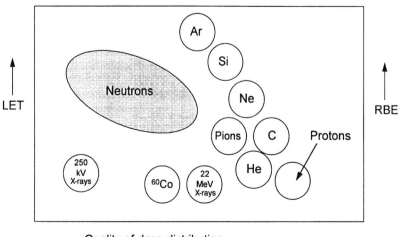

Figure 19.13 The radiations available for radiation therapy differ in the quality of beam that they produce, also in RBE. Based on Fowler (1981).

Key points

1. X- and γ-rays are sparsely ionizing radiations with a low linear energy transfer (LET). Some particle radiations (*e.g.* neutrons, α-particles or heavy ions) have a high LET.
2. High-LET radiations are *biologically* more effective per gray than low-LET radiations. This is measured by the relative biological effectiveness (RBE). For most high-LET radiations at therapeutic dose levels, RBE is in the range of 2 to 10.
3. RBE increases as the LET increases up to about 100 keV μm^{-1}, above which RBE decreases because of cellular overkill. The oxygen enhancement ratio (OER) also decreases rapidly over the same range of LET.
4. RBE increases as the dose is reduced *in vitro*, or the dose *per fraction* is reduced *in vivo*. In late-responding tissues, this increase occurs more rapidly than in early-responding tissues.
5. High-LET radiations may be clinically useful in selected cases.
6. Heavy charged particles such as He, C and Ne ions have a high LET and in addition they have improved physical depth–dose distributions.
7. Proton beams provide the best improvement in dose distribution for the lowest cost; their RBE is similar to low-energy photons.

BIBLIOGRAPHY

Barendsen GW (1968). Responses of cultured cells, tumours and normal tissues to radiations of different linear energy transfer. *Current Topics Rad Res Quart* 4:293–356.

Blakely EA (1982). Biology of Bevelac beams: cellular studies. In: *Pion and Heavy Ion Radiotherapy: Pre-clinical and Clinical Studies*, pp. 229–50. (Ed) Skarsgard LD. Elsevier, New York.

Britten RA, Warenius HM, Parkins C and Peacock JH (1992). The inherent cellular sensitivity to 62.5MeV neutrons of human cells differing in photon sensitivity. *Int J Radiat Biol* 61:805–12.

Castro JR (1995). Results of heavy ion radiotherapy. *Radiat Environ Biophys* 34:45–8.

Chapman JD (1980). Biophysical models of mammalian cell inactivation by radiation. In: *Radiation Biology in Cancer Research*, pp. 21–32. (Eds) Meyn RE and Withers HR. Raven Press; New York.

Goodhead DT (1989). The initial physical damage produced by ionizing radiation. *Int J Radiat Biol* 56:623–34.

Joiner MC (1988). A comparison of the effects of p(62)-Be and d(16)-Be neutrons in the mouse kidney. *Radiother Oncol* 13:211–224.

Joiner MC and Johns H (1987). Renal damage in the mouse: the effect of d(4)-Be neutrons. *Radiat Res* 109:456–68.

Raju MR (1980). *Heavy Particle Radiotherapy*. Academic Press; New York.

Wambersie A, Richard F and Breteau N (1994). Development of fast neutron therapy worldwide. *Acta Oncol* 34:261–74.

Withers HR, Thames HD and Peters LJ (1982). Biological bases for high RBE values for late effects of neutron irradiation. *Int J Radiat Oncol Biol Phys* 8:2071–6.

FURTHER READING

Alpen EL (1990). *Radiation Biophysics*. Prentice-Hall; London.

Bewley DK (1989). *The Physics and Radiobiology of Fast Neutron Beams*. Adam Hilger; Bristol.

Fowler JF (1981). *Nuclear Particles in Cancer Treatment*. Adam Hilger; Bristol.

Goodhead DT (1988). Spatial and temporal distribution of energy. *Health Physics* 55:231–40.

Hall EJ (1988). *Radiobiology for the Radiologist*, chapter 13. Lippincott; Philadelphia.

Nordic Conference on Neutrons in Research and Cancer Therapy (1994). *Acta Oncol* 33:225–327.

20

Combination of radiotherapy and chemotherapy: principles

G. Gordon Steel

20.1 The objectives of combined modality therapy

The biological problems presented by combined modality therapy are considerable. Radiotherapy by itself induces complex changes both in tumours and in the adjacent normal tissues. The response to chemotherapy is similarly complex: in addition to many of the factors that determine response to radiation treatment we also have problems of drug delivery, drug resistance and metabolism. The combination of radiotherapy and chemotherapy involves the combined complexity of *both* modalities, plus the interactions between them. This chapter seeks to provide a framework within which these processes can be examined. In general, what we seek is an improved therapeutic strategy, which may be defined as follows.

An improved therapeutic strategy – A combination of drugs and radiation that gives a greater tumour response than either of the component single agents, the combination and single-agent treatments being evaluated at comparable levels of overall toxicity.

This definition is illustrated by Figure 1.4. In experimental animals it is practicable to obtain dose–response curves for effects on tumours and on dose-limiting normal tissues. Using these dose–response curves it is possible to identify an improved therapy quite reliably. In clinical studies

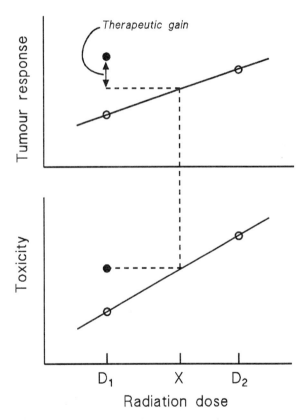

Figure 20.1 Scheme for the identification of an improved therapeutic strategy, in clinical research (see text). Open symbols show the tumour response and toxicity associated with radiation alone (doses D_1 and D_2). Solid symbols show the results with chemotherapy added to radiation dose D_1.

it is impracticable to examine dose–response curves over the range of responses used in experimental animals, although Figure 11.3 is an example. Restrictions on the size of clinical trials also limit the number of dose and treatment groups that can be compared simultaneously.

The strategy outlined in Figure 20.1 is the minimum scale of study necessary to show a therapeutic gain and to answer the question: 'Could an observed gain in tumour control by combined modality therapy also have been obtained merely by increasing the radiation dose?' A three-arm trial is envisaged, with two radiation doses (D_1 and D_2), the third arm consisting of the lower radiation dose plus chemotherapy. By 'dose' we here mean some measure of intensity of fractionated radiation treatment, for instance the total radiation dose. We also assume that 'toxicity' can be ascribed a numerical value, plotted on the vertical scale; this is a gross oversimplification, for as will be indicated below the addition of chemotherapy may not only enhance radiation-induced damage but will almost certainly introduce new toxicities as well. The higher radiation dose is expected to produce a higher level of toxicity than the lower dose. Adding chemotherapy is also expected to lead to greater toxicity. By drawing a horizontal (dashed) line we see that the toxicity of the combined arm is equivalent to a radiation dose X (lower panel). The tumour response that could have been *expected* from this dose (alone) can be found by interpolation as shown by the broken lines. If the *actual* tumour response is greater than this, then a therapeutic gain has been identified. Whether this gain is significantly different from zero will depend on the precision of all six experimental points in this analysis (three tumour, three toxicity). Clearly, to achieve sufficient precision will require a very large clinical trial. But if this cannot be done, or if this minimum three-arm study cannot be performed, then the *existence* of a therapeutic gain will be in doubt.

THE CONCEPT OF INTERACTIVE AND NON-INTERACTIVE COMBINATIONS

This distinction is central to the discussion of biological mechanisms in combined modality treatment. *Non-interactive* describes a situation where each modality appears to exert its own individual effect; *interactive* refers to a situation where there is evidence that one modality modifies the effect of the other.

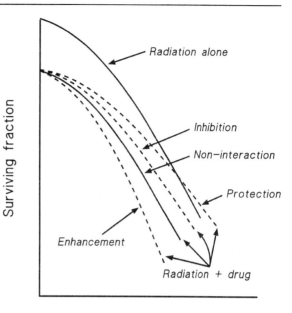

Figure 20.2 A terminology for the description of interactive processes between a cytotoxic drug and radiation. From Steel (1979), with permission.

A suggested terminology of interactive processes is described in Figure 20.2. If the result of adding a cytotoxic drug is to move the dose–response curve for radiation alone without changing its shape, then the combination is *non-interactive*. If the radiation dose–response curve is steepened, then this may be called *enhancement* of response; if it is made shallower, this is *inhibition*, and in the extreme case where the combination gives less effect than radiation alone, this is *protection*. For the reasons outlined below we avoid the use of the ambiguous terms *synergism* or *additive*.

20.2 Exploitable mechanisms in the combination of radiotherapy and chemotherapy

The processes that may lead to a therapeutic gain can be described under four headings (Steel and Peckham, 1979).

Spatial Co-operation

This term describes the use of radiotherapy and chemotherapy to hit disease in different anatomical sites. The commonest situation is where radiation is used to treat the primary tumour and chemotherapy is added as an adjuvant to deal with systemic spread (Figure 20.3). There is an analogous situation in the leukaemias where chemotherapy is the main-line treatment and radiotherapy is brought in to deal with disease in a 'seclusion site' such as the brain. This is an important aspect of combined modality treatment and there is some evidence for its practical efficacy (Rosen et al, 1974; Bleyer and Poplack, 1985). But it should be stressed that this combined use of drugs and radiation envisages no *interaction* between them: the two modalities are used separately to treat disease in spatially different anatomical locations. The optimum clinical approach will be to give the best radiotherapy and the best chemotherapy, seeking to *avoid* interactions between them.

The successful exploitation of spatial co-operation depends critically on the effectiveness of the available chemotherapy. As an illustration of this, consider a situation in which a small amount of disease has spread outside the radiation field. Radiotherapy alone will fail. If only 0.1 g of tumour is missed by radiation this could contain 10^8 cells of which upwards of 10^6 might be clonogenic. Recent evidence (Tannock, 1989) suggests that in the common solid tumours chemotherapy seldom achieves a surviving fraction better than 10^{-6} and in those cases this adjuvant use of chemotherapy will fail even to control such a very small amount of disseminated disease. For spatial co-operation to succeed more widely we need better drugs.

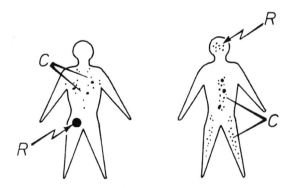

Figure 20.3 The concept of spatial co-operation. R = radiotherapy; C = chemotherapy.

Independent Cell Kill

This term describes the simple concept that if two therapeutic modalities can both be given at full dose, then *even in the absence of interactive processes* the tumour response should be greater than that achieved with either alone.

For the hypothetical example of treatment of a lung tumour, Table 20.1 illustrates the idea behind this approach. Radiation produces a (+++) tumour response and is limited, for example, by radiation pneumonitis. Chemotherapy also produces a tumour response (++) but is limited by intestinal or bone marrow damage. Even if the antitumour effects of these two treatments are *subadditive* (++++ instead of the expected +++++) the overall antitumour effect may well be greater than that of either drug or radiation alone. Clearly, *any* extra tumour response produced by the chemotherapy will improve on radiation alone. The cost of this advantage is that the patient has to tolerate a wider range of toxic reactions and a critical factor in the reasoning is the assumption that radiation lung damage is not enhanced by the addition of chemotherapy (still +++ for lung toxicity). If this is not the case and lung damage is increased, then to maintain an isotoxic treatment the radiation dose would have to be reduced and this might well lead to the loss of the extra small gain in tumour response. In order to exploit *independent cell kill* successfully it is thus vital to select effective antitumour drugs that do not exacerbate radiation damage to critical normal tissues within the radiation field.

To what extent is it possible to find effective chemotherapy that does not make radiation-induced normal-tissue damage worse? There is a large experimental literature on this, including some important reviews (Phillips and Fu, 1976; von der Maase, 1986; Steel, 1988). The extent to which chemotherapy enhances radiation damage varies

Table 20.1 Concept of independent toxicity leading to simple addition of antitumour effects as a mechanism in combined modality therapy

	Associated toxicity			Response of bronchial tumour
	Intestinal	Bone marrow	Lung	
Radiation	−	−	+++	+++
Drug	+++	+	−	++
Combination	+++	+	+++	++++

widely among drugs and among the normal tissues studied. In some cases there is little or no enhancement; in others there is severe enhancement and radiation dose would have to be reduced by up to perhaps 50% to keep the toxicity the same. The guidelines given by Phillips and Fu are probably still useful:

(i) Beware of the cytotoxic antibiotics (actinomycin, bleomycin, adriamycin, etc.) which enhance radiation damage in a variety of tissues.
(ii) Beware of drugs that have a recognized toxicity to the tissues that are being irradiated (examples are cyclophosphamide in the lung, and various proliferation-dependent agents in the intestine).
(iii) Avoid *concurrent* treatment with drugs and radiation (see below).

Most of the experimental evidence for drug-induced enhancement of radiation toxicity has concentrated on early-responding normal tissues and there are few studies on the enhancement of late reactions. If interactions are mediated by effects on cell proliferation, then they may be less prominent in late-responding tissues. But it would be a mistake to depend upon this, as evidenced by the enhancement of spinal cord damage in the rat when intravenous methotrexate was used with irradiation (van der Kogel and Sissingh, 1985).

Independent cell kill is an obvious and promising way in which chemotherapy may improve on the results of radiotherapy alone. Note that, as with *spatial co-operation*, this mechanism does not require interactive processes between drugs and radiation: in fact in critical normal tissues these are specifically to be avoided in order to retain tolerance to radiotherapy.

Protection of Normal Tissues

We now move to strategies in combined modality therapy that *do require* interaction, as defined above. First, can we identify chemotherapeutic drugs that, when used in conjunction with radiation, reduce the damage to dose-limiting normal tissues? This would seem unlikely, for two toxic agents usually tend to produce more damage than either alone. However, there are well-documented situations in experimental animals in which certain cytotoxic drugs increase the resistance of normal tissues to radiation or to a second cytotoxic treatment. W.W. Smith observed this with colchicine and the *vinca* alkaloids over 30 years ago, and Millar *et al* (1978) studied this phenomenon in detail. Cyclophosphamide, cytosine arabinoside, chlorambucil and methotrexate were found to be effective radioprotective agents. An important characteristic of this phenomenon is its dependence upon timing. Maximal radioprotection of animal survival was achieved when cytosine arabinoside was given 2 days before irradiation, but the optimum gap for cyclophosphamide was 3 days. A similar protection phenomenon has been found in which a priming treatment with one cytotoxic drug can protect against a large dose of another (Millar and McElwain, 1978).

Studies of the mechanism of this remarkable phenomenon have concentrated upon the bone marrow and intestinal epithelium. In the marrow it has been shown that the most effective of the 'protective' agents, cytosine arabinoside, did not modify stem cell radiosensitivity; it stimulated enhanced repopulation by surviving stem cells. In the small intestine (Phelps and Blackett, 1979) microcolony survival experiments have shown that cytosine arabinoside given 12 hours before irradiation greatly increased the survival of intestinal stem cells, perhaps by enhancing the repair of radiation damage.

Although attempts have been made to exploit this phenomenon in high-dose combination chemotherapy (Hedley *et al*, 1978), its critical dependence on timing has precluded its exploitation in fractionated radiotherapy.

Enhancement of Tumour Response: The Concept of Supra-additivity

This is often perceived to be the principal aim of adding chemotherapy to radiotherapy. Is it possible to obtain a greater tumour response than would be expected on the basis of simple addition of antitumour effects? This is, both conceptually and experimentally, a very difficult problem. If dose-response curves are linear, the additivity of two agents has a simple and unique meaning. When, as is usually the case in cancer therapy, dose-response curves are far from linear, the nature of an additive response is controversial.

It has been argued (Steel and Peckham, 1979) that when dose–response curves are non-linear there is no unique description of an additive response. Take, for example, the dose–response curve shown in Figure 20.4. The dose D gives an

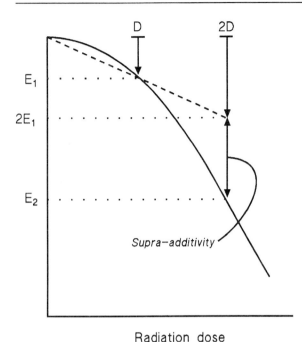

Figure 20.4 A non-linear dose–response curve can give a spurious impression of supra-additivity. In this example (two simultaneous doses of radiation given to mammalian cells in tissue culture) the effect plotted on the ordinate is log(surviving fraction).

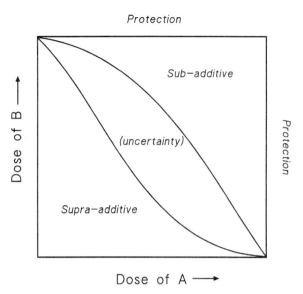

Figure 20.5 An isobologram is an isoeffect plot of the doses of two agents that together give a fixed biological effect. If dose–response curves are non-linear, there is a region of uncertainty about the existence of 'additivity'. From Steel and Peckham (1979), with permission.

effect E_1. On the assumption of *additivity* we would expect that if we give two such doses at the same time (dose $2D$) we would get twice the effect: $2E_1$. In fact, it can be seen from the dose–response curve that a dose $2D$ actually gives an effect E_2: there is an apparent *supra-additive* effect equal to $(E_2 - 2E_1)$. Such a result is therapeutically uninteresting: it derives purely from the shape of the survival curve. If one of the two doses were in fact a cytotoxic drug, what we would wish to know is whether the response is *greater than* E_2, which might be an interesting case of synergism. This example illustrates the fact that when dose–response curves are non-linear great care is needed in the description of additivity; there is an area of uncertainty whose magnitude depends upon the non-linearity of the responses. The extent of the uncertainty is best judged by the use of an isobologram (Figure 20.5). This is an isoeffect plot which indicates the separate doses of two agents that in combination (using any chosen timing) give the isoeffect. The lines in Figure 20.5 enclose the range of situations which *under some assumed analysis of the dose–response curves* might be regarded as additive (Steel and Peckham, 1979). Only a combination that gives an experimental point to the left of the left-hand boundary can confidently be described as *supra-additive*. The 'envelope of additivity' in this diagram should not be regarded as a new and more reliable definition of additivity: it is an expression of the *uncertainty* in this concept.

20.3 Possible mechanisms leading to interactions between the effects of drugs and radiation

A wide variety of biological mechanisms have been proposed to explain interactive processes between radiation and cytotoxic drugs. Some of these are as follows.

Inhibition of Repair of Radiation Damage

Many drugs have the property of inhibiting the repair of radiation damage (Kelland and Steel, 1988a). Some of these are antimetabolites that are of no interest as cytotoxic agents in their own right (3-aminobenzamide, cordycepin, caffeine, etc). Others have been used as anticancer agents (actinomycin-D, adriamycin, hydroxyurea, Ara-C, cis-platinum, etc). All of these agents are cytotoxic at sufficiently high drug concentrations. For experimental studies on repair inhibition it is usual to choose a non-toxic drug dose in order to simplify the analysis. In clinical treatment this is clearly unnecessary: provided the drug has some beneficial anticancer effect in its own right it will probably be best to employ it at its conventional dose level. Repair inhibition has been detected in a number of ways: removal of the shoulder on the cell survival curve, inhibition of split-dose recovery, inhibition of delayed-plating recovery, etc. Sensitization has been detected at low radiation doses and at low dose rate (Kelland and Steel, 1988b), which is of considerable therapeutic interest. However, there is little evidence so far that *selectivity* for effects on tumours rather than on normal tissues can be achieved. It must be borne in mind that as a result of local ischaemia, the tissue levels of a systemically administered drug are often lower in tumours, or in parts of tumours, than in normal tissues, which *depend for their tolerance of radiation on their ability to repair*. Selectivity is therefore essential.

Cell Synchronization

Many cytotoxic drugs show some degree of selectivity in killing cells at certain phases of the cell cycle (Mauro and Madoc-Jones, 1970). Some agents show maximum effect on cells that are undergoing DNA synthesis. Radiation also has a cell cycle dependence, often with peaks of *resistance* in the S-phase and in G1 (Section 6.8). There is therefore an attractive possibility of complementary action between drugs and radiation. As has been found with kinetically optimized drug-drug combinations, it is unfortunately the case that this approach to synergism only works well with rapidly cycling cells. Positive effects have been found in cell cultures and fast-growing experimental tumours but the existence of slowly growing or resting cells in human tumours could possibly explain why synchronization therapy has been disappointing (Tubiana *et al*, 1975; Tannock, 1989).

Recruitment

It follows from the above that response to therapy might be improved if non-proliferating cells could be stimulated to come into cycle. This has mainly been explored in combination chemotherapy. There is evidence (reviewed by Steel, 1977) that the growth fraction of some experimental tumours is increased by a suitable priming treatment. The resulting therapeutic benefit has not been large. There is also an academic point that tumours consist of dynamic cell populations and the non-dividing cells probably have a limited lifespan; as a result, if the growth fraction rises after treatment this may not indicate that some non-proliferating cells have begun to cycle but rather that the rate of production of non-proliferating cells has been reduced (*i.e.* a lower rate of decycling). If so, this response is still therapeutically useful, but *recruitment* is a misleading term for it.

Enhanced Repopulation

Any effective cytotoxic treatment reduces the total number of viable tumour cells and, perhaps by improving the nutritional status of the remainder, may lead to accelerated repopulation (Section 14.3). In the combined modality situation it is logical to use radiation to debulk the tumour and then seek to use proliferation-specific drugs to exploit the kinetic response. There is some evidence that chemotherapy given a few days after radiotherapy does lead to greater effects on experimental tumours (Steel, 1988). Unfortunately, this strategy is also the most damaging to normal tissues that repopulate rapidly after irradiation. It has also been suggested that enhanced repopulation may lead to a therapeutic *detriment* in combined modality therapy (Withers *et al*, 1988). If chemotherapy is given first it may switch on repopulation during a subsequent course of radiotherapy and thereby reduce its effectiveness.

Reduction of the Hypoxic Fraction

The debulking of a tumour by chemotherapy might lead to a reduction in the hypoxic fraction and improved response to radiotherapy. Although this has been postulated, there is little evidence for benefit being achieved in this way. It is also possible that the first few fractions of a course of radiation treatment will have a similar effect, which is the well-known phenomenon of reoxygenation (Section 15.5).

DEBULKING

This is probably the most promising basis for expecting a benefit from combined chemotherapy/radiotherapy. A tumour that has shrunk in response to one treatment may more easily be cured by a second, purely because there are fewer cells to be killed. In simple terms, this is what has been described above as *independent cell kill* and should not be confused with synergism. Debulking may also lead to improved oxygen supply or increased proliferation which, as indicated above, may lead to greater cell kill from subsequent radiotherapy. More promising, however, is the possibility that debulking by chemotherapy may allow the field size of subsequent radiotherapy to be reduced and the dose thus to be increased.

20.4 Time-dependence of interactive effects between drugs and radiation

The mechanisms described in the previous section are widely varied, difficult to relate to actual therapeutic response and in some cases hypothetical. The *time-line method* is an experimental approach that cuts through these difficulties. It is basically an empirical search for the optimum treatment with two agents. It can be applied to tumours or normal tissues in experimental animals, also to cells in tissue culture. The method was first described by Vietti *et al* (1971). Using mouse leukaemic cells growing *in vivo*, they selected fixed doses of 5-fluorouracil and radiation, then gave them both to groups of mice, changing only the time interval between them (Figure 20.6). In each group of mice they measured the survival of leukaemic cells, termed CFU in the figure. The data described an intricate response curve, with minimum CFU survival when 5-fluorouracil was given up to 8 hours *after* radiation. When the interval between the treatments (in either order) was over 40 hours, the level of cell kill was close to what would be expected by adding the log cell kill from the separate treatments. The form of this time-line has not been fully explained but from a therapeutic point of view its implications are obvious: it allows the optimum timing of the treatments to be identified.

Three further examples of time-line experiments are shown in Figure 20.7. Panel A shows a time-line for the interaction of cyclophosphamide and pelvic irradiation in mice. Radiation was given at time zero; each experimental point derives from a group of mice that were given cyclophosphamide (200 mg/kg) at some time up to 4 weeks before or 4 weeks after irradiation. Various radiation doses were used in order to derive the dose-enhancement factor (DEF: the ratio of radiation dose without drug to the radiation dose with drug, to give a fixed level of radiation damage). A DEF of 1.0 indicates no enhancement of radiation damage as a result of giving the drug. When DEF = 1.2, the radiation dose in the presence of the drug would have to be reduced by 17% (1/1.2 = 0.83) to produce the same amount of damage as with radiation alone. In this example there is massive enhancement of intestinal damage as a result of drug treatment. The peak DEF is 1.5 for cyclophosphamide given 3 days after irradiation. When the time interval between drug and radiation (or *vice versa*) was increased to 4 weeks, there was little enhancement of damage: the effect of the first treatment seems to have been forgotten. The minimum in DEF seen at -2 days is interesting. This may well reflect the protection phenomenon referred to in Section 20.2; in this case there is no protection (DEF > 1.0) but in the absence of this effect the DEF would have been around 1.3.

Figure 20.7B, from the same study, shows the results with three further chemotherapeutic agents. They gave similar responses: no enhancement of damage for drug given more than 5 days before radiation, a peak at +3 days, and a tendency to return towards no effect for drug given 14 days after irradiation.

Figure 20.7C is a rare example of a clinical time-line. This reports experience at the Royal Marsden Hospital (1976–1981) in treating testicular teratomas with combination chemotherapy (VB, PVB, or BEP) together with radiation (Yarnold *et al*, 1983). Normal-tissue damage was scored as the percentage of patients with subcutaneous fibrosis

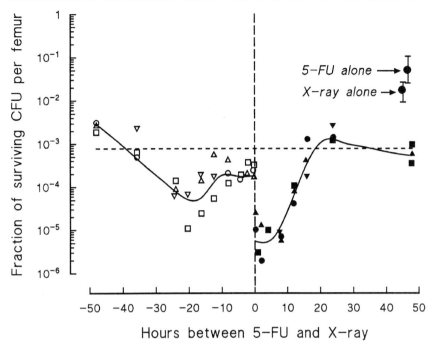

Figure 20.6 A time-line for the interaction of 5-fluorouracil (5-FU) and X-radiation in the treatment of mouse leukaemia. Radiation was given at time zero; points to the left of zero are for drug before radiation; points to the right are for drug after radiation. From Vietti *et al* (1971), with permission.

or gastrointestinal damage. The group that were treated with chemotherapy approximately 3 months after irradiation showed a high probability of damage and although the time-scale differs between man and mouse, the form of the time-line is roughly similar to those shown in the other two panels.

The time-line approach has been used extensively in studies on normal-tissue damage in mice (von der Maase, 1986; reviewed by Steel, 1988).

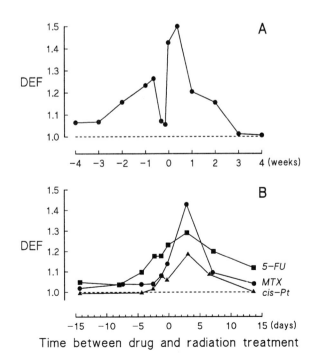

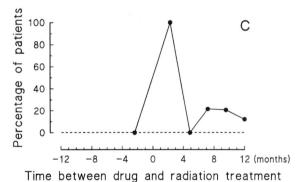

Figure 20.7 A: time-line for the interaction of cyclophosphamide (200 mg/kg) and pelvic irradiation in mice. B: time-lines for the interaction of 5-fluorouracil, methotrexate or *cis*-platinum and pelvic irradiation in mice. C: time-line for normal-tissue damage in patients treated for testicular teratoma with radiation and combination chemotherapy. A and B from Pearson and Steel (1984); C from Yarnold *et al* (1983), with permission.

The broad picture is that there is great variation in the enhancement of normal-tissue damage, but most chemotherapeutic drugs enhance damage in some normal tissue for some particular timing. Short time intervals between drugs and radiation most frequently lead to maximum damage. Increasing the time interval tends to reduce the interaction. There are some drugs, of which *cis*-platinum is an example (Figure 20.7B), for which no enhancement of damage occurs provided the time intervals are chosen carefully.

Most of the experimental experience with the time-line approach applied to normal-tissue damage has been with early-reacting tissues. Although one might anticipate that time-dependent phenomena would be less marked in late-reacting normal tissues, this is not the case in the mouse lung (Collis and Steel, 1983) nor in the spinal cord (van der Kogel and Sissingh, 1985), where proliferation-dependent cytotoxic drugs significantly enhanced radiation damage. Table 20.2 summarizes the results of two research groups in studying the enhancement by chemotherapy of lung damage in mice. It illustrates the large differences in enhancement due to different drugs, with cyclophosphamide and the cytotoxic antibiotics at the top of the list. The interactions generally fall off with increasing time between treatments, in either sequence.

The time-line method has also been applied in the search for the optimum timing of combined modality treatment in transplanted mouse tumours. It has not been possible to examine long time intervals between drug and radiation because of the rapid growth of these tumours. Overall, the results do not encourage the attempt to exploit time-dependent phenomena in the combined modality treatment of cancer: a peak of tumour response has seldom been seen and when a peak is seen in one tumour system it has not been confirmed in others (Steel, 1988).

The overall conclusion of these studies is therefore disappointing: time-dependent interactions are more commonly seen in the normal tissues of experimental animals than in experimental tumours. The take-home message for the 'mouse radiotherapist' is thus to avoid short time intervals between drugs and radiation; by doing so it will be possible to keep the drug and radiation doses high and thus to gain maximum benefit from *spatial co-operation* and *independent cell kill*.

Table 20.2 Lung damage in mice as a result of combined treatment with radiation and drugs

Drug	Drug administration*			Source**
	Before	Concurrent	After	
Cyclophosphamide	++	+++	++	C
	−	+++	−	M
Bleomycin	++	++	++	C
	−	+++	−	M
	++			S
Adriamycin	−	+++	++	C
	−	+++	−	M
	+			S
Actinomycin D	−	+		C
	+			S
Methotrexate	+	−	−	C
	−	−	−	M
	−			S
5-Fluorouracil	−	−	−	M
Vincristine	+	+	+	C
CCNU	−	−	+	C
cis-Platinum	−	−	−	C
	−	−	−	M
Mitomycin C	−	+	−	M

+++ severe enhancement of lung damage; ++ moderate enhancement; + slight enhancement (doubtful significance); − no enhancement.
* Before = 7 – 28 days before radiation; after = 7 – 28 days after radiation.
** C = Collis (1981), Collis and Steel (1983); M = von der Maase (1986); S = Steel *et al* (1979).

Table 20.3 Summary of evidence for exploitation of four mechanisms in the combination of radiotherapy and chemotherapy

	Evidence in mice	Evidence in man
Spatial co-operation	+++	++
Independent cell kill	+++	+
Protection of normal tissues	++	−
Enhancement of tumour response	+	−

20.5 The good news and the bad news

The four mechanisms in combined modality therapy that have been described here differ in their exploitability, both in laboratory and clinical studies. Table 20.3 is an attempt to summarize the general picture. *Spatial co-operation* and *independent cell kill* can easily be demonstrated in experimental animals. We have shown, for instance, that if Lewis lung tumours are implanted into the legs of mice and allowed to grow to a measurable size, the local tumour can be cured with high probability by irradiation (Steel *et al*, 1978); most of the mice, however, die of disseminated disease. Cyclophosphamide is very effective against this tumour, and when given in conjunction with local irradiation of the implant leads to cures in a high proportion of the mice. Increased cure and increased growth delay in primary implants as a result of *independent cell kill* by cytotoxic drugs and radiation also have been observed widely. Evidence has been cited above for the observation in mice of the phenomenon of protection of normal-tissue damage. In spite of the described difficulties in identifying supra-additive tumour responses, some examples of this do exist.

In clinical studies, the gains from adding chemotherapy to radiotherapy have been small and the scope for identifying mechanisms of improvement is therefore limited (Tannock, 1989; Tubiana, 1989). Those gains that have been made can be attributed either to *spatial co-operation* or to *independent cell kill*. There appear to be no clinical results that require supra-additive effects on the tumours to be invoked for their explanation. We can therefore summarize the clinical picture by saying that where gains have so far been observed by adding drug therapy to radiotherapy they have come from direct cell killing (rather than interactive processes), either within the irradiated field or outside it.

This review of concepts that have come out of the extensive laboratory studies that have been made on combined modality therapy thus leads to 'good news' and 'bad news'. The good news is that clear therapeutic benefit can be obtained by adding chemotherapy to radiotherapy. The benefits are probably obtained by simple addition of the antitumour effects of drugs and radiation, either in the tissues that are irradiated or in *spatial co-operation* between effects on irradiated and non-irradiated disease. The benefits are probably proportional to the effectiveness of the chemotherapeutic drugs themselves. Only when these can achieve multi-decade tumour cell kill is there likely to be a detectable clinical benefit. The bad news is that the large amount of time and effort that has been put into identifying sophisticated mechanisms of synergism between drugs and radiation has so far failed to give useful clinical benefits.

Key points

1. The processes by which the addition of chemotherapy to radiotherapy may lead to an improvement in therapeutic result may be classified as *interactive* or *non-interactive*.
2. Non-interactive processes include *spatial co-operation* and *independent cell kill*. These can be clearly demonstrated in experimental animals and are probably responsible for the modest gains that have so far been claimed for clinical combined modality therapy.
3. *Synergism* or *supra-additivity* is an intricate concept in the combined modality field; non-linear dose–response curves may give rise to spurious evidence for synergism.
4. Interactions between cytotoxic drugs and radiation are often time-dependent. This is especially the case in normal tissues, less so in tumours. The evidence from studies on experimental animals is therefore that concurrent treatment is bad, in terms of therapeutic index. Longer time intervals between drugs and radiation (or *vice versa*) are safer, and still allow potential benefit from the non-interactive processes.

Bibliography

Bleyer WA and Poplack DG (1985). Prophylaxis and treatment of leukemia in the central nervous system and other sanctuaries. *Sem Oncol* 12:131–48.

Collis CH (1981). The response of the lung to ionizing radiation and cytotoxic drugs. Thesis, University of London.

Collis CH and Steel GG (1983). Lung damage in mice from cyclophosphamide and thoracic irradiation: the effect of timing. *Int J Radiat Oncol Biol Phys* 9:685–9.

Hedley DW, Millar JL, McElwain TJ and Gordon MY (1978). Acceleration of bone-marrow recovery by pre-treatment with cyclophosphamide in patients receiving high-dose melphalan. *Lancet* 4 November:966–7.

Kelland LR and Steel GG (1988a). Inhibition of recovery from damage induced by ionizing radiation in mammalian cells. *Radiother Oncol* 13:285–99.

Kelland LR and Steel GG (1988b). Modification of radiation dose-rate sparing effects in a human carcinoma of the cervix cell line by inhibitors of DNA repair. *Int J Radiat Biol* 54:229–44.

Mauro F and Madoc-Jones H (1970). Age response of cultured mammalian cells to cytotoxic drugs. *Cancer Res* 30:1397–408.

Millar JL, Blackett NM and Hudspith BN (1978). Enhanced post-irradiation recovery of the haemopoietic system in animals pretreated with a variety of cytotoxic agents. *Cell Tissue Kinet* 11:543–53.

Millar JL and McElwain TJ (1978). Combinations of cytotoxic agents that have less than expected toxicity on normal tissues in mice. In: *Antibiotics and Chemotherapy*, pp. 271–82. (Ed) Schonfeld H. Karger; Basel.

Pearson AE and Steel GG (1984). Chemotherapy in combination with pelvic irradiation: a time dependence study in mice. *Radiother Oncol* 2:49–55.

Phelps TA and Blackett NM (1979). Protection of intestinal damage by pretreatment with cytarabine (cytosine arabinoside). *Int J Radiat Oncol Biol Phys* 5:1617–20.

Rosen G, Wollner N, Wu SJ et al (1974). Disease-free survival in children with Ewing's sarcoma treated with radiation therapy and adjuvant four-drug sequential chemotherapy. *Cancer* 33:384–93.

Steel GG (1977). *The Growth Kinetics of Tumours*. Oxford University Press; Oxford.

Steel GG (1979). Terminology in the description of drug-radiation interactions. *Int J Radiat Oncol Biol Phys* 5:1145–50.

Steel GG, Adams K and Peckham MJ (1979). Lung damage in C57Bl mice following thoracic irradiation: enhancement by chemotherapy. *Br J Radiol* 52:741–7.

Steel GG, Hill RP and Peckham MJ (1978). Combined radiotherapy-chemotherapy of Lewis lung carcinoma. *Int J Radiat Oncol Biol Phys* 4:49–52.

Tubiana M, Frindel E and Vassort F (1975). Critical survey of experimental data on *in vivo* synchronisation by hydroxyurea. In: *Recent Results in Cancer Research*, pp. 187–205. (Eds) Grundman E and Groos R. Springer-Verlag; Berlin.

van der Kogel AJ and Sissingh HA (1985). Effects of intrathecal methotrexate and cytosine arabinoside on the radiation tolerance of the rat spinal cord. *Radiother Oncol* 4:239–51.

Vietti T, Eggerding F and Valeriote F (1971). Combined effect of X-radiation and 5-fluorouracil on survival of transplanted leukemic cells. *J Nat Cancer Inst* 47:865–70.

von der Maase H (1986). Experimental studies on interactions of radiation and cancer chemotherapeutic drugs in normal tissues and a solid tumour. *Radiother Oncol* 7:47–68.

Withers HR, Taylor JMG and Maciejewski B (1988). The hazard of accelerated tumor clonogen repopulation during radiotherapy. *Acta Oncol* 27:131–46.

Yarnold JR, Horwich A, Duchesne G et al (1983). Chemotherapy and radiotherapy for advanced testicular non-seminoma. *Radiother Oncol* 1:91–9.

Further Reading

Phillips T and Fu KK (1976). Quantification of combined radiation therapy and chemotherapy effects on critical normal tissues. *Cancer* 37:1186–200.

Steel GG (1988). The search for therapeutic gain in the combination of radiotherapy and chemotherapy. *Radiother Oncol* 11:31–53.

Steel GG and Peckham MJ (1979). Exploitable mechanisms in combined radiotherapy-chemotherapy: the concept of additivity. *Int J Radiat Oncol Biol Phys* 5:85–91.

Tannock IF (1989). Combined modality treatment with radiotherapy and chemotherapy. *Radiother Oncol* 16:83–101.

Tubiana M (1989). The combination of radiotherapy and chemotherapy: a review. *Int J Radiat Biol* 55:497–511.

21

Combined radiotherapy and chemotherapy: clinical application and evaluation

Fiona A. Stewart and Michele I. Saunders

21.1 Clinical objectives

Virtually all classes of chemotherapeutic agents currently in use have been used in combined modality clinical trials involving one or more drugs plus a full course of radiotherapy. The aim of such trials is an improvement in local control and/or eradication of distant metastases. Substantial improvements in survival have been obtained using the combined modality approach in a few cancer types, notably childhood soft-tissue and bone sarcomas and Wilms' tumour (Figure 21.1). Hodgkin's disease and non-Hodgkin's lymphomas are other examples of tumours sensitive to both chemotherapy and radiotherapy, where combined modality treatment is standard and has proven to be effective. The majority of adult solid tumours are, however, not very chemosensitive and only modest increases in patient survival have resulted from a combined modality approach, although it is sometimes possible to avoid mutilating surgical procedures, for example by the combined use of 5-fluorouracil (5-FU) and mitomycin-C (MMC) with radiotherapy as an alternative to surgery for the treatment of anal cancers.

There are several possible ways of combining chemotherapy with radiotherapy and the choice of treatment protocol should be determined according to the aims and expectations of the trial.

Concomitant Chemotherapy. This is treatment with chemotherapy during a fractionated course of radiotherapy. The reasoning for this is based on the expectation of an interaction between the two modalities. This could be a direct influence of one modality on the intrinsic cellular sensitivity of the other modality (for instance through inhibition of repair of sublethal damage) or a more indirect interaction, such as physiological alterations involving oxygenation or pH status by one modality leading to increased sensitivity to the second modality. Advantages of concomitant chemotherapy are that it avoids delaying either the chemotherapy or the

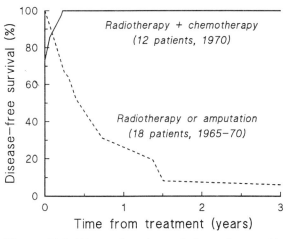

Figure 21.1 Disease-free interval in patients with Ewing's sarcoma treated with radiotherapy alone or in combination with adjuvant chemotherapy (dactinomycin, Adriamycin, vincristine and cyclophosphamide). Adapted from Rosen *et al* (1974) with permission.

radiotherapy and it affords the least chance for tumour repopulation during therapy. Improvements in both local tumour control and survival have been obtained using such schedules for head and neck and non-small-cell lung cancers. The use of concomitant chemotherapy, however, carries the greatest risk of increased normal-tissue complications. Examples are the high incidence of severe pulmonary complications and oesophageal stricture or mucositis after concurrent chemotherapy and radiotherapy for treatment of small-cell lung cancers or head and neck cancers (Fu et al, 1987). Concomitant chemotherapy schedules should, wherever possible, avoid the use of drugs that are known to produce specific toxicities in the organ which is to be irradiated, e.g. bleomycin with concomitant lung irradiation, cis-platinum with concomitant renal irradiation (Section 20.2).

Neoadjuvant Chemotherapy. Otherwise known as induction chemotherapy, this is the use of chemotherapy prior to radiotherapy. The rationale is that chemotherapy is used to debulk the primary tumour, thus permitting the use of a smaller radiotherapy treatment volume and higher radiation dose, and also to eradicate distant metastases. The disadvantage of such schedules is that they involve a delay in starting radiotherapy. Radiotherapy is usually the more effective of the two modalities and where this is the case it is unwise to delay it. There is also the possibility that chemotherapy will, by debulking the tumour, switch on tumour cell repopulation and this will inevitably be detrimental to tumour control by radiotherapy (Section 14.3).

Adjuvant Chemotherapy. This is the term usually given to the use of radiotherapy followed by chemotherapy. The aim is independent cell killing from both modalities, to achieve a greater total cell kill in the primary tumour, plus eradication of distant metastases. The disadvantage of such schedules is the possible increased metastatic burden (due to delaying chemotherapy) and possible emergence of drug-resistant clones if cross-resistance between irradiation and drugs develops, as has been demonstrated for vincristine and melphalan. Another problem with adjuvant chemotherapy, particularly for head and neck cancers, is poor compliance when drugs are scheduled in the post-irradiation period.

Alternating Chemotherapy and Radiotherapy. In such schedules the radiotherapy is given during breaks in the chemotherapy treatments. The rationale is that this allows early administration of both therapies and yet gives some temporal separation, thereby improving normal-tissue tolerance. Patient survival after alternating schedules for the treatment of small-cell lung cancer in Villejuif (Tubiana, 1989) are among the best ever reported (3- and 4-year survivals of 26% and 22%) but with less severe normal-tissue toxicity than in trials employing concomitant chemotherapy. Similar trials subsequently initiated in Great Britain were, however, abandoned due to unacceptable levels of leucopenia.

Alternating schedules have also been tested in randomized trials for advanced squamous cell carcinoma of the head and neck and significant increases in median survival were demonstrated for the combined modality group (Figure 21.2). The total radiation dose in the combined modality arm was only 60 Gy, compared with 70 Gy for radiation alone (Merlano et al, 1992). Similar protocols are now being evaluated by the EORTC.

21.2 Normal-tissue toxicity and the concept of therapeutic gain

Combined modality treatments always involve the risk of some extra normal-tissue toxicity compared

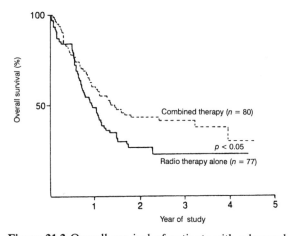

Figure 21.2 Overall survival of patients with advanced squamous cell carcinoma of the head and neck was significantly better in patients treated with combined chemotherapy (cis-platinum and 5-FU) and radiotherapy (60 Gy) in an alternating schedule, than for radiotherapy alone (70 Gy). From Merlano et al (1992), with permission.

with single-modality treatments. Greater acute toxicity is particularly common after concomitant schedules, for instance increased severity of oral mucositis, gastrointestinal toxicity and leucopenia. This extra toxicity should be taken into account when evaluating the success of combined modality therapy, particularly when the expected increased survival is a few months at best. Gelber and Goldhirsch (1986) proposed that estimation of TWIST (time without symptoms or toxic side-effects) would be a good way to compare the benefits of different treatment schedules used to treat breast cancer. This concept has recently been expanded and formalized to incorporate quality of life into survival analyses of post-menopausal breast cancer patients.

Combined modality trials which indicate a significant improvement in local control but also a significant increase in toxicity of normal tissues within the irradiated field are difficult to evaluate. This problem is addressed in Section 20.1 (*see* Figure 20.1). An example of this common situation is the randomized trial in head and neck cancer reported by Lo *et al* (1976). There was a significant improvement both in local control and survival for patients with oral cavity tumours treated with concurrent 5-FU and radiotherapy versus radiotherapy alone, although there was no trend for suppression of distant metastases in the combined modality group. Normal-tissue toxicity was, however, also increased after the combined modality treatment and major complications (including soft tissue or bone necrosis) were only seen in this group (Table 21.1). It is of course possible that a higher total irradiation dose without chemotherapy could have achieved the same therapeutic benefit. A reliable assessment of therapeutic gain requires that the combined modality treatment should either show an improved local control (and/or survival) for *the same normal-tissue toxicity* or reduced toxicity for the same level of tumour control. The total radiation doses in the radiation alone or combined therapy arms do not necessarily have to be equal: it is equivalent normal-tissue toxicity which is the important factor in assessing therapeutic gain.

21.3 Problems in assessing the efficacy of combined modality trials

The methodological issues which often make it difficult to interpret the value of combined modality treatments were clearly set out in a review by Tannock (1989). The vast majority of published papers on combined modality therapy report the results of small, non-randomized trials in which the outcome of treatment is compared with retrospective series where patients were treated with radiotherapy or surgery alone. It is virtually impossible to evaluate the therapeutic benefit of such studies, even if stringent efforts are made to match patients on the basis of various prognostic factors. The major factors identified as introducing bias in such comparisons are briefly as follows.

Patient selection often occurs because only those patients who are fit to receive the relatively toxic combined modality schedules will be entered into the trial. If this selected patient group is compared with all patients previously treated with radiotherapy alone, this will inevitably introduce bias in favour of the new treatment.

Stage migration is another problem associated with the use of historical controls, particularly when there is a long time interval between the studies. Improved diagnostic and detection methods such as CT or NMR scanning help to detect smaller amounts of metastatic disease and have the effect of shifting patients into higher-stage (*i.e.* more metastatic) categories. This removes from the lower-stage (*i.e.* low-risk) categories some patients who are least likely to respond and the result is a bias in favour of the newer treatment.

The duration of follow-up may be shorter for a new study than for the historical controls. The initial tumour response is often good in combined modality studies but early differences between control and combined modality arms tend to decrease with longer follow-up. Single-arm studies which make an early evaluation of tumour

Table 21.1 Tumours of the oral cavity treated with radiotherapy alone or in combination with chemotherapy: tumour response and complications

	RT alone	RT + 5-FU
Number of patients	33	33
Local control (2 yr)	9	20
Alive and well	5	11
Major complications	0	5

Adapted from Lo *et al* (1976).

response and compare actuarial survivals with historical controls after short follow-up are therefore very unreliable.

Small numbers of patients included in trials is another serious problem in evaluating many studies. In order to detect an improvement in survival of 15% with 90% confidence, approximately 300–400 patients randomized to control and test arms are required. Few of the published series contain enough patients to be able to detect such an improvement with statistical significance. Tannock (1989) concluded from his review of the literature that the methodology used in the majority of published trials did not allow conclusions to be made on the possible benefits of combined modality treatment, but only about toxicity and feasibility.

21.4 Current status of combined modality therapy

Head and Neck Cancer

A recent meta-analysis of published results from 54 randomized controlled trials of combined chemotherapy and radiotherapy in head and neck cancer (Munro, 1995) demonstrated a 6.5% improvement in survival (95%, confidence interval 3.1–9.9%) in favour of combined modality therapy. Single-agent chemotherapy concomitant with radiotherapy increased survival by 12.1% (CI 5–19%), whereas the survival benefit from neoadjuvant chemotherapy was less (3.7%, CI 0.9–6.5%). A subgroup analysis of these trials according to type of chemotherapy did not indicate that regimens containing 5-FU or *cis*-platinum were more effective than other drugs.

A previous meta-analysis of 28 combined modality trials for head and neck cancer had failed to demonstrate a benefit for combined modality (Stell, 1992). This may indicate a relative increase in efficacy of the more recent combined modality trials. The meta-analyses compared survival benefits for combined chemotherapy trials vs radiotherapy alone or radiotherapy with surgery. No information was given on relative risks for normal-tissue damage and morbidity. Several individual trials have, however, reported a significant increase in morbidity in combined modality head and neck cancer trials, particularly for regimens including bleomycin (Fu *et al*, 1987) and for concomitant chemotherapy trials (Jassem *et al*, 1992; Wibault *et al*, 1996).

The large randomized trials published so far generally indicate that concomitant (or alternating) chemotherapy and radiotherapy gives the greatest chance of improved local control but that this is very often at the expense of increased toxicity.

Lung Cancer

A meta-analysis of 52 randomized trials comparing chemotherapy with no chemotherapy for non-small-cell lung cancer has recently been published (NSCLC Collaboration Group, 1995). Older trials using alkylating agent chemotherapy combined with surgery for early-stage disease actually gave higher mortality rates than surgery alone. However, there were small but significant benefits for more modern combined modality regimens containing *cis*-platinum, compared with radiotherapy alone or surgery alone, both for early and

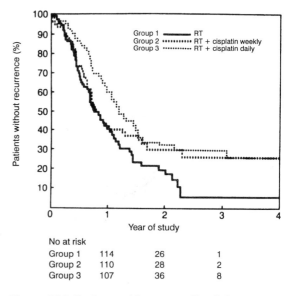

Figure 21.3 Patients with non-small-cell lung cancer treated with *cis*-platinum and radiotherapy (RT) had significantly longer disease-free survival than patients treated with radiotherapy alone ($p = 0.015$ overall). RT + weekly *cis*-platinum was not significantly different from RT alone ($p = 0.15$), but RT + daily *cis*-platinum gave significantly improved disease-free survival ($p = 0.003$) and overall survival ($p = 0.009$). From Schaake-Koning *et al* (1992), with permission.

locally advanced disease (4% and 5% increase in survival at 2 years). Some trials using *cis*-platinum-based regimens have demonstrated much larger improvement in local control and survival (Figure 21.3; Schaake-Koning *et al*, 1992). In a recent randomized trial where hyperfractionated radiotherapy was given with or without concurrent carboplatin and etoposide (Jeremic *et al*, 1996) there was also a significant improvement in 4-year recurrence-free survival (42% vs 19%), with no increase in acute or late toxicity.

For limited-stage small-cell lung cancer, irradiation of the chest has led to improved local control in many studies (relative to chemotherapy alone). A meta-analysis of 12 randomized trials, however, demonstrated an increased median survival of only 5 weeks (67 vs 62 weeks) (Arriagada *et al*, 1989).

Gastrointestinal Cancer

There have been many publications reporting 'positive' results for combined modality treatment of gastrointestinal cancers but the vast majority are uncontrolled single-arm studies which are difficult to evaluate properly. The most promising results have been obtained with concomitant irradiation and multidrug chemotherapy regimes including 5-FU and either MMC or *cis*-platinum. The RTOG has recently published results from a randomized study of oesophageal cancer which demonstrated a large benefit for treatment with 5-FU, *cis*-platinum and radiotherapy vs radiotherapy alone (30% vs 0% 5-year survival; $p < 0.001$; Al-Sarraf *et al*, 1996).

Regimes containing 5-FU have sometimes demonstrated improved local control and survival in rectal cancer when used post-operatively with radiotherapy (Gastrointestinal Tumor Study Group, 1985; Krook *et al*, 1991). Other trials have, however, not always confirmed this survival advantage, despite increased toxicity. The clearest indication for the use of combined modality in rectal cancer is for patients with Dukes' C stage tumours, where the chemotherapy appeared to delay metastatic spread. Concomitant radiotherapy and chemotherapy (5-FU infusion with MMC) is widely used as a successful alternative to surgery for anal cancer. This approach improves local control as well as avoiding mutilating surgery (Figure 21.4), although the overall survival remains unchanged.

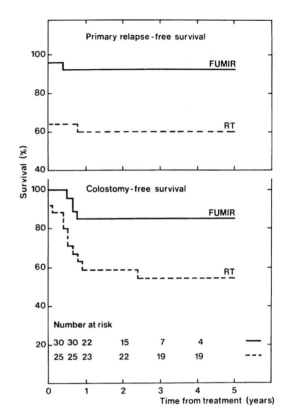

Figure 21.4 Actuarial primary tumour control and colostomy-free survival after treatment of anal cancer with radiotherapy alone (RT) or radiotherapy combined with 5-FU and MMC (FUMIR). The overall survival rates in these two groups of patients were not different (70% at 5 years, data not shown). From Cummings *et al* (1984), with permission.

Breast Cancer

Adjuvant chemotherapy (after mastectomy) in pre-menopausal women with positive axillary nodes significantly reduces the risk of distant metastases. Radiotherapy combined with post-operative chemotherapy reduces locoregional recurrence (Monyak and Levitt, 1989). A recent unpublished meta-analysis of Danish trials also indicates an improved survival. The role of combined chemotherapy and radiotherapy for early breast cancer after breast-conserving surgery is much less clear. Some studies have shown that adjuvant chemotherapy can be given after irradiation of the breast but there is no clear indication of benefit and the cosmetic results tend to be worse than for radiotherapy alone. Concomitant chemotherapy is rarely used because of increased toxicity and poor cosmetic results.

Combination chemotherapy and radiotherapy has been tested in numerous trials for advanced, inoperable breast cancer. The claims for positive results based on uncontrolled trials have generally not been confirmed in randomized trials, although delayed recurrence without improved survival has been demonstrated in some studies.

CANCER OF THE CERVIX

The outlook for patients with advanced (Stage IIIb) carcinoma of the cervix is extremely poor and attempts have been made to improve the results with combined chemotherapy and radiotherapy. On the whole the trials have included small patient numbers, but so far there has been no improvement in local tumour control or survival. Some randomized trials have actually demonstrated significantly worse survival in patients treated with combined modality therapy (Tattersall *et al*, 1995).

BLADDER CANCER

Survival rates for locally advanced bladder cancer are poor due to both local recurrence and metastatic spread. Combined radiation and chemotherapy has been tested with the aim of bladder conservation and reduced metastatic spread. Early pilot studies generally suggested benefits from the combined modality approach but results from the first large randomized trials were disappointing (Richards *et al*, 1983). Neither response rates nor survival were improved by the addition of doxorubicin and 5-FU to radiotherapy. More recent trials using *cis*-platinum-based regimens are now being analyzed. Preliminary results from a large EORTC/MRC trial indicate a 4% improvement in locoregional control and a 6% improvement in disease-free interval at 2 years, in favour of neoadjuvant chemotherapy (*cis*-platinum, methotrexate and vinblastine), although the overall survival rates were unchanged (Hall, 1996).

21.5 Future perspectives

Advances have been made in recent years with combined modality treatment but the gains are still modest for the majority of solid adult tumours. New drugs with increased activity against the major tumour types are needed, as well as methods for targeting these drugs to tumours. Current approaches include intraperitoneal or intrathecal or intra-arterial drug administration, conjugation of cytotoxic drugs to tumour-seeking carriers, and pH- or hypoxia-dependent cytotoxins. Alternatively, cytokines can sometimes be used to stimulate normal-tissue proliferation and recovery, for instance with high-dose chemotherapy and bone marrow transplant regimes for leukaemia. This approach allows drug escalation to levels which would otherwise produce life-threatening toxicity.

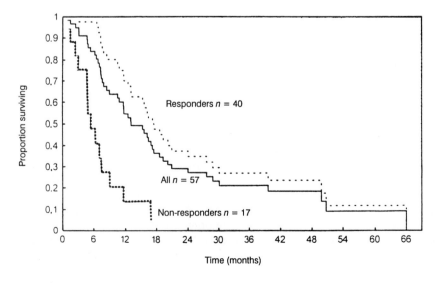

Figure 21.5 Advanced squamous cell carcinoma of head and neck. Survival of patients whose tumours responded (>50% size reduction) to combined chemotherapy and radiotherapy was significantly better than for non-responders. From Wibault *et al* (1996), with permission.

Prediction of those patients most likely to benefit from combined modality therapy is also a worthwhile goal (Section 25.3). Some combined modality trials have demonstrated significantly worse overall survival in patients whose tumours did not respond during the first 28 days after treatment (Figure 21.5). If patients who were resistant to the chemotherapy could be identified before treatment this would prevent unnecessary toxicity in this group and the benefits of combined modality therapy could be better evaluated in the remaining chemosensitive group. The development of predictive tests for this purpose has a long history and there have been examples of promising tests that have not yet been taken up by the clinical community (Baker et al, 1988). Current efforts to develop such tests involve the measurement of DNA damage and drug-DNA adducts, or chromosome damage.

Key points

1. The success of combined modality therapy depends on the availability of effective chemotherapy and the use of optimal radiotherapy and chemotherapy dose schedules. The development of new, more effective drugs and scheduling should lead to an improvement in combined modality therapy.
2. Concomitant chemotherapy usually gives the best chance of improved local control and survival but is also associated with the highest risk of toxicity and normal-tissue damage.
3. Increase in toxicity and decreased quality of life should be considered when assessing the value of aggressive combined modality schedules. This is especially true when little increase in survival is expected.
4 In order to demonstrate a real benefit from combined modality treatment, there should be a significant improvement in survival or local control, relative to radiotherapy alone, for *equivalent toxicity*.

Bibliography

Al-Sarraf M, Martz K, Herskovic A et al (1996). Superiority of chemo-radiotherapy (CT-RT) vs radiotherapy (RT) in patients with esophageal cancer. final report of an intergroup randomized and confirmed study. *Proceedings of ASCO Thirty-second Annual Meeting, Philadelphia*, vol. 15, p. 206 (abstract 464).

Arriagada R, Pignon JP and Le Chevalier T (1989). Thoracic radiotherapy in small cell lung cancer; rationale for timing and fractionation. *Lung Cancer* 5: 237–47.

Baker FL, Spitzer G, Ajani JA and Brock WA (1988). Drug and radiation sensitivity testing of primary human tumour cells using the adhesive-tumor-cell culture system (ATCCS). *Proc Clin Biol Res* 276: 105–17.

Cummings B, Keane T, Thomas G et al (1984). Results and toxicity of the treatment of anal canal carcinoma by radiation therapy or radiation therapy and chemotherapy. *Cancer* 54: 2062–8.

Fu KK, Phillips TL, Silverberg IJ et al (1987). Combined radiotherapy and chemotherapy with bleomycin and methotrexate for advanced inoperable head and neck cancer: update of a Northern California Oncology Group randomized trial. *J Clin Oncol* 5: 1410–18.

Gastrointestinal Tumor Study Group (1985). Prolongation of the disease-free interval in surgically treated rectal carcinoma. *N Engl J Med* 312: 1465–72.

Gelber RD and Goldhirsch A (1986). A new endpoint for the assessment of adjuvant therapy in postmenopausal women with operable breast cancer. *J Clin Oncol* 4:1772–9.

Hall RR (1996). Neo-adjuvant CMV chemotherapy and cystectomy or radiotherapy in muscle invasive bladder cancer. First analysis of MRC/EORTC intercontinental trial. (abstract 612) *Proceedings of ASCO* 15: 244.

Jassem J, Dewit L, Keus R and Bartelink H (1992). Concomitant chemotherapy and radiotherapy. In *Multimodality Therapy for Head and Neck Cancer*, pp. 126–46. (Eds Snow GB and Clark JR) Thieme Stuttgart–New York.

Jeremic B, Shibamoto Y, Acimovic L and Milisavljevic S (1996). Hyperfractionated radiation therapy with or without concurrent low-dose daily carboplatin/etoposide for stage III non-small-cell lung cancer: a randomized study. *J Clin Oncol* 14: 1065–70.

Krook JE, Moertel CG, Gunderson LL et al (1991). Effective surgical adjuvant therapy for high risk rectal carcinoma. *N Engl J Med* 324: 709–15.

Lo TC, Wiley AL, Ansfield FJ *et al* (1976). Combined radiation therapy and 5-fluorouracil for advanced squamous cell carcinoma of the oral cavity and oropharynx: a randomized study. *Am J Roentgenol* 126: 229–35.

Merlano M, Vitale V, Rosso R *et al* (1992) Treatment of advanced squamous-cell carcinoma of the head and neck with alternating chemotherapy and radiotherapy. *N Engl J Med* 327: 1115–21.

Monyak D and Levitt S (1989). The changing role of radiation therapy in the treatment of primary breast cancer. *Investigative Radiology* 24: 483–4.

Munro AJ (1995). An overview of randomized controlled trials of adjuvant chemotherapy in head and neck cancer. *Br J Cancer* 71: 83–91.

Non-small-cell Lung Cancer Collaborative Group (1995). Chemotherapy in non-small cell lung cancer: a meta-analysis using updated data on individual patients from 52 randomized clinical trials. *Br Med J* 311: 899–909.

Richards B, Bastable JR, Freedman L *et al* (1983). Adjuvant chemotherapy with dozorubicin (Adriamycin) and 5-fluorouracil in T3, NX, MO bladder cancer treated with radiotherapy. *Br J Urol* 55: 386–91.

Rosen G, Wollner N, Tan C *et al* (1974). Disease-free survival in children with Ewing's sarcoma treated with radiation therapy and adjuvant four-drug sequential chemotherapy. *Cancer* 33:384–93.

Schaake-Koning C, Van den Bogaert W, Dalesio O *et al* (1992). Effects of concomitant cisplatin and radiotherapy on inoperable non-small-cell lung cancer. *N Engl J Med* 326: 524–30.

Stell PM (1992). Adjuvant chemotherapy for head and neck cancer. *Semin Radiat Oncol* 2: 195–205.

Tannock IF (1989). Combined modality treatment with radiotherapy and chemotherapy. *Radiother Oncol* 16: 83–101.

Tattersall MHN, Lorvidhaya V, Vootiprux V *et al* (1995). Randomized trial of epirubicin and cisplatin chemotherapy followed by pelvic radiation in locally advanced cervical cancer. *J Clin Oncol* 13: 444–51.

Tubiana M (1989). The combination of radiotherapy and chemotherapy: a review. *Int J Radiat Biol* 55: 497–511.

Wibault P, Bensmaïne MEA, De Forni M *et al* (1996). Intensive concomitant chemoradiotheapy in locally advanced unresectable squamous cell carcinoma of the head and neck: a phase II study of radiotherapy with cisplatin and 7-week continuous infusional fluorouracil. *J Clin Oncol* 14: 1192–200.

22

Re-treatment tolerance of normal tissues

Fiona A. Stewart

22.1 Introduction

Curative radiotherapy usually involves treating a volume of normal tissue to the limit of tolerance. The intention will be to treat the tumour and a small volume of normal tissue around it to high dose; in addition there will be a larger volume of normal tissue, perhaps involving other organs, that receives a lower radiation dose. The composite effect of these differing radiation exposures to normal tissues will be to produce a level of morbidity which is the maximum that the therapist judges to be tolerable by the patient. This is what we mean by normal-tissue tolerance (Section 5.1). The actual value of tolerance dose will depend on the organ, fractionation, field-size and concomitant treatments, as well as the performance status of the patient (Section 11.5).

It is frequently necessary to consider re-treatment of radiation fields that have already received radiation therapy, either because of tumour recurrence or the appearance of a new primary within the exposed field. Decisions regarding re-treatment doses which can safely be given are very complex and depend on the timing and extent of any proliferative regeneration (in relation to the interval between treatments), as well as the extent of residual normal-tissue damage present after the regenerative process is complete.

This chapter reviews experimental data on re-treatment tolerance for various normal tissues. Only in laboratory animals has it been possible to perform well-controlled studies. Clinical experience with re-treatment is briefly reviewed in Section 22.9. In order to compare data on different normal tissues using various fractionation regimes, the results of published studies have been recalculated in terms of the biologically effective dose (BED) using linear-quadratic fractionation concepts (Chapter 13). For each study we have calculated the BED for some defined level of tolerance and we will call this the BED_t. The intensity of both the initial treatment and the re-treatment will be specified as a percentage of the BED_t.

22.2 Skin

Experimental studies using erythema and desquamation as end-points show that rodent skin can be re-treated beyond 8 weeks after initial treatment with a full tolerance dose (Figure 22.1). As shown by these data, recovery is faster after initial treatments that are below tolerance.

Fewer data are available for an analysis of *late* radiation effects after re-treatment of the skin but some results suggest a poorer re-treatment tolerance than for acute reactions. For example Brown and Probert (1975) demonstrated that 6 months after fractionated irradiation the re-treatment tolerance of mouse hind limbs was 85–90% of a full tolerance dose (BED_t) for acute skin reactions but when late deformity was scored in the same mice the re-treatment tolerance was only 50–70% of the corresponding BED_t (Figure 22.2). This is not a universal finding, however, and other studies have demonstrated very good re-treatment tolerance of both rat and pig skin for late deformity end-points.

Published data for re-treatment tolerance of acute skin reactions are consistent in demonstrat-

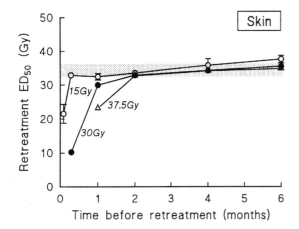

Figure 22.1 Re-treatment tolerance of mouse skin. The ordinate gives the re-treatment dose required for a specified level of skin damage (*i.e.* ED_{50} for desquamation) at different times after priming treatments with 15, 30 or 37.5 Gy. The shaded area shows the range of ED_{50} doses for the same level of skin damage for previously untreated mice. Adapted from Terry *et al* (1989), with permission.

ing a good recovery from the initial damage with restoration of almost full radiation tolerance within 8 weeks. This extent of recovery fits well with the known time-scale for the proliferative response of epithelial cells after irradiation

(Section 4.2). Accelerated proliferation in irradiated rodent skin occurs from about 8 days after treatment, depending on dose. A certain level of cell depletion is required to trigger this accelerated proliferation, which is then maintained until the original cell number is restored and tissue regeneration is complete. This occurs more rapidly after lower radiation doses (*i.e.* after less cell killing) than after high doses.

22.3 Intestine

Re-treatment tolerance of previously irradiated mouse jejunum follows a similar pattern to that of acute skin reactions, with full re-treatment tolerance being regained within 8 weeks of initial irradiation (Reynaud and Travis, 1984). The rapid recovery of the intestine and restoration of full re-treatment tolerance is again consistent with the rapid proliferation kinetics of this tissue. The cell turnover time for untreated crypt cells is about 15 hours in the mouse, with even faster proliferation at 3 days after irradiation.

There are currently no data available to determine the re-treatment tolerance for late bowel complications after previous irradiation.

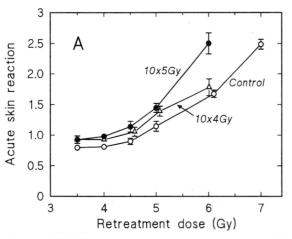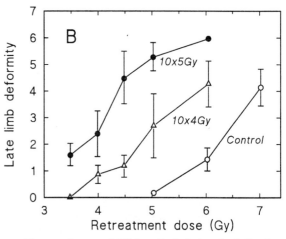

Figure 22.2 Differences in re-treatment tolerance for (**A**) acute skin reactions, and (**B**) late limb deformity following the irradiation of mouse hind limbs (expressed in arbitrary units). Re-treatment was at 6 months after 10 × 4 Gy or 10 × 5 Gy. Open circles show the dose response for damage without prior irradiation. Re-treatment curves show almost complete recovery of acute reactions but for deformity the curves are still well to the left of the control curve. Redrawn from Brown and Probert (1975), with permission.

22.4 Bone marrow

Acute depletion of the various cellular compartments of the bone marrow after irradiation are rapidly followed by a regenerative phase. After whole-body irradiation with single doses of up to 4 Gy, recovery of the stem cells (colony forming units, CFU-S) to near normal levels occurs in about 2–4 weeks. After higher doses, a sustained depletion of CFU-S may occur, although this is masked by their ability to maintain a normal content of peripheral blood cells through increased proliferation of committed precursor cells. However, repeated injury such as re-irradiation can reduce the stem cell pool to below a critical level. The CFU-S which survive repeated irradiations also have a reduced capacity for self-renewal. In such circumstances any additional insult could tip the balance towards severe bone marrow insufficiency. This has been dramatically demonstrated in patients who were re-irradiated with a dose of only 1 Gy 2–4 months after an initial dose of about 4 Gy whole-body irradiation. The second irradiation produced a severe, long-lasting bone marrow aplasia and resulted in the death of one patient (Tubiana *et al*, 1961). It can be concluded that some recovery of re-treatment tolerance does occur in bone marrow if sufficient time is allowed for stem cell regeneration, but estimates of the level of recovery based on mature peripheral blood cells can be very misleading.

22.5 Lung

Experimental studies in mice demonstrate remarkable recovery and re-treatment tolerance in the lung. Recovery is similar to that seen in rapidly dividing tissues such as the skin and intestine, despite the slow turnover times of most lung cell types. Mice given initial doses of 6 Gy to both lungs (*i.e.* 31% of the BED_t for 50% lethal pneumonitis) could be re-treated to full tolerance at 4 weeks (Terry *et al*, 1988). Recovery was slower after higher initial doses, and after initial doses in excess of 70% of the BED_t the full re-treatment tolerance was never achieved (Figure 22.3).

Two separate waves of lethality occur in mice after thoracic irradiation (Figure 5.5). The first, at 12–16 weeks, is characterized by histologically

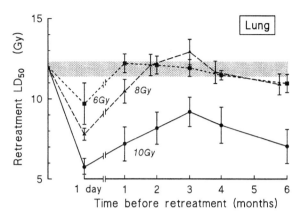

Figure 22.3 Re-treatment tolerance of the mouse lung. The ordinate gives the LD_{50} for re-treatment at the indicated times after priming treatment with 6, 8 or 10 Gy. The shaded area shows the range of LD_{50} values for previously untreated animals. Adapted from Terry *et al* (1988), with permission.

identifiable pneumonitis, whereas later deaths are associated with increased collagen deposition and fibrosis in the lungs of some strains of mice (Travis *et al*, 1980). The excellent re-treatment tolerance of irradiated lung which has been demonstrated in experimental studies probably only applies for the pneumonitis phase. It is likely that re-treatment tolerance for late lung damage may be poorer, although few quantitative data are available.

The re-treatment tolerance for radiation-induced pneumonitis is consistent with an early increase in proliferation of type II pneumocytes seen at 2–8 weeks after irradiation (Section 4.3). The proliferation of these cells is apparently able fully to compensate for cell loss after low radiation doses. Re-treatment after high doses leads to an earlier onset of pneumonitis, particularly for re-treatment after a short interval. This is consistent with the target-cell population being nearer to a critical level of cell depletion, and thus requiring longer for repopulation after higher initial doses.

22.6 Spinal cord

Re-irradiation of the spinal cord is not generally considered to be a realistic clinical option. Therapists are understandably cautious because of the disastrous consequences (myelopathy and paralysis) of exceeding tolerance. It has also been assumed that long-term regeneration will not be effective in such a slowly proliferating tissue. However, there is now substantial experimental evidence which indicates that long-term recovery does occur in the irradiated cord and that re-treatment, albeit with a reduced dose, is feasible.

Analysis of data obtained on spinal cord damage in rats and guinea-pigs allows three main conclusions to be made. First, re-treatment tolerance is inversely related to the size of the initial dose. Second, re-treatment tolerance increases between 2–26 weeks after irradiation and third, substantial residual injury remains, particularly after high-dose initial treatment. The maximum total dose (*i.e.* initial plus re-treatment doses) which could be given in the rat was about 140% of the BED_t for 50% paralysis (Figure 22.4).

These findings from rodent studies are supported by a study of re-treatment tolerance of monkey spinal cord (Figure 22.5). At 2 years after an initial treatment of 20 × 2.2 Gy, equivalent to about 60% of the BED_t, these monkeys could be re-treated with a further 66 Gy in 2.2-Gy fractions (85% BED_t), giving a total dose for both treatments equivalent to 145% of the BED_t.

The increase in total tolerance observed for the spinal cord with longer re-treatment intervals may be related to proliferation in the putative target cells (oligodendrocytes and endothelial cells).

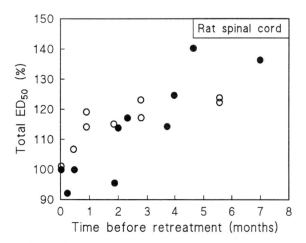

Figure 22.4 Long-term recovery after irradiation of rat cervical cord. The total (initial + re-treatment) isoeffective dose for 50% paralysis is expressed as a percentage of the tolerance dose (BED_t). Data are for 3-week-old (○) and adult (●) rats; adapted from Ruifrok *et al* (1992), White and Hornsey (1980) and van der Kogel *et al* (1982).

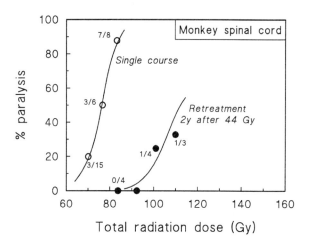

Figure 22.5 Re-treatment tolerance of the monkey spinal cord. At 2 years after an initial dose of 44 Gy (*i.e.* 60% of BED_t) a further 66 Gy (85% of BED_t) could be delivered for the specified level of damage (*i.e.* 50% paralysis). All treatments were given with 2.2 Gy per fraction. From Ang *et al* (1993), with permission.

Mature glial cells normally turn over very slowly, with a lifespan of approximately 140 days, but a marked increase in proliferation occurs at about 21 weeks after irradiation. This is followed by a rapid decline in the glial cell numbers, which precedes the onset of white matter necrosis. An early, transient increase in glial cell proliferation also occurs at 3 weeks after irradiation. This could be related to the early regenerative phase which has been identified in glial progenitor cells after irradiation of the rat optic nerve (van der Maazen *et al*, 1992). The main wave of increased labelling (beyond 20 weeks after irradiation) actually occurs rather later than the period over which dose-sparing effects are observed for re-treatment (4–14 weeks, see Figure 22.4). Indeed, it has been suggested that the late proliferative burst in glial cells merely serves to precipitate further damage as lethally damaged cells attempt to divide. The precise influence of stimulated proliferation of glial cells for regeneration of the spinal cord after irradiation remains unclear.

22.7 Bladder

Studies on re-treatment tolerance of mouse bladder have not demonstrated recovery from late,

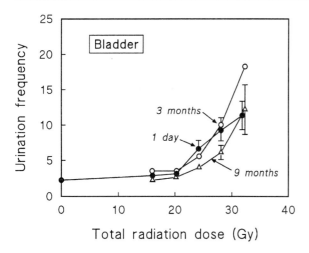

Figure 22.6 Dose–response curves for bladder damage after irradiation with two doses separated by 1 day (●), 3 months (○) or 9 months (△). The ordinate shows the number of urination events per ml of urine excreted. The total dose for a given amount of damage did not increase with increasing time from first treatment. From Stewart *et al* (1990), with permission.

functional damage (as indicated by increased frequency of urination or reduced bladder compliance) for re-treatment intervals of 12 or 40 weeks compared with short (1-day) intervals (Figure 22.6). The latent period before expression of permanent functional damage was also much shorter in animals that were re-irradiated, even at 40 weeks after low, subtolerance initial doses.

The normal, untreated bladder epithelium has a very slow cell turnover time (in excess of 200 days) but it is capable of rapid proliferation in response to injury (Section 4.3). The increase in proliferation is maximal at about 40 weeks after irradiation. Bladders re-irradiated at this time therefore have a urothelium in a state of rapid turnover and damage from the second treatment is expressed without the usual delay. Despite the evidence for extensive epithelial proliferation after bladder irradiation, the re-treatment studies failed to demonstrate any dose-sparing effect. Compensatory proliferation precipitated an early expression of damage without preventing or reducing the extent of late damage. This implies that epithelial damage, although certainly involved in the initial urination frequency response, is not the only factor that determines persistent bladder damage.

22.8 Kidney

The kidneys are among the most radiosensitive of organs, although the latent period before expression of functional damage may be very long, particularly after low doses. Progressive, dose-dependent development of functional damage, without apparent recovery, has been clearly demonstrated in rodents. This is consistent with clinical observations of slowly progressive renal damage which can occur over periods of at least 7 years after irradiation. Based on the known profiles for development of renal radiation injury, it is perhaps obvious that re-irradiation after large initial doses is unlikely to be feasible. However, it is less obvious what the situation would be after low, sub-tolerance doses to the kidneys.

In view of the slowly progressive nature of renal radiation injury, the absence of any measurable renal dysfunction at the time of re-treatment certainly cannot be interpreted as a sign that there is no latent or residual injury. Experimental studies demonstrate that doses of radiation too low to produce overt renal damage significantly reduce the tolerance to re-treatment; none of these studies has demonstrated any long-term functional recovery of the kidney. After an initial dose of only 6 Gy (25% of the BED) the tolerance to re-treatment actually *decreases* with time (Figure 22.7). This is consistent with there being continuous progression of occult damage in the interval between treatments.

22.9 Clinical experience with re-treatment

It is difficult to analyze clinical reports of re-treatment tolerance in the same way as the animal data since they are often anecdotal and always include patients treated with a wide range of initial and re-treatment doses at different times after initial treatment. There are, however, several reported series with at least 50 patients re-irradiated with curative intent for recurrent head and neck cancer (*e.g.* Wang, 1987; Pryzant *et al*, 1992). General conclusions from these and other smaller studies are that good local control (>50%) can be achieved with total re-treatment doses of 60 Gy but that lower doses are ineffective (local control <10%).

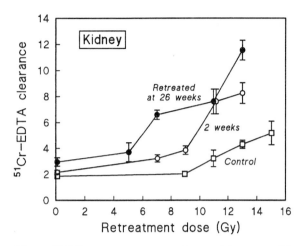

Figure 22.7 Dose–response curves for renal damage at 35 weeks after re-treatment. The initial treatment was 6 Gy (*i.e.* 25% BED$_t$) given at either 2 weeks (○) or 26 weeks (●) before re-treatment. The response of age-matched control animals (no initial dose) is also shown. Renal damage, as detected by ^{51}Cr-EDTA clearance (% injected dose per ml plasma) was worse for re-treatment with the longer 26-week interval than for a shorter interval, indicating progression of subthreshold damage rather than recovery. From Stewart *et al* (1989), with permission.

Serious complications in long-term survivors were seen in 6–40% of patients and this was clearly associated with the higher re-treatment doses and short re-treatment intervals. The best results (good local control with few complications) were achieved with re-treatment techniques involving both external-beam irradiation and intracavitary caesium.

An overview of results in head and neck cancer is given in Table 22.1. The largest study is that of Yan *et al* (1983) in nasopharyngeal cancer. The studies by Wang (1987), also in nasopharyngeal cancer, and Langlois *et al* (1985) clearly show the difference between a tumour dose greater and less than 60 Gy, but with significant necrosis.

Early experience of re-treatment of recurrent cervical cancer was not encouraging. Local control and survival rates were generally poor (10–20% long-term survival) and complication rates were high (30–50%). More recent studies, in which patients have been selected on the basis of volume and location of the cancer, demonstrate much better results, particularly for re-treatment using brachytherapy. In these studies long-term survivals of 60% with a severe complication rate of <15% were seen after full dose re-treatment (*e.g.* Prempree *et al*, 1984; Russell *et al*, 1982). Favourable conditions were small tumour volume, second malignancies and re-treatment with brachytherapy. Unfavourable conditions were recurrent cancer, large tumour volume and re-treatment with external-beam therapy.

Re-treatment of recurrent lung cancer with palliative intent (30 Gy/3 weeks) has also been reported to give good improvement of symptoms (cough, dyspnoea, pain and haemoptysis) and good tumour response (74%) if patients were selected on the basis of having achieved a good

Table 22.1 Re-irradiation of recurrent head and neck cancer

Author	No. of patients	Tumour response***	End-point	Severe complications
Fu *et al* (1975)	42	41%	5-yr survival	21% necrosis
Nisar Syed *et al* (1978)	64	49%	LC	30% necrosis
Skolyszewski *et al* (1980)	20	5% (<50 Gy) 65% (>50 Gy)	3-yr survival	0% necrosis 20% necrosis
Yan *et al* (1983)	219	18%	5-yr survival	13%
Langlois *et al* (1985)	35	8% (<60 Gy) 55% (>60 Gy)	CR	38% necrosis/ bleeding
Wang *et al* (1987)	51	0 (<60 Gy) 45% (>60 Gy)	5-yr survival	6 %
Pryzant *et al* (1992)	53*	21%	5-yr survival	4% <100 Gy** 39% >100 Gy**

* Of 53 patients 9 received 20–30 Gy external-beam irradiation plus 40–50 Gy intracavitary caesium at re-treatment: 5-yr local control for this subset was 67%, with no severe complications.
** Total dose (initial + re-treatment).
*** Re-treatment dose in parentheses.
LC = local control; CR = complete response.

response to initial therapy (Green and Melbye, 1982). The objective tumour response in cases not pre-selected for a good initial response was very limited, although palliation of symptoms was still achieved (Montebello *et al*, 1993). Re-treatment was associated with lower levels of oesophagitis than the initial treatment and no increase in pneumonitis. The re-treatment doses were of course much lower than the initial doses and most patients succumbed to their disease before late damage (*i.e.* lung fibrosis) had time to be expressed.

The limited clinical data available suggest that re-treatment can be carried out in some sites and provides either effective palliation without a high incidence of morbidity or, when a full dose is given, good local control and survival with an increased risk of morbidity. In some situations this risk is probably acceptable when weighed against the alternative option of no treatment.

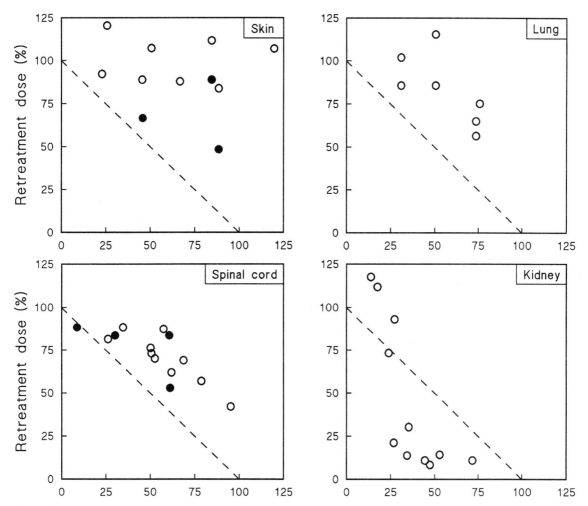

Figure 22.8 Summary of re-treatment tolerance in four normal tissues. The abscissa gives the size of the initial dose as a percentage of tolerance (*i.e.* BED_t). The ordinate shows the isoeffective re-treatment dose, again as a percentage of BED_t. The dashed line indicates the re-treatment dose which would be expected if there were full recovery of sub lethal damage but no long-term proliferative recovery. Sources: mouse skin desquamation (○) Brown and Probert (1975), Denekamp (1975), Terry *et al* (1989); skin late deformity (●) Brown and Probert (1975), Simmonds *et al* (1989); lung pneumonitis Field *et al* (1976), Terry *et al* (1988); lumbar cord paralysis (●) Hornsey *et al* (1982), Mason *et al* (1993); cervical cord paralysis (○) Ang *et al* (1993), van der Kogel *et al* (1982), White and Hornsey (1980), Wong *et al* (1993), Ruifrok *et al* (1992); renal function Stewart *et al* (1989, 1994), Stewart Oussoren, and Luts (1990), Stewart and van der Kogel (1994). Redrawn from Stewart and van der Kogel (1994), with permission

22.10 Summary

For four tissues where recovery following a range of initial treatments has been evaluated Figure 22.8 summarizes results for re-treatment tolerance. Both the initial and the re-treatment radiation exposures are shown as a percentage of the tolerance dose for a defined level of damage, calculated in terms of BED_t using the appropriate α/β ratio for each tissue. The dashed lines indicate the relationship that would be expected if there is full repair of sublethal injury but no proliferative recovery and no cell depletion between the treatments. Where data points for re-treatment beyond 4 weeks after initial irradiation fall above the dashed line (in skin, lung and cord) this indicates some long-term recovery in the tissue. Where the data points fall below the dashed line (kidney), this indicates damage progression and cell depletion rather than recovery. The general conclusion is that, on the basis of studies in experimental animals, several normal tissues are able to tolerate considerable re-treatment with radiation. The phenomenon is not, however, universal and care must be applied in translating this into the clinic. The relatively few substantial clinical studies confirm that provided the re-treatment interval is long, good responses can be achieved in a proportion of patients. But the therapeutic predicament is even greater than for initial radiotherapy: low doses can be given safely but with poor tumour response; when the re-treatment dose approaches the full tolerance dose the risk of severe morbidity is high.

Key points

1. Rapidly dividing tissues are capable of proliferative recovery to completely restore the original cell numbers and full re-treatment tolerance within a few months of irradiation. After repeated irradiations of bone marrow there may be a reduced density of stem cells leading to decreased re-treatment tolerance.
2. In terms of pneumonitis, the lung is capable of extensive long-term recovery after subtolerance irradiation. Re-treatment tolerance for late lung damage is probably less than for pneumonitis.
3. The spinal cord is capable of moderate long-term recovery and re-treatment doses of up to 80% of tolerance (*i.e.* BED_t) can be given after previous subtolerance irradiation.
4. Re-treatment tolerance for late bladder damage does not increase with increasing time between treatments. The onset of late bladder damage occurs much earlier after re-treatment than after initial irradiation.
5. Re-treatment tolerance of the kidney decreases with increasing interval between treatments, indicative of continuous progression of subthreshold damage.
6. Clinical experience confirms that re-treatment can usefully be given under some circumstances.

Bibliography

Ang KK, Price RE, Stephens LC et al (1993). The tolerance of primate spinal cord to re-irradiation. *Int J Radiat Oncol Biol Phys* 25:459–64.

Brown JM and Probert JC (1975). Early and late radiation changes following a second course of irradiation. *Radiology* 115:711–16.

Denekamp J (1975). Residual radiation damage in mouse skin 5 to 8 months after irradiation. *Radiology* 115:191–5.

Field SB, Hornsey S, Kutsutani Y (1976). Effect of fractionated irradiation on mouse lung and a phenomenon of slow repair. *Brit J Radiol* 49:700–7.

Fu KK, Newman H and Phillips TL (1975). Treatment of locally recurrent carcinoma of the nasopharynx. *Radiology* 117:425–31.

Green N and Melbye RW (1982). Lung cancer: re-treatment of local recurrence after definitive irradiation. *Cancer* 49:865–8

Hornsey S, Myers R and Warren P (1982). Residual injury in the spinal cord after treatment with X-rays or neutrons. *Brit J Radiol* 55:516–19.

Langlois D, Eschwege F, Kramar A and Richard JM (1985). Reirradiation of head and neck cancers. *Radiother Oncol* 3:27–33.

Mason KA, Withers HR and Chiang C-S (1993). Late effects of radiation on the lumbar spinal cord of guinea pigs: re-treatment tolerance. *Int J Radiat Oncol Biol Phys* 26(4):643–8.

Montebello JF, Aron BS, Manatunga AK *et al* (1993). The re-irradiation of recurrent bronchogenic carcinoma with external beam irradiation. *Am J Clin Oncol* 16:492–88

Nisar Syed AM, Feder BH, George FW III and Neblett D (1978). Iridium-192 afterloaded implant in the retreatment of head and neck cancers. *Brit J Radiol* 51:814–20.

Prempree T, Amornmarn R, Villasanta U *et al* (1984). Re-treatment of very late invasive squamous cell carcinoma of the cervix with irradiation. *Cancer* 54:1954–5

Pryzant RM, Wendt CD, Delclos L and Peters LJ (1992). Retreatment of nasopharyngeal carcinoma in 53 patients. *Int J Rad Oncol Biol Phys* 22:941–7

Reynaud A and Travis EL (1984). Late effects of irradiation in mouse jejunum. *Int J Radiat Biol* 46:125–34.

Ruifrok ACC, Kleiboer BJ and van der Kogel AJ (1992). Reirradiation tolerance of the immature rat spinal cord. *Radiother Oncol* 23:249–56.

Russell AH, Coh W-J, Markette K *et al* (1987). Radical re-irradiation for recurrent or second primary carcinoma of the female reproductive tract. *Gynecol Oncol* 27:226–32.

Simmonds RH, Hopewell JW and Robbins MEC (1989). Residual radiation-induced injury in dermal tissue: implications for retreatment. *Brit J Radiol* 62:915–20.

Skolyszewski J, Korzeniowski S and Reinfuss M (1980). The reirradiation of recurrences of head and neck cancer. *Brit J Radiol* 53:462–5.

Stewart FA, Luts A and Lebesque JV (1989). The lack of long term recovery and reirradiation tolerance in the mouse kidney. *Int J Radiat Biol* 56: 449–62.

Stewart FA and Oussoren Y (1989). Reirradiation of mouse kidneys: a comparison of re-treatment tolerance after single and fractionated partial tolerance doses. *Int J Radiat Biol* 58:531–44.

Stewart FA, Oussoren Y and Luts A (1990). Long-term recovery and reirradiation tolerance of mouse bladder. *Int J Radiat Oncol Biol Phys* 18:1399–406.

Stewart FA, Oussoren Y, Van Tinteren H and Bentzen SM (1994). Loss of reirradiation tolerance in the mouse kidneys with increasing time after single or fractionated partial tolerance doses. *Int J Radiat Biol* 66: 169–79.

Stewart FA and van der Kogel AJ (1994). Re-treatment tolerance of normal tissues. *Sem Radiat Oncol* 4: 103–11.

Terry NHA, Tucker SL and Travis EL (1988). Residual radiation damage in murine lung assessed by pneumonitis. *Int J Radiat Oncol Biol Phys* 14:929–38.

Terry NHA, Tucker SL and Travis EL (1989). Time course of loss of residual radiation damage in murine skin assessed by re-treatment. *Int J Radiat Biol* 55:271–83.

Travis EL, Down JD, Holmes SJ and Hobson B (1980). Radiation pneumonitis and fibrosis in mouse lung assayed by respiratory frequency and histology. *Radiat Res* 84:133–43.

Tubiana M, Lalanne C-M and Surmont J (1961). Total body irradiation for organ transplantation. *Proc R Soc Med* 54:1143–50.

van der Kogel AJ, Sissingh HA and Zoetelief J (1982). Effect of X rays and neutrons on repair and regeneration in the rat spinal cord. *Int J Radiat Oncol Biol Phys* 8: 2095–7.

van der Maazen RWM, Verhagen I, Kleiboer BJ and van der Kogel AJ (1992). Repopulation of O-2A progenitor cells after irradiation of the adult rat optic nerve analyzed by an *in vitro* clonogenic assay. *Radiat Res* 132:82–6.

Wang CC (1987). Reirradiation of recurrent nasopharyngeal carcinoma – treatment techniques and results. *Int J Rad Oncol Biol Phys* 13:953–6

White A and Hornsey S (1980). Time dependent repair of radiation damage in the rat spinal cord after X rays and neutrons. *Eur J Cancer* 16:587–962.

Wong CS, Poon JK and Hill RP (1993). Reirradiation tolerance in the rat spinal cord: influence of level of initial damage. *Radiother Oncol* 26:132–8.

Yan JH, Hu YM and Gu Z (1983). Radiation therapy of recurrent nasopharyngeal carcinoma. *Acta Radiol Oncol* 22:23–8.

23

Hyperthermia

Jens Overgaard and Michael R. Horsman

23.1 History and background

Knowledge of the effects of heat treatment on tumours is as old as the oldest written text in medicine. The Edwin Smith surgical papyrus, which was probably the first medical textbook, dating back more than 5000 years, gives a description of a patient with a tumour in the breast treated with hyperthermia (*i.e.* cautery). Similar procedures have subsequently been used throughout history. In the last decades of the nineteenth century hyperthermia underwent a renaissance, triggered by the observation that patients with high fever due to erysipelas in some instances demonstrated spontaneous regression of tumours. This led the New York surgeon William B. Coley to develop his 'Mixed Bacterial Toxin' and thereby he became the father both of the modern use of hyperthermia and non-specific immunotherapy for the treatment of cancer. Concurrent with Coley's interest, a more direct local application of hyperthermia was performed by others who demonstrated that moderately elevated temperatures (up to 45°C) could induce significant regression and even disappearance of tumours.

This treatment principle developed throughout the first part of the twentieth century, not only as a modality by itself, but also in combination with radiotherapy. The inspiration for using hyperthermia as an adjuvant to radiotherapy came indirectly from the observation by Gottwald Schwarz who in 1909 demonstrated the effect of *Kompressionsanämie*, which constricted the blood supply to tissues and reduced the effect of ionizing radiation. Attempts to *increase* the effect of radiation were subsequently performed, and one of the obvious methods was by heating the tissue in order to increase the blood supply. This principle, called 'thermopenetration', was proposed by Schwarz (1909) and subsequently carried out by Müller who in 1910 described the principle of the technique and a few years later presented the results of a phase I–II trial describing his first 100 cases (Müller, 1912). The biological mechanism of this combined effect was later found not to be by a change in blood flow.

Hyperthermia has also been applied in combination with chemotherapy, especially in association with whole-body heating or the regional perfusion of limbs. Most recently it has been applied in the treatment of advanced local tumours as a part of a multimodality strategy, or when other more conventional modalities have been ruled out.

The Achilles' heel of hyperthermia treatment of malignant tumours is the problem of achieving adequate heating of deep-seated tumours. Current technology only allows reasonably satisfactory heating of superficial lesions. The technology for superficial heating has become easily available and hyperthermia has increasingly been applied in the treatment of various superficial tumours, either alone or in combined approaches.

The present chapter gives an overview of the biological rationale and clinical experience with local hyperthermia, especially in combination with radiation. On the basis of present knowledge we will also outline the indications for hyperthermia in current oncological practice.

23.2 The biological effects of hyperthermia

HEAT ALONE

Moderate heat treatment (temperatures of 41 to 45°C) is able to destroy malignant and normal cells *in vitro*. The mechanisms of destruction by heat are complex and involve damage to cell nuclei, to membranes and to other cytoplasmic components. Sensitivity to hyperthermia varies greatly among different cell lines but as a whole there is no differentially enhanced sensitivity in malignant cells. Most experimental studies have been performed with rodent cell lines, which in general seem to be somewhat more sensitive to hyperthermia than human cells, but they do not seem to differ in a qualitative way.

Thermal damage to cells depends on two main factors: temperature and the heating time. This is illustrated in Figure 23.1, which shows that the growth delay of mouse mammary tumours increased linearly with heating time and that the slope of the relationship was highly dependent upon temperature in the range 41–44.5°C. Data of this type can be summarized in an Arrhenius plot, an example of which is given as an inset to Figure 23.1. This is an isoeffect plot in which the logarithm of the heating time for a given level of effect is plotted against temperature. Arrhenius plots are often linear up to around 42.5°, at which point the slope abruptly changes (in the direction of *longer* heating time at the higher temperature). It is possible to calculate from the slope of such plots the activation energy (μ) associated with the effect. The change in slope that occurs in Figure 23.1 implies that above 42.5° a different biological process is being inactivated: the activation energies correspond with damage to proteins and membranes rather than to DNA.

The sensitivity of cells to heat-induced cell killing varies through the cell cycle. A number of studies have shown that cells in the S-phase tend to be the most sensitive, in contrast to the situation with radiation cell killing. This has been proposed as one rationale for combining these two treatments.

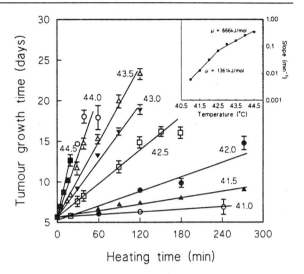

Figure 23.1 Tumour growth time as a function of heating time for a solid mouse mammary carcinoma heated *in vivo* in the range 41.0 – 44.5°C. Tumour growth time was the time taken for tumours to regrow to five times their treatment volume.
Inset: Time–temperature plot showing the slope of the dose–response curves as a function of heating temperature. From Lindegaard and Overgaard (1987), with permission.

THERMOTOLERANCE

Thermotolerance is the phenomenon by which one heat treatment induces temporary resistance against subsequent heating. If two heat treatments are separated by a short interval, thermotolerance may lead to the second having little effect. In the clinical application of hyperthermia, for instance in combination with radiotherapy, it may be necessary to deliver repeated heat treatments and thus thermotolerance is of considerable practical importance.

The magnitude and kinetics of thermotolerance depend on the intensity of the priming treatment and the effects differ among various tumours and normal tissues. An example is shown in Figure 23.2. Mouse mammary tumours were given a 30-minute pre-heating at a temperature between 42.5° and 44.5°. They were then reheated after various time intervals and the extent of protection (the

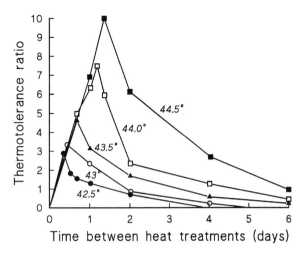

Figure 23.2 Kinetics and magnitude of thermal tolerance in C3H mammary carcinomas. An initial 30-minute heating was followed by treatment at the same temperature for various times, but with different time intervals between the two treatments. Thermotolerance ratio is the ratio of heating times for a fixed level of tumour response. From Overgaard *et al* (1987), with permission.

The main factor responsible for enhanced cell death within ischaemic tissues is probably the increased acidity (Figure 23.3). Since such a deprived environment usually exists only within solid tumours and not in normal tissues, this enhancement of sensitivity is a major factor in the rationale for applying hyperthermia in the treatment of cancer. Specific hyperthermic cytotoxicity is seen as direct heat killing of cells in a deprived microenvironment, characterized by insufficient blood supply, which leads to poor nutrition and increased acidity due to anaerobic metabolism and accumulation of lactic acid and other waste products. Cells situated in such a region are highly sensitive to hyperthermia and can be destroyed by a heat treatment that does not cause significant damage to cells in a normal environment.

A further and important effect of the reduced blood flow to ischaemic tissues is that they will experience less cooling and therefore for any given energy input they will rise to a higher temperature. As a result, heat treatment of superficial tumours often results in a higher temperature in the tumour than in the surrounding tissue. In experimental systems this has been shown to lead to a differential response between tumour and normal tissue. Whereas most tumours show a tendency for collapse of the blood vessels at a relatively low temperature, normal tissues sometimes react by *increasing* their blood flow and thus reaching a lower temperature than would otherwise be the case (Figure 23.4). This *physical* effect of blood flow is perhaps the most important factor leading to differential cell killing between tumours and normal tissues.

'thermotolerance ratio') was determined. The ratio increased linearly with time interval up to a peak, the height of which was very temperature-dependent and occurred between 12 and 48 hours after pre-heating; thereafter the protection declined and became negligible for intervals of 4–6 days.

Thermotolerance is a large effect which is important in determining the response to multiple heat treatments. It has been observed in all tissues where it has been sought and it must therefore be considered a general phenomenon. Its extent and timing vary from one tissue to another which makes it difficult to predict precisely.

Environmental and Vascular Effects

Heat-induced damage is also strongly influenced by the extracellular environment. Thus an environment characterized by deprived nutritional conditions including chronic hypoxia, increased acidity and starvation promotes hyperthermic damage. This is not, as with radiation, an oxygen effect. In fact, acute hypoxia or hypoxic conditions that persist for a limited time in an otherwise normal environment do not seem to influence the degree of hyperthermic damage.

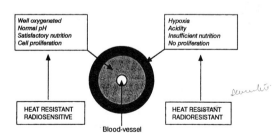

Figure 23.3 The microenvironment in tumours leads to regions that differ in a number of important respects. See also Figure 15.5.

Direct cell killing by heat occurs rapidly. Heated cells die within hours both *in vitro* and *in vivo*, in contrast to cell killing after irradiation which may take some days to appear. This rapid necrotic

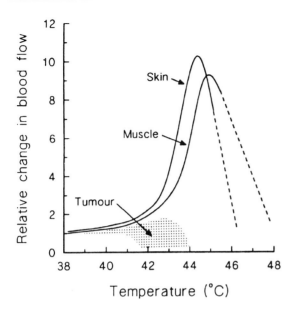

Figure 23.4 Relative changes in blood flow in tumours and normal tissues heated at different temperatures. The stippled area shows the range of values obtained for a variety of animal tumours heated at various temperatures for 30 – 40 minutes. The normal-tissue data are from blood flow measurements in skin and muscle of SD rats heated for the same time periods. Adapted from Song (1984), with permission.

process occurs even faster in tumours with a deprived environment and is probably due to accelerated cytoplasmic and membrane damage. Thermal damage is invariably *acute*: late effects are uncommon and where they have been observed they are thought to be the result of excessive acute tissue damage.

INTERACTION WITH RADIATION

In addition to direct cell killing, hyperthermia enhances the effect of radiation on cells. Hyperthermic radiosensitization occurs primarily when the radiation is delivered *during* hyperthermia, *i.e.* for simultaneous treatment. An interval between the two treatments reduces the effect. The enhancement is expressed as a quantitative increase in the radiation effect but without qualitative changes. The effect is complex, involving both an increased direct radiosensitivity, depending on temperature and heating time, and a reduced repair of damage. Interestingly, the oxygen effect does not seem to be affected, for the oxygenated and acutely hypoxic cells are often equally sensitized by hyperthermia.

The extent of hyperthermic radiosensitization is usually expressed as the thermal enhancement ratio:

$$\text{Thermal enhancement ratio (TER)} = \frac{\text{Radiation dose without heating}}{\text{Radiation dose with heating}}$$

to produce the same level of effect. The measurement of the TER value for mouse mammary tumours is illustrated in Figure 23.5.

TER values tend to be high for *simultaneous* heat and radiation treatment but of similar magnitude in tumour and normal tissues. Such a treatment will therefore not improve the therapeutic ratio unless the tumour is heated to a higher temperature than the normal tissue. An example of the time-dependence of hyperthermic radiosensitization is shown in Figure 23.6. Mouse mammary tumours were heated to 42.5°C and tumour control was documented as a function of radiation dose. This allowed TER values to be calculated for each selected time interval between the treatments. In parallel experiments skin damage was also measured. It can be seen that effects on both tumour and skin peaked at zero time interval (*i.e.* for simultaneous treatment), the TER values reaching 2.5 in each case. For intervals of up to 24 hours between the treatments the tumour data gave a

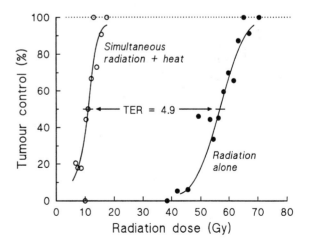

Figure 23.5 Dose–response curves for the control of C3H mammary carcinomas by radiation alone or combined with hyperthermia (43.5°C for 60 minutes), illustrating the determination of a thermal enhancement ratio. From Overgaard et al (1987), with permission.

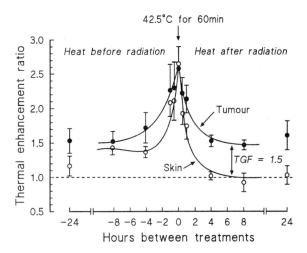

Figure 23.6 Thermal enhancement ratio (TER) as a function of time interval and sequence between hyperthermia and radiation treatment of C3H mammary carcinomas and the surrounding skin. Maximum TER values were observed after simultaneous treatment and any time interval reduced the thermal sensitization. From Overgaard et al (1987), with permission.

symmetrical curve, falling to a TER of 1.5. In the skin, there was evidence for sensitization with heat before radiation, but not *vice versa*. In this experimental situation there was therefore a therapeutic advantage in post-radiation heating. This advantage may be expressed as:

Therapeutic gain factor = $\dfrac{\text{TER for tumour response}}{\text{TER for normal-tissue damage}}$
(TGF)

In experimental systems it is possible to distinguish the effects of direct hyperthermic cytotoxicity from hyperthermic sensitization. This is illustrated in Figure 23.7. Mouse mammary tumours were either heated simultaneously with radiation treatment or the treatments were given sequentially. Thermal enhancement ratios were determined for the two situations. Whereas the simultaneous treatment resulted in a linear increase in TER with heat dose, the sequential treatment led to a plateau at a TER of approximately 2.0, with no increase by further heating.

In a few clinical studies it has been possible to obtain dose–response curves for tumour response with and without heat treatment, and thereby to derive TER values. Examples are shown in Figure 23.8 yielding TER values in the range 1.5–1.6 (Overgaard, 1989).

RATIONALE FOR COMBINED HEAT AND RADIATION *IN VIVO*

The biological rationale for combining hyperthermia and radiation has been described in detail by Overgaard (1989). Briefly, it relies on two different mechanisms, namely hyperthermic radiosensitization and direct hyperthermic cytotoxicity.

Hyperthermic radiosensitization is expressed as increased damage from radiation when hyperthermia and radiation are applied simultaneously. This effect does not qualitatively alter the radiation response, but only gives a quantitative enhancement. Hyperthermic radiosensitization occurs to an equal extent in normal tissues and in tumours and is most pronounced for a truly simultaneous application of the modalities.

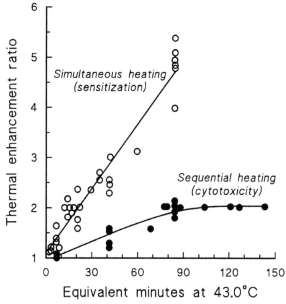

Figure 23.7 Thermal enhancement ratio as a function of heat treatment in a mouse mammary carcinoma exposed either to simultaneous *(radiosensitization)* or sequential *(cytotoxicity)* radiation and hyperthermia (heat 4 hours after radiation). From Overgaard (1989), with permission.

Hyperthermic cytotoxicity is seen as direct heat killing of cells in a deprived microenvironment, characterized by insufficient blood supply with subsequent poor nutrition and increased acidity due to anaerobic metabolism with accumulation of lactic acid and other acidic waste products. Cells situated in such an area are highly sensitive to

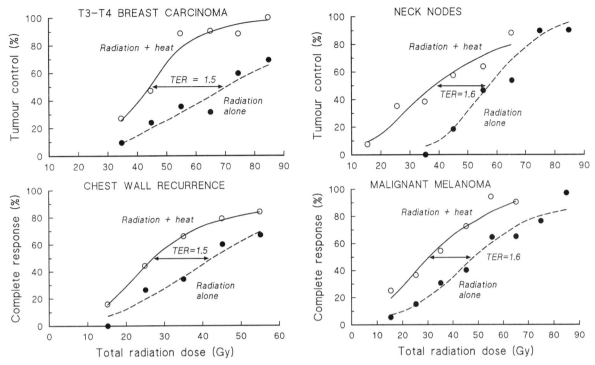

Figure 23.8 Dose–response relationships for advanced breast carcinoma, recurrent breast carcinoma, advanced neck nodes and malignant melanoma treated with radiation or combined radiation and hyperthermia. The total radiation doses were normalized to daily fractions of 2 Gy using an α/β ratio of 25 Gy for breast and head and neck carcinomas and 2.5 Gy for malignant melanoma. From Overgaard (1989), with permission.

hyperthermia and can be destroyed by a heat treatment which does not cause significant damage to cells in a physiological environment. Furthermore, the environmental factors which enhance the hyperthermic damage are typically those that reduce the radiation response (*e.g.* hypoxia). Thus, the areas of a tumour which, due to insufficient vascularization, are resistant to radiation are likely to be the same as those sensitive to hyperthermia. On the other hand, the well-vascularized tumour areas (characterized by a normal physiological environment) are sensitive to radiotherapy, but relatively resistant to hyperthermia.

Strategy for Clinical Application

Two strategies have been derived on that basis for combining heat and irradiation. One is as a radiosensitizing agent enhancing the effect of ionizing radiation in heated tissues. This strategy is only advantageous if hyperthermia and radiation are applied simultaneously. The other strategy utilizes the specific cytotoxic effect of heat against radioresistant cells, which results in damage to radioresistant areas of the tumours and is achieved by sequential treatment. Although hyperthermic radiosensitization with simultaneous treatment is able to yield the highest thermal enhancement ratio (TER), the clinical applicability of this protocol is dubious due to technical problems and it is likely that the dominating effect is due to the hyperthermic cytotoxicity. This biological rationale has been documented clinically (Overgaard, 1989).

The quantitative difference between simultaneous and sequential treatments appears to be less for clinical studies than would be expected from animal studies. However, recent studies have indicated that this is probably again due to the fact that truly simultaneous treatments have not been used in the clinical situation. As shown in Figure 23.9, an interval of only a few minutes between heat and radiation, or even placing the radiation early or late in the heating session, reduces the amount of hyperthermic radiosensitization. Therefore, if up to half an hour is allowed between radiation and

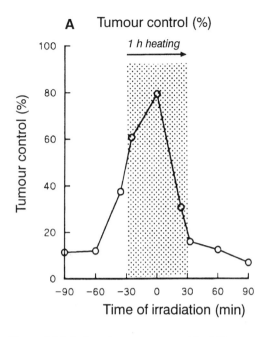
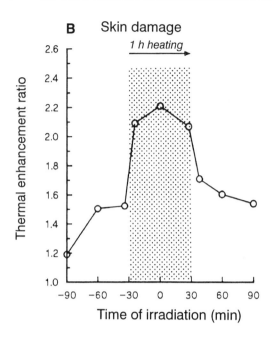

Figure 23.9 A: Local tumour control in C3H mouse mammary carcinomas, and **B**: damage to surrounding skin as a function of sequence and interval between hyperthermia and radiation. Tumours were irradiated with 26 Gy at various times before, during or after heating to 42.5°C for 60 minutes; tumour control was determined 120 days after treatment. From Overgaard *et al* (1987), with permission.

hyperthermia, the major effect of such treatment will be dominated by direct hyperthermic cytotoxicity, although some radiosensitization may persist both in tumour and normal tissues.

THE PROBLEM OF FRACTIONATED TREATMENT

The clinical use of radiation and heat is most likely to be in fractionated schedules. This may be a problem in view of thermotolerance, as described above. Thermotolerance not only reduces the effect of hyperthermic cytotoxicity: also the effects of simultaneous heat and radiation treatment may be less pronounced if thermotolerance is present. Table 23.1 shows the results of giving C3H mouse mammary tumours a sequence of five daily treatments. When each treatment consisted of simultaneous heat and radiation, a good response was observed (TER = 2.59). When each treatment allowed a gap between radiation and heat the TER fell to 1.34. The kinetics of this response is identical to that seen for thermotolerance developing after heat alone. Thus, for fractionated combined treatment we have

a 'Catch 22 situation': the optimal heat plus radiation interaction requires a long radiation fractionation interval in order to avoid thermotolerance, but that will decrease the radiation effect. Thus we cannot avoid a second-best option: to use a large single or a few widely separated heat treatments, preferably given as a boost simultaneously with a small dose of radiotherapy (see below).

Table 23.1 Fractionated heat and radiation treatment of a C3H mammary carcinoma

Treatment	TCD_{50} (Gy)	TER
Radiation alone	60	
R+R+R+R+R		
Simultaneous		
RH+RH+RH+RH+RH	23	2.59
R+R+R+R+RH	41	1.46
Sequential		
RH+RH+RH+RH+RH	45	1.34
R+R+R+R+RH	46	1.31

Treatment was 42.5°C / 60 min applied sequentially or simultaneously with all or only one of five equal-sized daily radiation fractions.

23.3 Clinical hyperthermia

INDICATIONS FOR HYPERTHERMIA IN RADIOTHERAPY

The indications for hyperthermia in clinical radiotherapy are twofold. First, hyperthermia may be included in the primary treatment of locally advanced tumours where improved local control is expected to result in improved cure rate and survival. Second, hyperthermia has an evident role in palliation, especially of recurrent tumours in previously irradiated areas. Both in the curative and in the palliative situations clinical experience is limited by the difficulties of heat delivery and it remains a major technical problem how to deliver homogeneous heating to a defined tumour volume. This is especially the case for other than superficial tumours.

Patients with a relatively long life expectancy are particularly in need of optimal primary treatment. Hyperthermia should here be considered as an adjuvant to radiation treatment schedules that are already known to be as effective as possible. It is essential that the adjuvant hyperthermia should not sensitize the normal tissues to radiation treatment, for in this situation the radiation dose will already be at tolerance. The treatment principle should therefore be based on optimal radiotherapy using a standard daily fractionation schedule and adding a heat treatment once or twice a week. The argument for this approach has been discussed in detail, especially with regard to the development of thermal tolerance and the interval between the hyperthermic fractions (Overgaard and Nielsen, 1983).

If normal-tissue heating can be avoided, the heat treatment should be given immediately after radiation, as some degree of radiosensitization may add further to the tumour-destructive effect. Otherwise, an interval of approximately 3–4 hours is recommended between radiation and hyperthermia to avoid any sensitization of radiation damage to normal tissues.

CLINICAL STUDIES WITH COMBINED HYPERTHERMIA AND RADIATION

Hyperthermia has been used extensively in the treatment of cancer. The use of heat as an *adjuvant* to radiation in the treatment of local and regional disease is the area where it currently seems to offer the greatest advantages. There is now abundant evidence from clinical phase I and II studies of combined heat and radiation showing that heat may enhance radiation effects on tumours to a significant degree (Figure 23.8; Vernon et al, 1996).

Even though heating techniques are still far from ideal, the multicentre evaluation of hyperthermia in Europe has resulted in a number of randomized clinical trials (Table 23.2). The data suggest that treatment has been most beneficial for chest-wall recurrent breast cancers and malignant melanoma. Pelvic tumours, to which more sophisticated heating techniques can be applied, have also shown benefit but this was not the case with neck nodes or primary breast cancer. The poor results with advanced breast cancers may have been due to insufficient heating. This should be further explored when improved techniques are available, because these results are in disagreement with previous studies, especially in neck nodes where a substantial gain with hyperthermia has been reported (Overgaard and Andersen, 1995).

The overall effect of hyperthermia in the trials reviewed in Table 23.2 is very substantial. This clearly indicates that this modality has a sound biological rationale and when properly applied may result in a substantial therapeutic benefit. One should, however, be aware that the heating technique is still poor and few tumours achieved the prescribed heating level. Further studies should therefore explore better heating techniques.

In addition to the European trials a number of other randomized studies have been completed and although most have involved relatively small numbers of patients they show a consistent trend towards a beneficial effect of adjuvant hyperthermia. Table 23.3 shows the results of a meta-analysis of these studies. Overall there is evidence for a highly significant improvement in tumour response.

PROGNOSTIC PARAMETERS

The effect of heat combined with radiotherapy depends on a number of factors that have been shown to influence the response. These include tumour volume, dose of radiation, the site of the tumour (differing in heating feasibility), histopathology, tumour cell environment and blood flow, primary or recurrent nature of the

Table 23.2 Overview of European randomized multicentre trials evaluating hyperthermia in combination with radiotherapy

Protocol and tumour type	Treatment	No. of tumours	CR rate (%)	Local control (%)	Odds ratio* (95% CL)
ESHO-1/MRC	Rad. alone	67	64	51	1.37
Advanced breast	Rad. + heat	76	71	53	(0.67 – 2.07)
ESHO-2	Rad. alone	32	53	33	0.88
Neck nodes	Rad. + heat	30	50	31	(0.00 – 1.88)
ESHO-3	Rad. alone	65	35**	28**	2.81**
Melanoma	Rad. + heat	63	62**	46**	(2.09 – 3.53)
ESHO-5/MRC	Rad. alone	88	32**	20**	3.43**
Chest wall recurrence	Rad. + heat	117	62**	52**	(2.85 – 4.01)
DDHK/AMC	Rad. alone	70	37**	29**	2.38**
Pelvic tumours	Rad. + heat	77	58**	50**	(1.72 – 3.04)
All trials	Rad. alone	322	42**		2.18***
	Rad. + heat	363	61**		(1.87 – 2.49)

* Odds ratio on complete remission (CR) rate.
** $p < 0.05$, *** $p < 0.00001$.
See Overgaard et al (1995) for sources and further details.

Time–Temperature Relationship ('Thermal Dose')

Long heating time, high temperature and large heated volume are all parameters of good prognostic value. Thermal dose is a means of expressing a biological heat trauma in equivalent minutes at a given temperature (e.g. 43°C), based on a well-defined Arrhenius relationship for a time–temperature isoeffect (Field and Morris, 1983). Although the concept is theoretically attractive, problems of uniformity of heating and thermotolerance make it difficult to apply in practical clinical therapy. Nevertheless, many attempts have been made to establish a generally acceptable thermal isoeffect dose concept. A variety of correlations have been made: with minimal and maximal temperature,

Table 23.3 Meta-analysis of trials of radiotherapy combined with hyperthermia

Tumour site	No. of trials	Total pts.	Local control RT + heat* (%)	Local control RT alone* (%)	Odds ratio (95% CL)
Advanced breast	2	143	68	69	0.99 (0.49–2.01)
Chest wall	4	276	59	38	2.37 (1.46–3.86)
Cervix	3	146	72	50	2.52 (1.27–5.01)
Rectum	2	164	17	5	4.01 (1.26–12.8)
Bladder	1	46	79	41	5.49 (1.49–20.2)
Prostate	1	49	81	79	1.16 (0.28–4.77)
Melanoma	1	128	56	31	2.81 (1.36–5.80)
Head & neck	5	274	51	33	2.08 (1.28–3.39)
Mixed	3	442	39	34	1.24 (0.84–1.82)
All trials	22	1668	51*	37*	1.77 (1.46–2.15)

* $p < 0.00001$
See Overgaard et al (1995) and Overgaard and Andersen (1995) for details.

mean temperature, in a single treatment or all treatments, or with the area of the tumour heated above a certain temperature, etc. These attempts have not yet led to a uniform approach.

23.4 Optimizing the interaction with radiotherapy

It has been supposed that the most serious therapeutic limitation on clinical hyperthermia is that simultaneous radiotherapy and heat treatment will lead to an unacceptable enhancement of normal-tissue damage. This concept should be re-evaluated in the light of work with boost treatments. As an example, a 3-Gy boost given during a 43.5°C heating for 1 hour will yield a therapeutic enhancement ratio in the order of 5, thus resulting in a radiation damage equivalent to a single dose of 15 Gy to the tumour or approximately 30 Gy when given in 2 Gy per fraction (assuming $\alpha/\beta = 10$ Gy). In addition the benefit of hyperthermic cytotoxicity will be maintained. Technically such treatments are possible, for instance by the use of interstitial infra-red radiation or hot-water needles (Horsman and Overgaard, 1992). An illustration of the magnitude of benefit is given in Table 23.4, which shows the thermal enhancement ratios which have been achieved in C3H mouse mammary carcinomas.

23.5 Interaction with chemotherapy

Interaction between hyperthermia and chemotherapy also promises useful clinical gain. The indications for heat and drug therapy are in situations where an increased locoregional drug concentration may be beneficial, such as treatment of locally refractory soft tissue sarcomas. Hyperthermia has

Table 23.4 Effect of sequential or simultaneous boost treatment

Treatment*	TCD_{50} (Gy)	TER
Rad. alone (no boost)	68	–
Rad. + heat	47	1.44
Rad. + (3 Gy–4 h-heat)	39	1.74
Rad. + (3 Gy–sim-heat)	13	5.23

* Heat treatment: 43.5°C for 60 min.
To avoid influence of reoxygenation, the large radiation dose was given under clamped conditions.

been shown to enhance the effect of some chemotherapeutic drugs. Thus, if there exists an indication for locoregional drug therapy which may be improved by an increased 'biological' drug dose, hyperthermia does have potential. It is important to stress that the effect envisaged here is an enhancement of drug damage rather than a heat effect by itself. Such a drug-enhancement effect may be considerably less temperature-dependent than the hyperthermic radiosensitization. The exact temperature level may thus be of less importance and enhancement may occur at relatively low temperatures (*e.g.* 41°C). Homogeneous heating of the tissue in question will therefore not be so critical, and this gives promises for the use of regional hyperthermia combined with systemic drug therapy. It must, however, be stressed again that the clinical indication for such a treatment must exist. This is the case with locally recurrent or refractory soft-tissue sarcomas, as seen in the interesting early results presented by Issels and co-workers (1990). At a time where bone marrow transplantation has become an accepted tool to increase the dose of chemotherapy it is interesting to know that a similar or better effect can be achieved by combining drugs and hyperthermia. It is of course essential that all the tumour cells can be satisfactorily heated and this approach is probably applicable to cases where the disease is a locoregional problem alone. The therapeutic effect will depend on the type of drug and the sequence and interval between the two modalities and the drug should be present at the time of heating in order to achieve optimal enhancement.

> **Key points**
>
> 1. Tissue damage by hyperthermia depends on two main factors: temperature and heating time. Cell killing rises steeply for temperatures above 41°C and is increased by an acidic environment.
> 2. One heat treatment reduces the effect of a second, given up to 48 hours later: the phenomenon of *thermotolerance*.
> 3. In ischaemic tissues the reduced blood flow leads to less cooling and therefore to higher temperatures for a given energy input. This is an important factor in tumour effects.
> 4. Hyperthermia enhances the effects of radiotherapy by two mechanisms: direct cell killing and radiosensitization.
> 5. The interaction between hyperthermia and radiation is very time-dependent. Maximum cell killing is produced by simultaneous treatment.
> 6. Significant clinical benefits from combined hyperthermia and radiotherapy have been reported but until large-scale clinical trials have been made, this should be considered to be an experimental therapy.

BIBLIOGRAPHY

Field SB and Morris CC (1983). The relationship between heating time and temperature: its relevance to clinical hyperthermia. *Radiother Oncol* 1:179–86.

Horsman, MR and Overgaard J (1992). Simultaneous and sequential treatment with radiation and hyperthermia: a comparative assessment. In: *Interstitial Hyperthermia*, pp. 11–33. (Ed) Handl-Zeller L. Springer-Verlag; Wien–New York.

Issels RD, Prenninger SW, Nagele A *et al*. (1990). Ifosfamide plus etoposide combined with regional hyperthermia in patients with locally advanced sarcomas: a phase II study. *J Clin Oncology* 8: 1818–29.

Lindegaard JC and Overgaard J (1987). Factors of importance for the development of the step-down heating effect in a C3H mammary carcinoma *in vivo*. *Int J Hyperthermia* 3:79–91.

Müller C (1912). Therapeutische Erfahrungen an 100 mit Kombination von Röntgenstrahlen und Hochfrequentz, resp. Diathermie behandelten bösartigen Neubildungen. *Medizinische Wochenschrift* 59:1546–9.

Overgaard J (1985). *Hyperthermic Oncology 1984*. Vol. 1–2. Taylor & Francis; London and Philadelphia.

Overgaard J (1989). The current and potential role of hyperthermia in radiotherapy. *Int J Radiat Oncol Biol Phys* 16:535–49.

Overgaard J and Bach Andersen J (1995). Hyperthermia. In: *Textbook of Oncology*, pp. 823–35. (Eds) Peckham M, Pinedo H. and Veronesi U). Oxford University Press; Oxford.

Overgaard J, Gonzalez Gonzalez D, Hulshof MCCM *et al* (1996). Hyperthermia as an adjuvant to radiation therapy of recurrent or metastatic malignant melanoma (ESHO 3–85). A multicenter randomized trial by the European Society for Hyperthermic Oncology. *Int J Hyperthermia* 12: 3–20. (see also *Lancet* 345: 540–3, 1995).

Overgaard J and Nielsen OS (1983). The importance of thermotolerance for the clinical treatment with hyperthermia. *Radiother Oncol* 1:167–78.

Overgaard J, Nielsen OS and Lindegaard JC (1987). Biological basis for rational design of clinical treatment with combined hyperthermia and radiation. In: *An Introduction to the Practical Aspects of Clinical Hyperthermia.* (Eds) Field SB and Franconi C. NATO ASI Series E: Applied Sciences, No 127. Nijhoff; Dordrecht and Boston.

Overgaard J, van der Zee C, Vernon C, Gonzalez Gonzalez D and Arcangeli G (1995). Thermoradiotherapy of malignant tumors. European randomized multicenter trials evaluating the effect of adjuvant hyperthermia in radiotherapy. In: *Progress in Radio-Oncology V.*, pp. 507–13. (Ed) Kegelnik HD. Monduzzi; Bologna.

Schwarz G (1909) Ueber Desensibilisierung gegen Röntgen- und Radiumstrahlen. *Munchener Medizinische Wochenschrift* 24:1–2.

Song CW (1984). Effect of local hyperthermia on blood flow and microenvironment: A review. *Cancer Res* 44:4721s-4730s.

Further Reading

Hand JW, Lagendijk JJW, Bach Andersen J and Bolomey JC (1989). Quality assurance guidelines for ESHO protocols. *Int J Hyperthermia* 5: 421–8, 1989.

Seegenschmiedt MH, Fessenden P Vernon CC (Eds) (1995). *Thermoradiotherapy and Thermoradiochemotherapy.* Vol. 1. Biology, Physiology and Physics. Springer-Verlag; Berlin.

Seegenschmiedt MH, Fessenden P and Vernon CC (Eds) (1996). *Thermoradiotherapy and Thermoradiochemotherapy.* Vol. 2. *Clinical Applications*, Springer-Verlag, Berlin.

Vaupel P and Kallinowski F (1987). Physiological effect of hyperthermia. *Recent Results Cancer Res* 104:71–109.

Vernon CC, Hand JW, Field SB *et al* (1996). Radiotherapy with or without hyperthermia in the treatment of superficial localised breast cancer: results from five randomised controlled trials. International Collaborative Hyperthermia Group. *Int J Radiat Oncol Biol Phys* 35:731–44.

24

Targeted radiotherapy

Thomas E. Wheldon

24.1 Introduction

Targeted radiotherapy means using biological differences between tumour cells and normal tissues to deliver radionuclide atoms selectively, so that the tumour receives a higher radiation dose than normal tissues. The ultimate goal is to sterilize all the tumour cells by delivering to them a sufficiently high radiation dose, without exceeding the tolerance of any critical tissue. As with other forms of radiotherapy, its effectiveness will be determined both by physical and by radiobiological factors. Targeted radiotherapy has some similarities to low dose rate brachytherapy. Both modalities achieve preferential cell kill by placing radionuclides close to tumour cells. In brachytherapy this is done by physical implantation, while with targeted radiotherapy it is the biological properties of the tumour that provide the basis for selective irradiation. In principle, targeted radiotherapy has the advantage that it may be capable of sterilizing metastases or micrometastases anywhere within the body, whether the individual tumours can be visualized or not. In practice, targeted radiotherapy is still at an early stage of development and has a long way to go before it achieves its full potential.

24.2 Targeting agents

The most immediate problem is to identify targeting agents or 'vehicles' that can be used to convey radionuclides selectively to tumour cells. Historically, the first clinical example of the targeted radiotherapy principle in oncology was the use of iodine-131 in the treatment of well-differentiated thyroid carcinomas. Unfortunately, no other tumour types exist for which so simple a targeting agent can be used.

In recent years a major impetus for both laboratory and clinical work on targeted radiotherapy has been the development of monoclonal antibody technology (Vaeth and Meyer, 1990). Monoclonal antibodies hold out the possibility of selectively targeting quite small molecular groups (*i.e.* epitopes) on the surface of cancer cells which, it is hoped, will be sufficiently different from corresponding epitopes on normal cells. Disappointingly, no really dramatic differences between normal and cancer cell surfaces have yet been found. Differences do exist, but these are usually quantitative rather than qualitative, so that monoclonal antibodies directed against the corresponding epitopes often show some degree of cross-reaction with histologically similar normal tissues. It is also recognized that antibodies, which are large molecules, may penetrate rather poorly into some tumours and also that not every cell of a tumour will have equal avidity for the antibody: the tumour heterogeneity problem. Radiolabelled antibodies seem promising for the treatment of only a few tumour types at present; one of these is lymphoma and this application is described below.

There is considerable current interest in other kinds of molecules as potential targeting agents. One of the most promising of these is the catecholamine precursor analogue meta-iodo-benzylguanidine (MIBG), which is preferentially taken up by the catecholamine-synthesizing cells of certain tumour types such as neuroblastoma and phaeochromocytoma. The application of radiolabelled MIBG to the treatment of neuroblastoma is described below. The hormone somatostatin is also attracting attention as a way of targeting some neuroendocrine tumours. These are smaller molecules than antibodies and it is likely that problems of tumour penetration will be less. Problems resulting from normal tissue cross-reaction and tumour heterogeneity of uptake are likely to remain. Many other putative targeting agents are now being studied. These include a variety of growth factors that might be used against tumours whose cells have been found to overexpress the receptor for the appropriate growth factor (Capala and Carlsson, 1991; Mairs *et al*, 1991). Some actual and potential targeting agents are listed in Table 24.1.

24.3 Radionuclides for targeted therapy

Many radionuclides have been proposed or explored for use in targeted therapy but only a few have reached the point of clinical use. Important considerations in choosing a radionuclide include the radiobiological characteristics of the particle emission (*e.g.* its RBE), its energy and mean range, its physical half-life and the ease and feasibility of conjugating it to appropriate targeting agents. Table 24.2 summarizes some properties of radionuclides of current interest. Generally, the radionuclides may be divided into long-range β-emitters, short-range β-emitters and emitters of α-particles and

Table 24.1 Targeting agents for radionuclide therapy

Agent	Appropriate tumour type	Clinical use?
Monoclonal antibodies	Wide variety	Yes
Molecular precursors of tumour-associated proteins (MIBG, somatostatin)	Neuroblastoma, phaeochromocytoma, neuroendocrine tumours	Yes
Growth factors (EGF, NGF, TGF-α or -β)	Tumours overexpressing receptor (*e.g.* glioma, melanoma, neuroblastoma, squamous cell carcinomas)	No

Table 24.2 Properties of radionuclides of current interest in targeted therapy

Radionuclide	Half-life	Emission	Mean particle range
^{90}Y	2.7 days	β	5 mm
^{131}I	8 days	β	0.8 mm
^{67}Cu	2.5 days	β	0.6 mm
^{199}Au	3.1 days	β	0.3 mm
^{211}At	7 hours	α	0.05 mm
^{212}Bi	1 hour	α	0.05 mm
^{125}I	60 days	Auger	~1 μm
^{77}Br	2.4 days	Auger	~1 μm

Auger electrons of still shorter range. Generally, clinical experience to date has been confined to the long-range β-emitters (^{131}I and ^{90}Y). Ultra-short range Auger electrons are capable of direct cell kill only when incorporated into DNA itself. Physical half-lives of a few days, comparable with the biological half-lives of most vehicles, are thought to be ideal. Very short half-lives (as with the α-emitters ^{211}At and ^{212}Bi) may present considerable logistic difficulties. Chemical conjugation to carrier molecules is relatively easy for the halogens (F, Br, I, At) and becomes progressively more difficult with radiometals which tend to detach from the biological carrier molecule. Work is now in progress to find new ways to bind radiometals more firmly. It is possible that these compounds will further extend the list of usable radionuclides.

Long-range β-emitters have the property of irradiating untargeted cells by *cross-fire* from adjacent targeted cells. This may be advantageous when tumour heterogeneity leads to non-uniformity of dose delivery (Section 24.4), but it has the disadvantage that adjacent normal tissues may also be irradiated. An up-to-date review of radionuclides and carrier molecules for therapy has been published by Zweit (1996).

24.4 Radiation dosimetry and microdosimetry

In contrast to external-beam irradiation, for which dosimetry is a precise science, the dosimetry of radiation delivered by radionuclides involves considerable uncertainties. At present, doses both to tumour and normal tissues can only be estimated approximately. Uncertainties arise from the non-uniformity of radionuclide distribution, its changing pattern with time, and its dependence on individual patient pharmacokinetics. The usual approach to dose estimation is to employ the *MIRD Schema* (Howell, 1994), a set of data published by the US Nuclear Medicine Society which provide tables of estimated absorbed doses in various tissues as a function of the radionuclide distribution and its variation with time. Because of the approximate nature of the dose estimates obtained by the MIRD approach, considerable margins of error must be allowed on the tolerances of normal tissues, for which dose rate effects may also have to be considered (see below).

Despite the imprecision of existing dosimetric methods, sufficient information has been gathered to allow approximate estimates of how effective targeted radiotherapy may be expected to be, in terms of tumour cell kill.

Dosimetry and Radiocurability

Vaughan *et al* (1987) collated published data on the uptake of radiolabelled antibodies into human tumours and on the kinetics of antibody clearance. They observed for systemically injected radioactivity that a tumour uptake of 0.0005% per gram of tumour was typical, and that the concentration of bound antibody usually declined exponentially with a half-life of 2–3 days. These authors then used these parameters in a dosimetric model to calculate the doses that could be delivered to tumours by monoclonal antibody-targeted ^{131}I or ^{90}Y. They concluded that if the total whole-body radiation dose had to be limited to around 2 Gy (in the absence of bone marrow rescue) then it would not be possible to deliver a curative dose (say of 60 Gy) to macroscopic solid tumours. To achieve such a dose, better targeting agents would be needed, or targeted therapy would have to be given by some strategy which was more effective in terms of therapeutic differential than simply injecting radiolabelled antibodies into the bloodstream.

Microdosimetry and Radiocurability

A major difference between external-beam and radionuclide-targeted irradiation relates to microdosimetry, especially the absorption of radiation dose in small tumours where dimensions are less than the mean path length of the particles emitted by the radionuclide (see Table 24.2 for particle path lengths). In the case of ^{131}I, the most commonly used radionuclide in clinical practice, microdosimetric problems arise in tumours whose diameter is less than 1–2 mm. If such microtumours have radionuclide atoms uniformly distributed throughout their volume, the absorption of the emitted energy will be relatively inefficient, for a substantial proportion of nuclear particles will escape into the surrounding tissues (Figure 24.1). Microdosimetric calculations by Humm (1986) have provided estimates of the fraction of emitted energy which would be absorbed by a spherical microtumour as a function of its diameter, for several different radionuclides. However, although the smaller microtumours absorb less dose, they also contain fewer clonogenic cells and therefore less dose is needed to sterilize them. These factors result in a relationship between tumour curability and tumour size which is more complex for targeted radiotherapy than for external-beam irradiation. With conventional radiotherapy, it is well

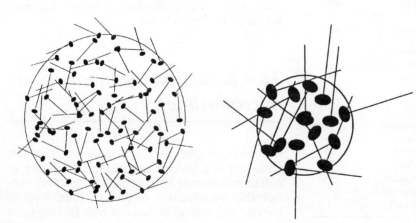

Figure 24.1 The proportion of energy deposited *outside* a radiolabelled metastasis becomes considerable when its diameter is comparable with the mean range of the emitted radiation. On the left is a large lesion, on the right a small one. The particle range is considered constant. Reproduced with permission of Dr J.A. O'Donoghue.

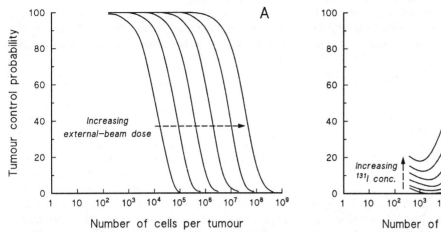

Figure 24.2 Relationship between tumour cure probability and tumour size: for external-beam irradiation at six different dose levels (**A**) and for spherical tumours labelled uniformly with various ^{131}I concentrations (**B**). The tumour size for maximum cure rate is a property of the isotope (Table 24.3) but independent of dose.

known that the larger a tumour, the lower will be the chance of its being sterilized by any given dose of radiation (Section 17.4). This is illustrated in Figure 24.2A which shows how at each of several radiation dose levels, tumour cure probability decreases steadily as a function of increasing tumour size. In contrast, Figure 24.2B shows how tumour cure probability changes with tumour size for several activity levels of ^{131}I. Unlike the external irradiation situation, the family of curves shows a rising component (increasing tumour size leading to more efficient absorption of emitted energy), a peak value and a falling component where the increasing number of cells to be sterilized dominates the response. As a result, for each radionuclide there is an optimal tumour diameter for radiocurability (Table 24.3); both smaller and larger tumours will be less radiocurable than those at the optimal size. The lower radiocurability of very small micrometastases is a unique feature of radionuclide therapy and is one of the reasons why the combination of targeted radiotherapy with external-beam irradiation may be an effective strategy (Wheldon and O'Donoghue, 1990).

24.5 Radiobiological factors in targeted radiotherapy

As with other forms of therapeutic irradiation, the biological effectiveness of targeted radiotherapy will depend on radiobiological factors as well as on the physics of dose absorption. The *5 Rs* of radiobiology (Section 6.9) will be involved to a greater or lesser extent. *Intrinsic radiosensitivity* is certainly an important factor. Because of the limited uptake and specificity of targeting agents now available, only tumour types with high intrinsic sensitivity (*e.g.* neuroblastoma, myeloma, seminoma, leukaemia and lymphoma) will have any prospect of being sterilized by targeted radiotherapy alone. *Repair* assumes importance because of the usually low and exponentially declining dose rate at which most of the treatment will be given. For many tumours, especially those for which the α/β ratio exceeds 10 Gy, the sparing effect of low dose rate irradiation will probably not be great. Fowler (1990) has calculated that for such a tumour an extra 10–20% dose would have to be given, compared with conventionally fractionated external-

Table 24.3 Tumour size for optimal radiocurability by uniformly targeted radionuclides

Radionuclide	Optimal tumour diameter	Equivalent cell number
^{90}Y	~2 cm	~10^{10}
^{131}I	~2 mm	~10^6
^{211}At	~60 μm	~10
^{125}I (Auger component)	~1 μm	~1

beam radiotherapy. However, some tumour types (*e.g.* glioma, melanoma, osteosarcoma) may have lower α/β ratios and for these considerable sparing could occur. Similarly, for normal tissues, the magnitude of the sparing will be modest for acute-responding tissues (typically with high α/β ratios, Section 13.5) but considerably greater for late-responding tissues (with low α/β ratios). Wheldon and O'Donoghue (1990) have calculated that for biological isoeffect in late-responding tissues, the effectiveness of targeted radiotherapy may be reduced several-fold relative to single-dose irradiation at high dose rate.

Other radiobiological factors are probably of lesser importance. *Redistribution* could be of some significance, because of a possible inverse dose rate effect. Theoretically, this could result from a gradual recruitment of radiation-inhibited cells into more sensitive phases of the cell cycle, as the dose rate falls. This mechanism would be expected to apply to rapidly proliferating tumour cells and would probably not occur in late-responding tissues, most of whose cells are out of cycle. Hypoxia and lack of complete *reoxygenation* could be an important cause of resistance, since most of the radiation dose delivered by targeted radiotherapy is usually given within a week, as a result of the biological clearance of the agent and radioactive decay. A week may not be sufficient for complete reoxygenation. However, the clinical efficacy of some forms of brachytherapy (for which the time-scale is similar) and the reduced OER for low dose rate irradiation argue against this being a serious problem. Nevertheless, hypoxic cell radiosensitizers or bioreductive drugs should be explored in conjunction with targeted radiotherapy in laboratory systems.

Finally, it is unlikely that *repopulation* will be a major factor for a treatment scheme that delivers most of the radiation dose over a few days. It is worth noting, however, that tumour cell proliferation will counteract the modest cell kill achieved by radiation at low dose rates. As a result, when the dose rate falls below 2–3 cGy/hour the treatment is likely to be ineffective (Fowler, 1990).

24.6 Strategies for improved clinical results

It is generally agreed that the specificity of existing targeting agents is insufficient to enable sterilization of macroscopic solid tumours. The number of possible targeting agents is growing, however, and protein engineering techniques are continually being used to shape antibody molecules to give them superior properties.

Several strategies have been developed to improve the therapeutic efficacy of targeted radiotherapy. One approach, pioneered by Epenetos and colleagues (Bamias *et al*, 1991), is the use of regionally targeted therapy: radiolabelled antibodies or other agents are injected into a body region or compartment instead of into the general circulation. Obviously, this will not work for widespread cancer but it might be effective in controlling more localized spread. Regional targeting is now being explored clinically for the treatment of carcinoma of the ovary (Stewart *et al*, 1989) and also of tumour cells within the cerebrospinal fluid (Lashford *et al*, 1988), and it may be used in other tumour sites.

Fractionation is also being explored. There is no strong radiobiological case for this, since sparing would not be expected in the fractionation of low dose rate irradiation (Figure 18.6C). However, there are some non-radiobiological arguments, such as the possibility of improving tumour uptake of the agent if all cellular targets had been saturated during the first treatment. Fractionated treatment using antibodies sometimes leads to the development of an immunological response to antibodies that are usually of murine origin. Attempts are being made to modify mouse antibodies to make them less antigenic in humans, or to replace portions of the antibody molecule with a human equivalent (*humanization* of antibodies) to overcome the problem of human anti-mouse antibody generation by the host immune system.

There are also theoretical arguments in favour of two-step targeting strategies, to improve the kinetics of the targeting reaction. For example, unlabelled antibodies conjugated to avidin might be injected and allowed to concentrate in the tumour (perhaps over 12–48 hours) before injecting a radiolabelled small molecule (such as biotin) which rapidly targets the avidin component of the conjugate. Two-step targeting strategies could be especially useful with radionuclides with short half-lives, for instance α-emitters such as ^{211}At. Another strategy is to seek to reduce normal-tissue reactions to the targeted therapy, thus to allow higher doses to be given. At present, myelosuppression due to irradiation of bone marrow is the most important side-effect of targeted radiotherapy.

This could be reduced, in selected patients, by the use of bone marrow or stem cell rescue or, perhaps more widely, by administration of haemopoietic growth factors. Bone marrow rescue is an important component of the strategy of combined modality therapy described below.

24.7 Microdosimetric rationales for combined modality treatment

The future of targeted radiotherapy may be in combination with some other treatment modality. Since targeted radiotherapy is ineffective in treating bulky disease, there is a clear case for combining it with local radiotherapy to sites of known disease, the approach described as *spatial co-operation* (Section 20.2). Combination with chemotherapy might also exploit non-overlapping toxicities. It was suggested some years ago on the basis of macroscopic dosimetry that the combination of targeted radiotherapy with external-beam irradiation may be therapeutically advantageous (Wessels and Rogus, 1984). Combination treatments are already being employed clinically, mostly on an *ad hoc* basis (Order, 1990). However, the development of microdosimetric modelling has provided new rationales for combination therapy, which now make this seem almost obligatory if cure of disseminated malignant disease is to be achieved.

There are two distinct microdosimetric rationales, one based on the inefficient dose absorption by uniformly targeted microtumours of dimension less than the emitted particle mean range, the other derived from considerations of tumour heterogeneity and consequent non-uniformity of dose distribution within the tumour. Fortunately, these rationales lead to the same conclusion.

As indicated in Table 24.3, the most efficient absorption of radiation from ^{131}I occurs in tumours whose diameter is about 2 mm, which are therefore the most radiocurable by this isotope. Wheldon *et al* (1992) have used these microdosimetric findings in exploring theoretically the clinical implications for treatment of radiosensitive tumours like neuroblastoma or lymphoma. First, note that for local external-beam irradiation it is necessary that a tumour be large enough to be imaged and localized. Generally, this means that the tumour must be at least a few millimetres in diameter. Total-body irradiation (TBI) in conjunction with bone marrow rescue can be given in 2-Gy fractions up to a total dose of perhaps 14 Gy. This is sufficient to sterilize single cells and small micrometastases of radiosensitive tumours but it is insufficient to treat micrometastases approaching millimetre dimensions. Similar conclusions hold when chemotherapy rather than TBI is used to provide systemic treatment for disseminated micrometastases. Therefore it seems that microtumours which lie just below the threshold of detectability (1 mm to 1 cm diameter) are especially likely to provide foci of recurrence following systemic treatment. This size range corresponds to optimal curability for ^{131}I which might therefore be a useful component of the treatment. It would not be ideal as a single modality treatment, however, because of inefficient energy absorption by smaller micrometastases. These considerations thus predict benefit from combining external-beam TBI and targeted ^{131}I in treating disseminated malignant disease. Larger tumours should be treated using localized modalities such as high-dose external-beam irradiation or surgery.

The second microdosimetric rationale derives from the likelihood that dose to cellular targets will be non-uniform, as a result of limited diffusion of targeting agent or intra-tumour gradients due to local absorption of the targeting molecule if not present in excess. In all these circumstances, dose non-uniformity is highly likely, and has been observed in practice (Yorke *et al*, 1987; Griffith *et al*, 1988; Moyes *et al*, 1989). O'Donoghue (1991) has explored theoretically the implications of heterogeneity in targeted radiotherapy. He assumed that a small proportion of cells receive no dose from targeted radiotherapy, a reasonable assumption for targeting by short-range emitters but an approximation for long-range β-emitters where untargeted cells receive some dose from adjacent targeted cells. This led to an important general principle. If a small proportion of cells (say 1 in 10^4) receive no dose from targeting, then it is not useful to apply targeted therapy at treatment intensities that give more than 4 logs of tumour cell kill (*i.e.* a survival of below 10^{-4}). Otherwise, *overkill* of targeted cells will result, with no advantage in the killing of untargeted cells, yet with an increase in normal-tissue toxicity. This optimal level of targeted treatment would be better combined with sufficiently high levels of some other treatment modality (such as external-beam irradiation or chemotherapy) to sterilize the cells that have escaped the targeting treatment. In a series of model calculations on targeted ^{131}I, TBI and bone marrow rescue, O'Donoghue (1991) showed that the TBI component is obligatory for cure and often forms a high proportion of the total dose given to normal tissues.

These microdosimetric rationales, together with earlier arguments by Wessels and Rogus (1984), make a strong case for the incorporation of targeted radiotherapy as a component of combined modality treatment. Clinical studies are now under way to determine whether these theoretical predictions are borne out in practice.

24.8 Radiolabelled MIBG therapy of neuroblastoma

Neuroblastoma is a paediatric tumour of the sympathetic nervous system which is usually disseminated at presentation and has a generally poor prognosis. Neuroblastoma cells often retain the catecholamine-synthesizing property of sympathetic nerve cells which allows them to be targeted using the precursor molecule MIBG. It is important that tumour-selective MIBG uptake requires active transport and occurs only in metabolically competent cells. Clinical studies using single-agent ^{131}I-MIBG have achieved promising responses (Lashford et al, 1992), especially when this treatment is given at first presentation (de Kraker et al, 1995). The micro-dosimetric rationales above suggest that ^{131}I-MIBG should preferably be given as part of a combined modality regimen. An extension of these rationales implies that ^{131}I-MIBG should be the first treatment modality given; this is because uptake will be maximal in undamaged cells and the cross-fire effect will help to nullify the consequences of an untargeted subpopulation (Gaze and Wheldon, 1996). New clinical studies organized by the UK Children's Cancer Study Group (UKCCSG) conform to these requirements. Initial ^{131}I-MIBG therapy in selected poor-prognosis patients will be followed by multiagent chemotherapy. Dose escalation will proceed until stem cell rescue is required. If the first studies prove sufficiently promising, it is intended that this treatment strategy will eventually be used for the majority of neuroblastoma patients.

24.9 Radiolabelled antibody therapy of lymphoma

Lymphoma is a radiosensitive neoplastic disease which, like neuroblastoma, usually presents when already disseminated. This is one of the few tumour types which can be treated successfully using radiolabelled antibodies. A high-dose strategy, pioneered in Seattle, involves the administration of large activities of an ^{131}I-labelled monoclonal antibody which has moderate selectivity for B-lineage lymphocytes or lymphoma cells. The doses given are ablative and bone marrow rescue is required. This approach has given impressive results in treating B-cell lymphoma patients in terms of long-duration remissions and possible cures (Press et al, 1993). However, the high activities of ^{131}I present logistic difficulties in relation to radiation protection. Theoretically, a combined modality strategy with reduced ^{131}I activity, but including TBI or chemotherapy, might be advantageous. If the Seattle approach can be optimized, the treatment of B-cell lymphomas might become one of the most important clinical applications of targeted radiotherapy.

24.10 Boron neutron capture therapy (BNCT)

This is a specialized approach to targeted therapy that has been under development for many years; it is nevertheless one that holds considerable promise (Barth et al, 1990). Its title is somewhat misleading in that this is not a form of neutron therapy but a way of generating α-particles within the tumour site; it is this secondary radiation that is responsible for any tumoricidal effect. The approach relies on the physical properties of the non-radioactive boron isotope ^{10}B. When exposed to 'slow' neutrons (i.e. with thermal or epithermal energies) this isotope has a large cross-section for neutron capture. A fission reaction ensues in which the boron nucleus splits into ^{7}Li plus ^{4}He, also emitting a pulse of γ-rays. The ^{4}He nucleus is an α-particle which produces a densely ionizing track over a range of a few cell diameters. In the absence of boron, fission

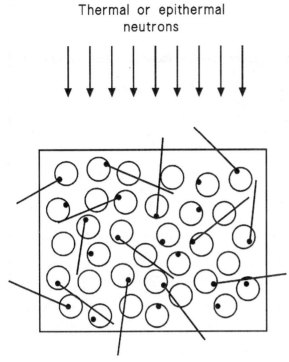

Figure 24.3 The local production within a small volume of tissue of isotropic alpha-particle radiation by neutron-induced fission of boron atoms. (○ cells; ● ^{10}B atoms; — α-tracks).

neutrons at the intensities required for this process produce little biological damage.

BNCT is a two-step strategy: targeting of ^{10}B to tumour cells, followed by fission of the isotope by neutron irradiation (Figure 24.3). The use of an external neutron beam and the fact that ^{10}B itself is a 'cold' isotope gives the therapist more control over the process than is usual in targeted therapy. Boron may accumulate in organ other than in the tumour sites (*e.g.* the liver), but if these sites are known it may be possible to avoid irradiating them with neutrons. The effectiveness of this treatment requires sufficient difference in uptake between the tumour and *local* normal tissues. α-particle radiation is very damaging to targeted cells and even the most resistant cell types will be efficiently sterilized. Research effort is being directed at the production of filtered beams of epithermal neutrons which have superior penetration properties compared with thermal neutrons. In Europe, this effort is concentrated at the Nuclear Reactor Facility at Petten in the Netherlands with the initial objective of treating intracranial disease.

As with any form of targeted therapy, success depends critically on the availability of a suitable targeting agent, in this case one that will carry boron into tumour cells to a higher concentration than in local normal tissues. The search for such compounds is under way (Barth *et al*, 1990; Nguyen *et al*, 1993). Monoclonal antibodies are an obvious candidate, but whether these are capable of carrying sufficient boron is yet to be ascertained. The use of boron-conjugated EGF to target tumours that overexpress the EGF receptor is also being considered (Andersson *et al*, 1992). Oligonucleotides conjugated to boron are further potential candidates for exploiting specific abnormal DNA sequences in malignant tumours and these may have the advantage of requiring less boron incorporation for cell killing by BNCT.

Although unique in some respects, BNCT may well be limited by factors that are common to all radioactive targeting methods. Its therapeutic effectiveness will depend on the selectivity of uptake of boron into the tumour. The α-particles generated by BNCT have a short range and the level of cell killing will depend on the homogeneity of boron uptake among tumour cells. For this reason, the combination of BNCT and external-beam treatment, as described in Section 24.7, may be necessary.

Key points

1. Targeted radiotherapy is the irradiation of tumours by means of radionuclides that are selectively delivered by tumour-seeking molecules (*i.e.* by *targeting agents*).
2. Targeting agents may be monoclonal antibodies, molecular precursors of tumour-associated proteins, hormones or growth factors. Many radionuclides are under consideration for targeting but only the β-emitters ^{131}I and ^{90}Y have been used clinically to any extent.
3. The most important radiobiological factors governing the success of targeted radiotherapy will be intrinsic radiosensitivity of the tumour cells and the repair capacity of cells in tumours and in critical normal tissues.
4. The major problems of targeted radiotherapy are insufficient selectivity of tumour uptake, limited penetration of the targeting agent into solid tumours and heterogeneous expression of cellular targets leading to non-uniformity of absorbed dose. When long-range β-emitting radionuclides are used, microtumours whose dimensions are less than the β-particle mean range will be spared due to low absorption of the radiation energy. Some killing of untargeted cells will however result from *cross-fire* from adjacent targeted cells.
5. Clinical strategies to overcome these problems include the use of targeting agents of low molecular weight, regional targeting, two-stage targeting procedures and the combination of targeted radiotherapy with other treatments.
6. Radiobiological modelling studies suggest that the combination of targeted radiotherapy with external-beam irradiation (local and systemic) may be the optimal strategy for those targeting agents which are presently available.
7. Boron neutron capture therapy is an aspect of targeting in which non-radioactive boron atoms selectively incorporated into tumour cells are induced to emit α-particles locally within the tumour.

Bibliography

Andersson A, Andersson J, Burgman JO *et al* (1992). Program for BNCT with accelerator keV neutrons and related chemical and biological studies. In: *Progress in Neutron Capture Therapy for Cancer*, pp. 41–52. (Eds) Allen BJ, Moore DE and Harrington BV. Plenum Press; New York.

Bamias A, Keane P, Krausz T, Williams G and Epenetos AA (1991). Intravesical administration of radiolabeled antitumour monoclonal antibody in bladder carcinoma. *Cancer Res* 51(2): 724–8.

Capala J and Carlsson J (1991). Influence of chloroquine and lidocaine on retention and cytotoxic effects of ^{131}I-EGF: studies on cultured glioma cells. *Int J Radiat Biol* 60: 497–510.

de Kraker J, Hoefnagel, CA, Caron H *et al* (1995). First line targeted radiotherapy, a new concept in the treatment of advanced stage neuroblastoma.*Eur J Cancer* 31A: 600–2.

Gaze MN and Wheldon TE (1996). Radiolabelled MIBG in the treatment of neuroblastoma. *Eur J Cancer* 32A: 93–6.

Griffith MH, Yorke ED and Wessels BW (1988). Direct dose confirmation of quantitative autoradiography with micro-TLD measurements for radioimmunotherapy. *J Nuc Med* 29: 1795–1809.

Howell RW (1994). The MIRD Schema. From organ to cellular dimensions. *J Nucl Med* 35(3): 531–3.

Lashford LS, Davies AG, Richardson RB *et al* (1988). A pilot study of ^{131}I monoclonal antibodies in the therapy of leptomeningeal tumours. *Cancer* 61:857–68.

Lashford LS, Lewis, IJ, Fielding SL *et al* (1992). Phase I/II study of 131-iodine metaiodobenzylguanidine in chemoresistant neuroblastoma: a United Kingdom Children's Cancer Study Group investigation. *J Clin Oncol* 10: 1889–96.

Mairs RJ, Angerson W, Gaze MN *et al* (1991). The distribution of alternative agents for targeted radiotherapy within human neuroblastoma spheroids. *Br J Cancer* 63: 404–9.

Moyes JSE, Babich JW, Carter R *et al* (1989). A quantitative study of meta-iodo-benzyl-guanidine (mIBG) in children with neuroblastoma: correlation with tumour histopathology. *J Nuc Med* 30: 474–80.

Nguyen T, Brownwell GL, Holden SA *et al* (1993). Subcellular distribution of various boron compounds and implications for their efficacy in boron neutron capture therapy by Monte Carlo simulations. *Radiat Res* 133: 33–40.

O'Donoghue JA (1991). Optimal scheduling of biologically targeted radiotherapy and total body irradiation for the treatment of systemic malignant disease. *Int J Radiat Oncol Biol Phys* 21: 1587–94.

Press OW, Eary JF, Appelbaum FR *et al* (1993). Radiolabeled-antibody therapy of B-cell lymphoma with autologous bone marrow support. *N Engl J Med* 329(17): 1219–24.

Stewart JSW, Hird V, Snook D *et al* (1989). Intraperitoneal radioimmunotherapy for ovarian cancer: pharmacokinetics, toxicity and efficacy of ^{131}I labelled monoclonal antibodies. *Int J Radiat Oncol Biol Phys* 16: 405–13.

Wessels BW and Rogus KD (1984). Radionuclide selection and model absorbed dose calculations for radiolabelled tumour associated antibodies. *Med Phys* 11: 638–44.

Wheldon TE and O'Donoghue JA (1990). The radiobiology of targeted radiotherapy. *Int J Radiat Biol* 58: 1–21.

Yorke ED, Griffith ML and Wessels BW (1987). Quantitative autoradiography and micro-TLD measurements in radioimmunotherapy. *J Nuc Med* 28: 617.

Further Reading

Barth RF, Soloway AH and Fairchild RG (1990). Boron neutron therapy of cancer. *Cancer Res* 50: 1061–70.

Fowler JF (1990). Radiobiological aspects of low dose rates in radioimmunotherapy. *Int J Radiat Oncol Biol Phys* 18: 1261–9.

Humm JL (1986). Dosimetric aspects of radiolabelled antibodies for tumour therapy. *J Nuc Med* 27: 1490–7.

Order SE (1990). Perspectives in radioimmunoglobulin therapy. In: *The Present and Future Role of Monoclonal Antibodies in the Management of Cancer*. (Eds) Vaeth JM and Meyers JL. *Frontiers of Radiation Therapy and Oncology, Vol 24*. Karger; Basel.

Vaeth JM and Meyer JL (Eds) (1990). *The Present and Future Role of Monoclonal Antibodies in the Management of Cancer. Frontiers of Radiation Therapy, Vol 24*. Karger; Basel.

Vaughan ATM, Anderson P, Dykes P and Bradwell AR (1987). Limitations to killing of tumours using radiolabelled antibodies. *Br J Radiol* 60: 567–78.

Wheldon TE, Amin AE, O'Donoghue JA and Barrett A (1992). Radiocurability of disseminated malignant disease by external beam irradiation and targeted radionuclide therapy. In: *Advances in the Applications of Monoclonal Antibodies in Clinical Oncology*, 8th International Hammersmith Meeting. (Ed) Epenetos AE. John Wiley; London

Zweit J (1996). Radionuclides and carrier molecules for therapy. *Phys Med Biol* 41: 1905–14.

25

Individualization of radiotherapy

Adrian C. Begg

25.1 Introduction

The most important goal of a predictive assay is to obtain information that can be used to choose a treatment protocol, so that each individual patient will receive optimal treatment. At the present time, the treatment plan is usually based on parameters such as tumour site, histological type, tumour stage and performance status. Within these broad categories, some tumours show a greater response to radiotherapy than others. If these could be identified before treatment, alternative therapies might be selected which may give a better chance of cure than the standard conventional therapy. The choices may include chemotherapy or surgery in preference to radiotherapy, or non-standard radiotherapy schedules. Predictive assays should be simple, quick and reliable tests for parameters that will predict response to a particular treatment.

It is also widely believed that individual cancer patients differ in their tolerance of radiation therapy. Among a group of patients given the same treatment protocol, some suffer more severe normal-tissue reactions than others. It is these severe reactors that limit the dose of radiation that can be prescribed to that group of patients and if they could be identified prior to therapy then it might be possible to improve their management (*e.g.* by reducing their dose) as well as that of the rest of the patient group (*e.g.* by increasing their dose). This chapter will deal with these two aspects of the individualization of radiation therapy.

25.2 Assay types

A wide variety of assays are currently being explored for predicting either the natural history of individual tumours or their response to therapy. Some of these are listed in Table 25.1. Cytogenetic changes and oncogene expression allow subgroups of patients to be identified who potentially may have a different prognosis. These aspects may also provide an indication of treatment response; *ras* oncogene expression and mutations in the *p53* tumour suppressor gene, for example, have been

Table 25.1 Potential predictors of natural history and treatment response

Parameter	Predictor of	
	Natural history	Treatment response
Proliferation cell kinetics	+	+
Ploidy/cytogenetics	+	?
Surface receptors/antigens	+	?
Proto-oncogene expression	+	?
NK-cell activity	+	−
Intrinsic cell sensitivity	−	+
pO_2/pH	−	+
O_2-and heat-regulated proteins	−	+
Endogenous thiols	−	+
p-glycoprotein	−	+
DNA-drug adducts	−	+
DNA repair enzymes	−	+
Tumour regression/necrosis	?	+
Host cell infiltrates	?	+
Magnetic resonance imaging or spectroscopy:		
^{32}P	?	+
^{1}H	?	?
Positron-emission tomography	?	?

associated with both drug and radiation resistance in some cell lines. Other assays are purely designed to indicate treatment response; for instance, measurement of DNA-drug adducts in tumour cells after chemotherapy can be an indicator of the likelihood of response to treatment (more adducts = greater response). Some assays are direct measures of cell killing (*e.g.* colony assays) and are applicable to most forms of treatment, while others give an indirect measure of sensitivity, for example tissue oxygen tension (pO_2) and thiol content for radiation response, DNA adducts and *p*-glycoprotein for drug response. Direct methods are preferable to indirect ones, and measurements directly on biopsy material are preferable to those in which the tumour cells must be treated *in vitro* with radiation or drug, since the response may be different from that *in vivo*. This is not always possible, and *in vitro* treatments can give useful predictive information, as will be illustrated below.

The ideal predictive assay should have the following characteristics:

- Harmless
- Quick and cheap
- Few false positives or negatives
- Independent predictive power

Within the range of predictive tests that may be of value to clinical radiotherapy there are three that stand out at the present time: the prediction of intrinsic radiosensitivity, of tumour hypoxia, and of repopulation rate during radiotherapy.

25.3 Intrinsic sensitivity to drug or radiation treatment

RADIOSENSITIVITY

Intrinsic radiosensitivity of tumour cells can be measured by colony formation after irradiation *in vitro* (Section 6.4). For these studies, a cell suspension from the biopsy must first be made and the cells grown in dishes or culture tubes. One widely used measure of radiosensitivity is the surviving fraction after 2 Gy (SF_2, Section 17.5). This dose is low enough to indicate the initial slope of the survival curve, although it is not identical with it, especially in radioresistant cell lines. It is also a commonly used dose per fraction in clinical radiotherapy. Although there are many reasons why *in vivo* and *in vitro* responses may differ (*e.g.* the presence or absence of hypoxic cells, or cell-contact effects), some studies have shown a correlation between the *in vitro* radiosensitivity of human tumour cell lines and their clinical response (see below).

Assays for cell survival can take up to 4 weeks to complete and it is normally inappropriate to delay the start of radiotherapy for this length of time. There is therefore a need for more rapid assays of cellular sensitivity. One approach is to measure growth inhibition of tumour cells plated out *in vitro*: the so-called growth assays (Section 6.4). In this type of assay, the change in total cell number in dishes given different treatment doses is measured; the higher the dose the fewer cells are present after a given growth period. This is similar to a growth delay experiment for tumours *in vivo*. A variety of methods have been used to rapidly determine the number of cells in the cultures at the end of the growth period, including cell counting with the aid of a vital stain for viable cells (the dye-exclusion method) and automated colorimetric assays using the tetrazolium (MTT) method (Carmichael *et al*, 1987; Mitchell, 1988) or its variants.

An alternative approach is to measure some parameter of radiation-induced DNA damage as an indicator of radiosensitivity. This is still an area of active laboratory research and it is not yet clear whether a clinically useful assay can be found. The methods described in Section 8.3 are being employed, especially pulsed-field gel electrophoresis (PFGE) and the comet assay. These can be used to measure either the initial level of induced DNA double-strand breaks or the speed and extent of strand-break rejoining. The level of initial damage shows a tendency to be higher in the more radiosensitive cell lines that have been studied, but there are significant exceptions to this. Some investigators have reported a better correlation of cell kill with the level residual double-strand breaks after repair, while others have reported that radioresistant cells show faster rejoining of double-strand breaks (Schwartz *et al*, 1988). It is clear that strand breaks, particularly residual double-strand breaks, lead to cell kill, but this has not yet provided a practicable and reliable predictive assay for routine use.

The measurement of chromosomal damage is another possible approach. Conventional techniques of counting radiation-induced chromosome aberrations in metaphase, the scoring of micronuclei, and newer techniques including fluorescence in situ hybridization (FISH) and premature chromosome condensation (PCC) are being tested (Section 8.4). There are encouraging reports of correlations with radiosensitivity but these methods have yet to be applied and validated in the clinic. The biggest problem at present is to obtain sufficient numbers of metaphases from small tumour biopsies in the short time required for a predictive assay (*i.e.* less than 2 weeks). Insufficient cell numbers mean bad statistics and consequently low resolving power for distinguishing between different radiosensitivities.

Radiosensitivity Genes

The rapid growth in knowledge of oncogenes and tumour suppressor genes, and techniques for rapidly assessing gene expression and mutations (Chapter 9), have stimulated an increasing number of reports correlating these molecular parameters with prognosis. In addition, in the area of chemotherapy, expression of genes known to be involved in drug resistance, *e.g.* the multidrug resistance *mdr* genes, can be measured and correlated with treatment outcome. This raises the question of whether it would be possible, now or in the future, to screen for expression of, or mutations in, genes known to be involved in radioresistance. This would offer the possibility of a rapid test of radiosensitivity in individual tumours which could easily be carried out in the time between diagnosis and the start of treatment.

Problems with this approach arise from the multiplicity of factors known to influence cell survival after irradiation. Complementation analysis with radiosensitive mutant cell lines shows that there are at least ten genes involved (Jeggo *et al*, 1994). As more studies are done, this could increase by several fold. Processes involved in the handling of radiation damage include recombination, excision repair, post-replication repair, replication, apoptosis, and cell cycle checkpoints, each of which is no doubt controlled by a number of genes. Some of these processes may dominate but no consistent picture has so far emerged. Recently, at least five genes influencing ionizing radiation sensitivity have been cloned. The *Ku70*, *Ku80* and *SCID* gene products together form the DNA-dependent protein kinase complex and all are important in strand break repair. In addition, *XRCC4*, a gene also critical for double-strand break repair (Li *et al*, 1995), and the ataxia telangiectasia (*ATM*) gene (Savitsky *et al*, 1995) have recently been cloned. Mutation in any of these genes leads to an increase in radiosensitivity. This opens the possibility in the future of screening for mutations in these genes, together with cell cycle checkpoint and apoptosis genes, to help predict radiosensitivity, but this is still some way off at present.

Chemosensitivity

Similar techniques can in principle be used to measure cell killing by drugs. Figure 25.1 shows results of a cell population growth assay to assess sensitivity of two human tumours to four different drugs. The tumour cells were primary explants, *i.e.* grown direct from biopsy material and tested immediately after removal from the patient. Data on two patients are compared and their responses to each of these drugs differed considerably. This illustrates the fact that differences among tumour cell lines to a given cytotoxic drug are usually larger than is the case with radiation. For a given cell line there are also wide differences in the effectiveness of various chemotherapeutic agents (Steel *et al*, 1983). Perhaps as a result of their more specific biochemical modes of action, drugs are more selective in their cell killing effects. For this reason, prediction of tumour response should be easier to achieve than is the case with radiation. A number of assays have been developed for this purpose but none has yet come into general use. The difficulties in achieving a successful assay are illustrated by the 'human tumour stem-cell assay' (Salmon *et al*, 1978). A suspension of tumour cells was exposed to a range of drug concentrations, after which the cells were grown in tissue culture medium thickened with agar (to keep the colonies compact and to inhibit growth of fibroblasts). After an appropriate culture period, colonies of more than 40 cells were counted. Although widely used at the time, this assay has not persisted, in part because clumps of cells in the original tumour cell suspensions were sometimes as frequent as the colonies that grew in this system (Selby *et al*, 1983).

A general feature of chemosensitivity assays so far described has been that whilst they are variably successful in predicting sensitivity they have a high success rate in predicting drug resistance (KSST, 1981; Bellamy, 1992). In principle, this should be a significant benefit to cancer chemotherapy since it will allow patients with resistant disease to be spared the toxicity of ineffective treatment. This potential advance has not, however, been taken up,

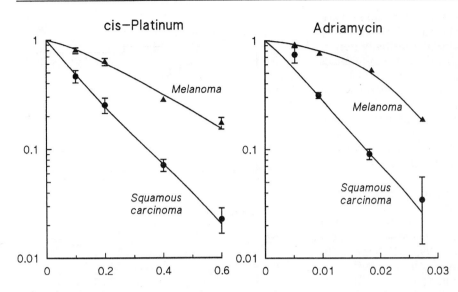

Figure 25.1 Survival curves for melanoma cells (triangles) and cells from a squamous carcinoma of the lung (circles) exposed continuously for 5 days to each of four chemotherapeutic agents and assayed in primary culture using the 'adhesive tumour cell culture' system. The abscissa is drug concentration in mg/ml. From Baker *et al* (1986), with permission.

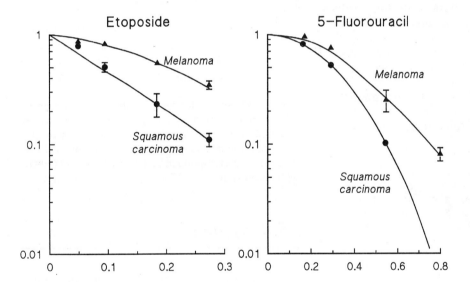

presumably because it is often the view of medical oncologists that, even if the likelihood of response is low, the patient should be given the chance of benefiting from chemotherapy.

Cell survival or cell growth assays can also be used to examine the individuality of response to other treatments such as hyperthermia or high-LET radiotherapy. Their main disadvantage remains that the treatment takes place in a culture dish and not in the patient.

CORRELATION WITH OUTCOME

Intrinsic radiosensitivity data for human tumours using a cell population growth assay are shown in Figure 25.2 (Brock *et al*, 1987). Cell suspensions made from tumour biopsies were irradiated *in vitro* and assayed for survival by cell population growth on specially prepared plates. The cumulative plots of the surviving fraction at 2 Gy were obtained from complete *in vitro* survival curves. The slope of each line indicates the extent of inter-tumour variation (flatter = more variable) and the lateral position represents the overall radiosensitivity (left = more radiosensitive). Significant differences between tumour types can be seen which are broadly consistent with clinical experience of radioresponsiveness of these tumours. In a series of head and neck tumours, a trend was observed that tumours eventually recurring were more radioresistant, although this was not significant, probably due to the addition of surgery in this patient group.

A similar type of study is shown in Figure 25.3. The *in vitro* radiosensitivity of cervical tumours was measured with a soft-agar colony assay and compared with the clinical response to radical radiotherapy (West *et al*, 1993). Patients with local recurrence were found to have significantly more radioresistant tumours (panel A). The survival of patients with low SF_2 values was also significantly better than those with high values (panel B). This illustrates the predictive potential of *in vitro* radiosensitivity measurements. The radiotherapy in this study was given over a relatively short time (3–4 weeks), thus minimizing the influence of proliferation rate differences between tumours (see below). The predictive power of SF_2 was observed with all tumour stages and was independent of tumour grade (West *et al*, 1995a).

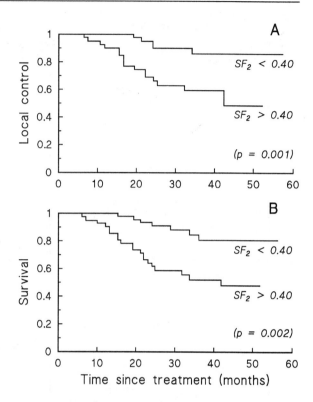

Figure 25.3 SF_2 values as a predictor of outcome of radiotherapy in carcinoma of the cervix. Both survival and local control are significantly correlated with SF_2, such that patients with radiosensitive tumour cells *in vitro* ($SF_2 < 0.4$) responded best. From West *et al* (1993), with permission.

25.4 Hypoxia

The evidence that tumour hypoxia can reduce radiotherapy cure rates in some sites, particularly in head and neck cancer, is set out in Chapter 16 where the need to identify patients most in need of 'anti-hypoxia' therapy is pointed out. The development of new and more reliable methods to measure hypoxia have made this a more feasible option.

METHODS TO MEASURE HYPOXIA

Many methods to detect tumour hypoxia have been described (Chapter 15), including the use of fine glass oxygen electrodes, autoradiographic detection of radiolabelled misonidazole or ana-

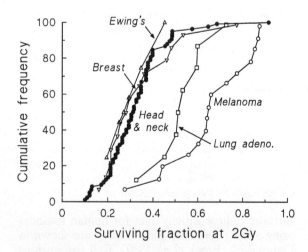

Figure 25.2 Cumulative frequency curves for SF_2 values in five tumour types, measured by a population growth assay. Differences in average radiosensitivity can be seen, with melanoma being the most resistant. From Brock *et al* (1987), with permission.

logues which selectively bind to hypoxic cells, the detection of fluorine-labelled misonidazole or its derivatives using positron-emission tomography (PET), and the detection of ^{123}I-labelled nitroimidazoles by external scanning. Recent studies have shown that nitroimidazole compounds, after reduction and binding in hypoxic cells, can also be detected immunologically, opening up the possibility of detection by both immunocytochemistry and flow cytometry. Most of these studies have been carried out in animals, although they are now being tested in the clinic. No large-scale studies have yet been done. Their disadvantage is the need to administer a potentially toxic drug to the patient. Other methods involve measuring vascular parameters, such as blood-vessel density or intercapillary distance. The advantage of these is that they can be done on routine histological sections of biopsies and are amenable to image analysis for rapid and objective scoring. They are indirect measures of hypoxia, however, and can be confounded by such factors as variations in oxygen consumption and haemoglobin saturation, and the existence of non-functional blood-vessels.

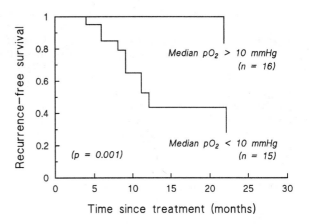

Figure 25.4 Correlation of intra-tumoral oxygen tension (pO$_2$) with outcome after radiotherapy in cervical cancer. Tumours with high median pO$_2$ values, *i.e.* less hypoxic, responded significantly better than the more hypoxic tumours. Data of Höckel *et al* (1993), with permission.

CORRELATION WITH OUTCOME

A few studies have been done with oxygen microelectrodes in which tumour hypoxia has been measured prior to treatment and correlated with outcome after radiotherapy. The results of a study on tumour-involved lymph nodes are shown in Figure 15.8, showing that patients with more hypoxic tumours at the start of treatment did worse. Similar results were reported by Höckel *et al* (1993, 1994) on primary cervix tumours (Figure 25.4). Hypoxia may not be a problem in all tumours, however, and ways to overcome hypoxia, *e.g.* chemical hypoxic cell sensitizers, carbogen, hypoxic cytotoxins or high-LET radiation, should only be given to patients where hypoxic tumour cells can be demonstrated to present a therapeutic problem.

25.5 Tumour cell repopulation

As described in Section 14.3, tumour cell repopulation during fractionated radiotherapy can reduce the probability of local tumour control. Evidence for the importance of repopulation has come from two main sources: the analysis of clinical data and measurements of the kinetics of cell proliferation in human tumour biopsies.

The clinical evidence has come from the analysis of data on the correlation between local control and overall treatment time. The review by Withers *et al* (1988) gave the first clear indication that clinical data can yield estimates of the doubling time for tumour cell repopulation (Figure 14.4A). As indicated in Section 14.3, they summarized published estimates of the tumour control dose (TCD$_{50}$) for head and neck tumours and plotted these against the duration of radiotherapy treatment. After an initial lag, they observed a positive correlation with a slope that was equivalent to a required dose increase of around 0.6 Gy/day for isoeffect. Based on assumptions about the radiosensitivity of the cells, Withers *et al* deduced a clonogen doubling time for head and neck cancers of about 4 days. The subsequent analysis of these data by Bentzen and Thames is shown in Figure 14.4B. Similar analyses have been made for other tumour types and some of these are listed in Table 25.2. If assumptions are made about tumour cell radiosensitivity, it is possible to calculate effective doubling times during radiotherapy; these values are remarkably short (3–8 days) compared with pre-treatment volume doubling times of perhaps 3–15 weeks (Section 2.4).

240 Individualization of radiotherapy

Table 25.2 Effective proliferation rates calculated from clinical radiotherapy results

Tumour	Source	Change in overall time (days)	Result	Effective doubling time (days)
Hodgkin's disease	Friedman	20	TCD increased by 6 Gy	3
Burkitt lymphoma	Norin	7	CR increased by 63%	3
Head and neck, T2/3	Parsons	21	TCD increased by 16%	6
Larynx	Niederer	15	TCD increased by >5 Gy	6
Larynx, T3/4	Maciejewski	16	TCD increased by 8 Gy	4
Skin carcinoma	Hliniak	8	TCD increased by 51%	8

TCD=local tumour control dose; CR=complete response.
From Fowler (1986).

PROLIFERATION MEASUREMENTS IN HUMAN TUMOURS

The simplest way to obtain proliferation information is to stain tumour sections with proliferation-dependent antibodies such as Ki67 and PCNA. The fractions of stained cells are related to, but not equal to, the growth fraction and the thymidine (or BrUdR) labelling index. They give no rate or time information but can still be useful to rank tumours approximately according to their proliferation potential. The three principal kinetic rate parameters in tumours are cell cycle time, potential doubling time and volume doubling time (Section 3.3). Of these, only the potential doubling time (T_{pot}) can be measured in a patient from a single biopsy. With the relative movement method (Section 3.4), a non-toxic dose of IUdR or BrUdR is given intravenously and a tumour biopsy is taken several hours later for flow cytometric analysis. The DNA histogram (Figure 25.5B) shows the presence of both normal and malignant cells in the biopsy. Only the aneuploid tumour cells are analysed. Cells labelled with the anti-IUdR antibody can be seen clearly distinguished from the unlabelled cells (Figure 25.5A), from which T_{pot} can be calculated using Equation 3.1. A wide range of T_{pot} values have been found by this method in clinical head and neck tumours. Figure 25.6 shows some of these results in head and neck tumours: the values range from 1.5 to over 11 days. It is remarkable that the average value of 4.5 days from this study is similar to that calculated from the clinical tumour control data shown in Figure 14.4A. This provides support for the view that pre-treatment T_{pot} values may predict for clonogen repopulation during radiotherapy.

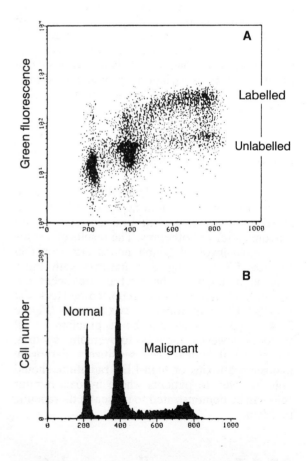

Figure 25.5 Flow cytometry analysis of a human tumour. **A**: a scatter diagram of IUdR content (vertical scale) and DNA content (horizontal scale). **B**: the distribution of DNA contents. From Begg, Hofland and Horiot, unpublished.

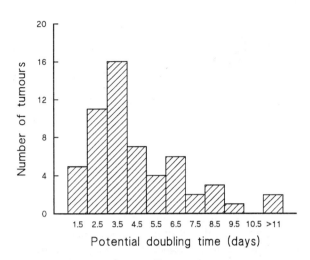

Figure 25.6 Distribution of T_{pot} values for head and neck tumours, derived from flow cytometry analysis. The mean potential doubling time is 4.5 days. From Begg, Hofland and Horiot, unpublished.

CORRELATION OF CELL KINETICS WITH
TREATMENT OUTCOME

Pre-treatment cell kinetics (T_s, LI, T_{pot}) have now been measured by *in vivo* BrUdR or IUdR labelling in many studies on head and neck tumours, including at least one randomized phase III trial. Most but not all of these studies show a trend for the faster-growing tumours (*i.e.* those with higher LI and shorter T_{pot} values) to respond less well to conventional radiotherapy than more slowly growing tumours. Some of these trends have been statistically significant. A recent meta-analysis, pooling the data from nine of these head and neck trials involving radiotherapy alone, showed that labelling index, but not T_{pot} or T_s, significantly correlated with local control in a survey of over 440 patients. Results for 210 aneuploid tumours, in which cell kinetics can be more accurately measured, are shown in Figure 25.7. The worse performance of T_{pot} (a combination of LI and T_s) may be due in part to the greater technical difficulties in measuring T_s. These data provide support for the hypothesis that pre-treatment kinetic measurements can predict repopulation rates during fractionated radiotherapy.

25.6 Prediction of normal-tissue tolerance

It is well recognized by radiotherapists that among a group of patients who are given the same radiation treatment some will show more severe reactions than others. The underlying causes of these differences include:

(i) *Dosimetry*. Differences in the actual radiation dose delivered to the target cells of the normal tissue in question. These may arise by technical factors in dose delivery and from differences in patient anatomy.

(ii) *Treatment volume*. Tumours differ in size, and prescription practices require that the volume of normal tissue irradiated will vary with tumour size (Section 5.4).

(iii) *Poisson statistics*. When irradiation has reduced the number of 'tissue-rescuing units' to critically low levels the failure or continued integrity of the tissue depend upon the survival of one or more of these entities. This is dealt with in detail in Section 10.2.

(iv) *Physiology*. Inter-patient differences exist in tissue physiology or tissue biology, for instance in blood flow to the tissue and proliferation rate. Some of these could have a genetic basis, others may depend on the age, environment or lifestyle of the patient. They may also be tissue-dependent, in the sense that the physiological parameters of various tissues may rank differently among a group of patients.

(v) *Genetics*. Here we think particularly about genetically based differences in radiation sensitivity. Predisposition to enhanced radiation sensitivity is known to be associated with a number of rare syndromes (Section 9.1). There are also individuals in the population who are heterozygotes for the genes that give rise to these forms of enhanced sensitivity; they also, but perhaps to a lesser extent, may be more sensitive to radiation treatment than individuals who do not carry those genes. Other predisposing genes may as yet be unrecognized.

All of these sources of variation will contribute to the observed differences in normal-tissue damage observed in radiotherapy patients. If methods of predicting their susceptibility to radiation damage can be developed, then either or both of two

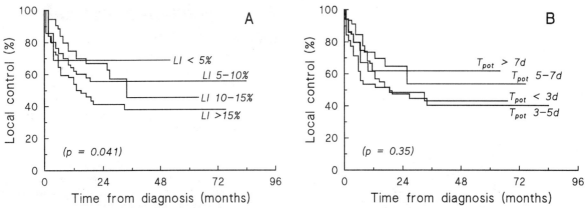

Figure 25.7 Correlation of outcome with cell kinetic data for 210 aneuploid tumours from a total of 441 head and neck cancer patients given conventional radiotherapy (6 weeks or longer). **A**: local control in relation to labelling index (rapid proliferation = higher LI). **B**: local tumour control in relation to potential doubling time (faster proliferation = shorter T_{pot}). The data in each case have been split into four categories for analysis of statistical trends. Data taken from pooled data of nine separate clinical trials (Begg AC, Wilson GW, Hart AM *et al*, unpublished).

courses of action are open: to reduce radiation dose for the susceptible patients, thus sparing them the risk of severe tissue damage, or to increase radiation dose for the non-susceptible patients, with the likelihood of improved tumour response. It is causes (iv) and (v) above that are likely to be predictable by this approach.

Evidence for the feasibility of predicting normal-tissue damage comes originally from the work of Burnet *et al* (1992), and subsequently Geara *et al* (1993) and Johansen *et al* (1994). All these groups obtained skin biopsies from untreated areas in patients who had been, or were going to be, treated with radiotherapy. Fibroblasts were grown out of the biopsies and their radiosensitivity determined by colony formation. Results have shown a significant correlation with the severity of late radiation damage to skin. Two of these data sets with the largest number of patients are shown in Figure 25.8. The studies have tended to show a better correlation for late effects than for early effects.

The measurement of fibroblast cell survival is too slow to be used as a predictive assay and efforts are being made to identify rapid and reliable alternatives. Chromosome damage (Russell *et al*, 1995) and the micronucleus test (Shibamoto *et al*, 1991) are rapid methods for detecting radiation cell killing and in normal tissues have potential as predictive assays. The extent of induced differentiation of fibroblasts, relevant to collagen production and subsequently fibrosis, is another possibility. Efforts are also being made to develop tests based on radiation DNA damage. These include the use of single-cell electrophoresis (the 'comet' assay, Section 8.3). Making cell suspensions from biopsies rather than waiting for cell outgrowth is also a faster procedure for obtaining fibroblasts. Blood lymphocytes are more readily available than skin biopsies and their damage can be assessed soon after a test dose of radiation with one or other of the rapid methods mentioned above (West *et al*, 1995b). However, lymphocytes may be more susceptible to the extraneous influence of infection *etc*. and their reproducibility for predictive purposes has to be demonstrated.

Bibliography

Baker F, Spitzer G, Ajani J *et al* (1986). Drug and radiation sensitivity measurements of successful primary monolayer culturing of human tumor cells using cell-adhesive matrix and supplemented medium. *Cancer Res* 46:1263–74.

Begg AC, Hofland I, van Glabekke M *et al* (1992). Predictive value of potential doubling time for radiotherapy of head and neck tumour patients: results from the EORTC cooperative trial 22851. *Semin Radiat Oncol* 2:22–5.

Bellamy WT (1992). Prediction of response to drug therapy of cancer: a review of *in vitro* assays. *Drugs* 44:690–708.

Brock WA, Bhadkamkar VA, Williams M and Spitzer G (1987). Radiosensitivity testing of primary cultures derived from human tumors. In: *Progress in Radio-Oncology*, Vol. III, pp. 300–6. (Eds) Karcher KH,

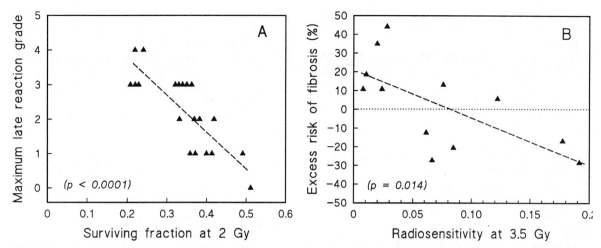

Figure 25.8 Correlation between late skin reactions to radiotherapy and the *in vitro* radiosensitivity of skin fibroblasts. A significant correlation was found in both studies. Data of **A**: Geara *et al* (1992), **B**: Johansen *et al* (1994), with permission.

Key points

1. The three most widely investigated predictive assays for tumour response to radiotherapy are those for intrinsic radiosensitivity, hypoxia and clonogen repopulation.
2. Intrinsic radiosensitivity can be measured by cell growth or colony assays after irradiation *in vitro* and there are encouraging signs of a correlation between SF_2 and outcome. A disadvantage of the colony assay is its approximately 4-week duration. Indirect techniques such as DNA damage, chromosome damage and gene mutation/expression are being developed as more rapid alternatives.
3. Pre-treatment levels of tumour hypoxia have been shown to correlate with outcome in several studies, detected mainly with the use of oxygen electrodes. New methods for measuring hypoxia are being developed, primarily involving the selective hypoxic binding of nitroimidazole compounds.
4. Analysis of clinical data indicates that repopulation during radiotherapy limits cure in several tumour types. Tumour potential doubling times measured before treatment using *in vivo* labelling with thymidine analogues are similar to values for repopulation rate calculated from clinical radiotherapy response data. Current clinical trials support the predictive value of pre-treatment labelling index measurements and favour accelerated radiotherapy for the more rapidly proliferating tumours.
5. Prediction of normal-tissue sensitivity to radiotherapy is also a potentially feasible option which has yet to be fully evaluated.

Kogelnik HD and Szepesi T. Int. Club for Radio-Oncology; Vienna.

Brown JM, Evans J and Kovacs MS (1992). The prediction of human tumour radiosensitivity *in situ*: an approach using chromosome aberrations detected by fluorescence in situ hybridization. *Int J Radiat Oncol Biol Phys* 24:279–86.

Burnet NG, Nyman J, Turesson I *et al* (1992). Potential for improving radiotherapy cure rates by predicting normal tissue tolerance from *in vitro* cellular radiation sensitivity. *Lancet* 339:1570–1.

Carmichael J, DeGraff WG, Gazdar AF *et al* (1987). Evaluation of a tetrazolium-based semiautomated colorimetric assay: assessment of radiosensitivity. *Cancer Res* 47:943–6.

Fowler JF (1986). Potential for increasing the differential response between tumours and normal tissues: can proliferation rate be used? *Int J Radiat Oncol Biol Phys* 12:641–5.

Geara FB, Peters LJ, Ang KK, Wike JL, Sivon SS, Guttenberger R, Callender DL, Malaise EP and Brock WA (1992). Intrinsic radiosensitivity of normal human fibroplasts and lymphocytes after high- and low-dose-rate irradiation. *Cancer Res* 52:6348–52.

Geara FB, Peters LJ, Ang KK, Wike JL and Brock WA (1993). Prospective comparison of *in vitro* normal cell radiosensitivity and normal tissue reactions in radiotherapy patients. *Int J Radiat Oncol Biol Phys* 27:1173–79.

Höckel M, Knoop C, Schlenger K *et al* (1993). Intratumoral pO2 predicts survival in advanced cancer of the uterine cervix. *Radiother Oncol* 26:45–50.

Höckel M, Terris DJ and Vaupel P (1994). The role of oxygen tension distribution on the radiation response of human breast carcinoma. *Adv Exp Med Biol* 345:485–92.

Hodgkiss RJ, Webster L and Wilson GD (1995). Development of bioreductive markers for tumour hypoxia. *Acta Oncol* 34:351–5.

Jeggo PA, Carr AM and Lehmann AR (1994). Cloning human DNA repair genes. *Int J Radiat Biol* 66:573–7.

Johansen J, Bentzen SM, Overgaard J and Overgaard M (1994). Evidence for a positive correlation between *in vitro* radiosensitivity of normal human skin fibroblasts and the occurence of subcutaneous fibrosis after radiotherapy. *Int J Radiat Biol* 66(4):407–12.

KSST: Group for Sensitivity Testing of Human Tumours (1981). *In vitro* short-term test to determine the resistance of human tumours to chemotherapy. *Cancer* 48:2127–35.

Li Z, Otevrel T, Gao Y *et al* (1995). The XRCC4 gene encodes a novel protein involved in DNA double-strand break repair and V(D)J recombination. *Cell* 83:1079–89.

Lord EM, Harwell L and Koch CJ (1993). Detection of hypoxic cells by monoclonal antibody recognizing 2-nitroimidazole adducts. *Cancer Res* 53:5721–6.

Mitchell JB (1988). Potential applicability of nonclonogenic measurements to clinical oncology. *Radiat Res* 114:401–14.

Olive PL, Banath JP and Durand RE (1990). Heterogeneity in radiation-induced DNA damage and repair in tumour and normal cells measured using the 'comet' assay. *Radiat Res* 122:86–94.

Rodemann HP and Bamberg M (1995). Cellular basis of radiation-induced fibrosis. *Radiother Oncol* 35:83–90.

Russell NS, Artlett CF, Bartelink H and Begg AC (1995). Use of fluorescence in situ hybridization to determine the relationship between chromosome aberrations and cell survival in eight human fibroblast strains. *Int J Radiat Biol* 68(2):185–96.

Salmon SE, Hamburger AW, Soehnlen B *et al* (1978). Quantitation of differential sensitivity of human-tumour stem cells to anticancer drugs. *New England J Med* 298:1321–7.

Savitsky K, Bar-Shira A, Gilad S *et al* (1995). A single ataxia telangiectasia gene with a product similar to PI-3 kinase. *Science* 268:1749–53.

Schwartz JL, Rotmensch J, Giovanazzi S *et al* (1988). Faster repair of DNA double-strand breaks in radioresistant human tumour cells. *Int J Radiat Oncol Biol Phys* 15:907–12.

Selby P, Buick RN and Tannock I (1983). A critical appraisal of the 'human tumour stem-cell assay'. *New England J Med* 308:129–34.

Shibamoto Y, Streffer C, Fuhrmann C and Budach V (1991). Tumor sensitivity prediction by the cytokinesis-block micronucleus assay. *Radiat Res* 128(3):293–300.

Steel GG, Courtenay VD and Peckham MJ (1983). The response to chemotherapy of a variety of human tumour xenografts. *Brit J Cancer* 47:1–13.

Stone HB, Brown JM, Phillips TL and Sutherland RM (1993). Oxygen in human tumours: correlations between methods of measurement and response to therapy. Summary of a workshop held 19–20 November 1992, at the National Cancer Institute, Bethesda, Maryland. *Radiat Res* 136:422–34.

Shibamoto Y, Streffer C, Fuhrmann C and Budach V (1991). Tumour radiosensitivity prediction by the cytokinesis-block micronucleus assay. *Radiat Res* 128:293–300.

West CM, Davidson SE, Roberts SA and Hunter RD (1993). Intrinsic radiosensitivity and prediction of patient response to radiotherapy for carcinoma of the cervix. *Br J Cancer* 68:819–23.

West CM, Davidson SE, Burt PA and Hunter RD (1995a). The intrinsic radiosensitivity of cervical carcinoma: correlations with clinical data. *Int J Radiat Oncol Biol Phys* 31:841–6.

West CM, Elyan SA, Berry P, Cowan R and Scott D (1995b). A comparison of the radiosensitivity of lymphocytes from normal donors, cancer patients, individuals with ataxia-telangiectasia (A-T) and A-T heterozygotes. *Int J Radiat Biol* 68:197–203.

Withers HR, Taylor JMG and Maciejewski B (1988). The hazard of tumour clonogen repopulation during radiotherapy. *Acta Oncol* 27:131–46.

Further Reading

Chapman JD, Peters LJ and Withers HR (Eds) (1989). *Prediction of Tumour Treatment Response*. Pergamon Press; Oxford.

Larsson R and Nygren P (1993). Laboratory prediction of clinical chemotherapeutic drug resistance: a working model exemplified by acute leukaemia. *Eur J Cancer* 29A:1208–12.

Mattern J and Volm M (1992). Prediction of drug resistance in human tumours using immunohistochemical techniques. *Anticancer Res* 12:413–8.

Paliwal BR, Fowler JF, Herbert DE *et al* (1989). *Prediction of Response in Radiation Therapy: Part I, the Physical and Biological Basis.* American Institute of Physics; New York.

Trott KR and Kummermehr J (1985). What is known about tumour proliferation rates to choose between accelerated fractionation or hyperfractionation? *Radiother Oncol* 3:1–9.

Tucker SL and Chan KS (1990). The selection of patients for radiotherapy on the basis of tumour growth kinetics and intrinsic radiosensitivity. *Radiother Oncol* 18:197–282.

West CM (1995). Intrinsic radiosensitivity as a predictor of patient response to radiotherapy. *Br J Radiol* 68:827–37.

Glossary of terms in radiation biology

α/β ratio: The ratio of the parameters α and β in the linear-quadratic model; used to quantify the fractionation sensitivity of tissues.

Accelerated fractionation: Reduction in overall treatment time; a schedule in which the average rate of dose delivery exceeds the equivalent of 10 Gy per week in 2-Gy fractions.

Additive: A situation in which the effect of a combination is the sum of the effects of the separate treatments (= 'independent cell kill').

Apoptosis: A mode of rapid cell death after irradiation in which the cell nucleus displays characteristic densely staining globules and some at least of the DNA is subsequently broken down into internucleosomal units. Sometimes postulated to be a 'programmed' and therefore potentially controllable process.

Autoradiography: Use of a photographic emulsion to detect the distribution of a radioactive label in a tissue specimen.

Biologically effective dose (BED): In fractionated radiotherapy, the total dose that would be required in very small dose fractions to produce a particular effect, as indicated by the linear-quadratic equation. Otherwise known as 'extrapolated total dose' (ETD). BED values calculated for different α/β ratios are not strictly comparable.

Brachytherapy: Radiotherapy using radioactive sources inserted into a body cavity or through needles into tissues.

Cell cycle time: The time between one mitosis and the next.

Cell loss factor (ϕ): The rate of cell loss from a tumour, as a proportion of the rate at which cells are being added to the tumour by mitosis. Usually calculated by the relation: $\phi = 1 - T_{pot}/T_d$, where T_{pot} is potential doubling time and T_d is the cell population doubling time.

Chromosomal instability: An effect of irradiation in which chromosomal aberrations continue to appear through many cell generations.

Chronic hypoxia: Persistent low oxygen concentrations such as exist in viable tumour cells close to regions of necrosis.

Clonogenic cells: Cells that have the capacity to produce an expanding family of descendents (usually at least 50). Also called 'colony-forming cells' or 'clonogens'.

Colony: The family of cells derived from a single clonogenic cell.

Complementation: Identification of whether a (radiosensitive) phenotype in different mutants is due to the same gene. Studied by means of cell fusion.

D_0: A parameter in the multitarget equation: the radiation dose that reduces survival to e^{-1} (*i.e.* 0.37) of its previous value on the exponential portion of the survival curve.

Direct action: Ionization or excitation of atoms within DNA leading to free radicals, as distinct from the reaction with DNA of free radicals formed in nearby water molecules.

Dose-modifying factor (DMF): When a chemical or other agent acts as if to change the dose of radiation, DMF indicates the ratio: (dose without/dose with) the agent for the same level of effect. Similarly: **Dose-reduction factor (DRF)** or **sensitizer enhancement ratio (SER)**

Dose rate effect: Decreasing radiation response with decreasing radiation dose rate.

'Double trouble': A hot spot within a treatment field receives not only a higher dose but also a higher dose per fraction, which means that the biological effectiveness of the dose is also greater.

Doubling time: Time for a cell population or tumour volume to double its size.

'Early' responses: Radiation-induced normal-tissue damage that is expressed in weeks to a few months after exposure. Generally due to damage to parenchymal cells. α/β ratio tends to be large.

Elkind repair: Recovery of the shoulder on a survival curve when irradiation follows several hours after a priming dose.

ED$_{50}$: Radiation dose that produces a specified effect in the normal tissues of 50% of animals ('Effect-Dose-50%').

Exponential growth: Growth according to an exponential equation: $V = V_0 \exp(kt)$. The volume doubling time is constant ($= (\log_e 2)/k$).

Extrapolated total dose (ETD): Calculated isoeffect dose when the dose rate is very low, or when fraction size is very small (see **Biologically effective dose**).

Extrapolation number: A parameter in the multitarget equation: the point on the survival scale to which the straight part of the curve back-extrapolates.

Field-size effect: The dependence of normal-tissue damage on the size of the irradiated area; also known as 'volume effect'.

FISH: Fluorescence in situ hybridization. Fluorescent dyes are attached to specific regions of the genome, thus aiding the identification of chromosomal damage.

'Flexible' tissues: Non-hierarchical cell populations in which function and proliferation take place in the same cells.

Flow cytometry: Analysis of cell suspensions in which a dilute stream of cells is passed through a laser beam. DNA content and other properties are measured by light scattering and fluorescence following staining with dyes or labelled antibodies.

Free radical: A fragment of a molecule containing an unpaired electron, therefore very reactive.

gray (Gy): Unit of absorbed dose. 1 Gy = 1 joule per kg (= 100 rad).

Growth delay: Extra time required for an irradiated tumour to reach a given size, compared with an unirradiated control.

Growth fraction: The proportion of cells in a population that are cycling.

'Hierarchical' tissues: Cell populations comprising a lineage of stem cells, proliferating cells and mature cells. The mature cells do not divide.

Hyperbaric oxygen (HBO): The use of high oxygen pressures (2–3 atmospheres) to enhance oxygen availability in radiotherapy.

Hyperfractionation: Increase in number of fractions and reduction in dose per fraction below a conventional level of 1.8–2.0 Gy.

Hyperthermia: The use of heat treatments in excess of 42°C to treat cancer.

Hypofractionation: The use of dose fractions substantially larger than the conventional level of ~2 Gy.

Hypoxia: Low oxygen tension; usually the very low levels that are required to make cells maximally radioresistant. Sometimes used to mean **Anoxia** (= literally, the complete absence of oxygen).

Incomplete repair: Increased damage from fractionated radiotherapy when the time interval between doses is too short to allow complete recovery.

Indirect action: Damage to DNA by free radicals formed through the ionization of nearby water molecules.

Inducible response: A response to irradiation that is modified by a small dose of radiation given shortly before.

Initial slope: The steepness of the initial part of the oxic cell survival curve, sometimes indicated by the surviving fraction at 2 Gy.

Interphase death: The death of irradiated cells before they reach mitosis.

Isoeffect plots: Graphs of the total dose for a given effect (*e.g.* ED_{50}) plotted, for instance, against dose per fraction or dose rate.

$LD_{50/30}$: Radiation dose to produce lethality in 50% of animals by 30 days; similarly $LD_{50/7}$ etc.

Labelling index: Proportion or percentage of cells within the S-phase, and therefore labelled by ^{3}H-thymidine or other precursors such as bromodeoxyuridine.

'Late' responses: Radiation-induced normal-tissue damage that in humans is expressed months to years after exposure. Generally due to damage to connective tissue cells. α/β ratio tends to be small (< 5 Gy).

Latent period or latency interval: Time between irradiation and expression of injury.

Linear energy transfer (LET): The rate of energy loss along the track of an ionizing particle. Usually expressed in keV/μm.

Linear-quadratic (LQ) model: Model in which the effect (E) is a linear-quadratic function of dose (d): $E = \alpha d + \beta d^2$. For cell survival: $S = \exp(-\alpha d - \beta d^2)$.

Log-phase culture: A cell culture growing exponentially.

Mitotic death: Cell death associated with a post-irradiation mitosis.

Mitotic delay: Delay of entry into mitosis, or accumulation in G2, as a result of treatment.

Mitotic index: Proportion or percentage of cells in mitosis at any given time.

Multitarget equation: Model which assumes the presence of a number of critical targets in a cell, all of which require inactivation to kill the cell. Survival is given by: $S = 1 - [1 - \exp(D/D_0)]^n$.

Non-stochastic effect: An effect where the severity increases with increasing dose, perhaps after a threshold region.

NSD: Nominal standard dose in the Ellis formula.

Oxygen enhancement ratio (OER): The ratio of dose given under anoxic conditions to the dose resulting in the same effect when given under oxic conditions.

Photodynamic therapy: Cancer treatment using light to activate a photosensitizing agent, thereby releasing cytotoxic free radicals.

Plateau-phase cultures: Cell cultures grown to confluence so that proliferation is markedly reduced (= 'stationary phase').

Plating efficiency: The proportion or percentage of *in vitro* plated cells that form colonies.

Potential doubling time (T_{pot}): The predicted cell population doubling time in the assumed absence of cell loss.

Potentially lethal damage (PLD): Cellular damage that is recovered during the interval between treatment and assay, especially under suboptimal growth conditions.

Prodromal phase: Signs and symptoms in the first 48 hours following irradiation of the central nervous system.

Quasi-threshold dose (D_q): Point of extrapolation of the exponential portion of a multitarget survival curve to the level of zero survival: $D_q = D_0 \ln(n)$.

Radioresponsiveness: A general term, indicating the overall level of clinical response to radiotherapy.

Radiosensitizer: In general, any agent that increases the sensitivity of cells to radiation. Most commonly applied to electron-affinic chemicals that mimic oxygen in fixing free-radical damage.

Radiosensitivity: The radiation dose required to produce a defined level of cell inactivation. Usually indicated by the surviving fraction at 2 Gy (*i.e.* SF_2) or by the parameters of the linear-quadratic or multitarget equations.

Reassortment or **Redistribution:** Return towards a more even cell-age distribution, following the selective killing of cells in certain phases of the cell cycle.

Recovery: An increase in cell survival as a function of time during or after irradiation (see **Repair**).

Regression rate: The rate at which a tumour shrinks during or after treatment.

Relative biological effectiveness (RBE): Ratio of dose of a reference radiation quality (usually 250-keV X-rays) and dose of a test radiation that produce equal effect.

Reoxygenation: The process by which surviving hypoxic clonogenic cells become better oxygenated during the period after irradiation of a tumour.

Repair: Restoration of the integrity of damaged macromolecules (see **Recovery**).

Repair saturation: Explanation of the shoulder on cell survival curves on the basis of the reduced effectiveness of repair after high radiation doses.

Reproductive integrity: Ability of cells to divide many times and thus be 'clonogenic'.

Sievert (Sv): Dose equivalent in radiation protection. Dose in grays is multiplied by a radiation quality factor.

SF_2: Surviving fraction at 2 Gy.

Slow repair: Long-term recovery which takes place on a time-scale of weeks to months.

Spatial co-operation: The use of radiotherapy and chemotherapy to hit disease in different anatomical sites.

Spheroid: Clump of cells grown together in tissue culture suspension. Not usually a colony.

Split-dose recovery (SLD recovery): Decrease in radiation effect when a single radiation dose is split into two fractions separated by times up to a few hours (= Elkind recovery, or recovery from sublethal damage).

Stathmokinetic method: Study of cell proliferation using agents that block cells in mitosis.

Stem cells: Cells capable of self-renewal and of differentiation to produce all the various types of cells in a lineage.

Stochastic effect: An effect where the incidence, but not the severity, increases with increasing dose (*e.g.* carcinogenesis).

Sublethal damage: Non-lethal cellular injury that can be repaired, or accumulated with further dose to become lethal.

Supra-additivity or **Synergism:** A biological effect due to a combination that is greater than would be expected from the addition of the effects of the component agents.

Target cell: A stem cell whose death contributes to a reduction in growth or tissue function.

Targeted radiotherapy: Treatment of disseminated cancer by means of drugs that localize in tumours and carry therapeutic amounts of radioactivity.

Target theory: The idea that the shoulder on cell survival curves is due to the number of unrepaired lesions per cell.

Therapeutic index: Tumour response for a fixed level of normal-tissue damage.

Thermal dose: A function of temperature and heating time that is thought to relate well to biological effect.

Thermotolerance: The observation that an initial heat treatment reduces the effect of a second heat treatment given shortly afterwards.

Time–dose relationships: The dependence of isoeffective radiation dose on the duration (and number of fractions) in radiotherapy.

Tolerance: The maximum radiaton dose or intensity of fractionated radiotherapy that the therapist judges to be acceptable. Usually expressed in dose units. Actual values will depend on fractionation, field size, concomitant treatments etc.

Transient hypoxia: Low oxygen concentrations associated with the transient closing and opening of blood-vessels. Sometimes called *acute* or *cyclical* hypoxia.

Tumour bed effect (TBE): Slower rate of tumour growth after irradiation due to stromal injury in the irradiated 'vascular bed'.

Tumour cord: Sleeve of viable tumour growing around a blood capillary.

Volume doubling time: Time for a tumour to double in size.

Volume effect: Dependence of radiation damage to normal tissues on the volume of tissue irradiated.

Xenografts: Transplants between species; usually applied to the transplantation of human tumours into immune-deficient mice and rats.

Index

Page numbers in **bold** refer to the Glossary.

α/β ratio   106, **246**
   values, for experimental tumours   110
   values, for human tissues and tumours   111
   values, for normal tissues   110
Accelerated radiotherapy   125–30, **246**
   and proliferation rate   125–6
   clinical evaluation   128–9
Accelerated repopulation   27, 157–8
Acute effects, *see* Early reactions
Acute hypoxia in tumours   134, 148
Additivity of drug-radiation effects   187–8, **246**
Alkali-labile sites   63
Apoptosis   25, 66, 73–4, **246**
Arrhenius plot   213
Assays, clonogenic   41–5
Ataxia-telangiectasia   66, 70–6
   ATM gene   75
Autoradiography   15, **246**

Biologically effective dose (BED)   112–14, 122, **246**
   for continuous irradiation   117, 122, 167–8
Bladder, tolerance and α/β ratio   32, 110
   proliferation after irradiation   26–7
   radiation pathology   36
   retreatment tolerance   206–7
Bladder cancer, combined radiotherapy/chemotherapy   200
Blood flow, effect of hyperthermia   214–15
Bone marrow damage, dose-rate effect   165
   retreatment tolerance   205
Boron neutron capture therapy   230–1
Brachytherapy   163–71, **246**
   distribution of biological effect   169–70
   isoeffect calculations   166–8
   pulsed   171
   radiobiological advantages   170
Bragg peak   180–1
Breast cancer, combined radiotherapy/chemotherapy   199
   SF_2 values   238
Bromodeoxyuridine (BrUdR)   17–22

Cancer predisposition, and radiosensitivity   76
Cell cycle control, and radiosensitivity   74
Cell cycle progression   49, 163
Cell cycle times, of human tumours   15
Cell cycle, variation in radiation cell killing   48–9
Cell death, sequence of events   66–7
   see also Apoptosis
Cell loss factor, definition   15, **246**
Cell loss from tumours   14–16
Cell proliferation, *see* Proliferation
Cell survival, concept   40–50, 156
   relation to gross tumour response   45–6
Cell survival curves, initial slope   158–9
Cell synchronization, by drugs and radiation   189–91
Central nervous system, pathogenesis of radiation damage   33–5
   see also Spinal cord
Cervix cancer, combined radiotherapy/chemotherapy   200
CHART   128–30
Chemotherapy, combination with radiotherapy   184–93
   prediction of drug response   236
   role in cancer management   1–2
   tumour size effect   157
Chromatin structure   67
Chromosome aberrations, radiation-induced   64–5
Chromosome instability   68, **246**
Chronic hypoxia   134, **246**
Clinical normal-tissue damage   87–96
   effect of patient age   95
   endpoints   87
   reporting systems   89–91
Clonogenic cells   40–50, **246**
   assays   4, 41–2
Clustering of ionization events   58–9
CNS, *see* Central nervous system
Cohen model of fractionation   103
Colony, definition   40, **246**
Combination of radiotherapy and chemotherapy   93, 184–93
   clinical evaluation   195–201

    exploitable mechanisms   185–90
    objectives   184
    terminology   185
    time-dependence   193
Comet assay   63
Compensatory cell proliferation   26
Complementation analysis   72–3, **246**
Cumulative radiation effect (*CRE*)   100
Cyclins   20–1
Cysteamine   59
Cytokeratins   20
Cytotoxic drugs, resistance mechanisms   139

D_0 52–3, **246**
Debulking   190
Delayed-plating experiments   46–7
Direct and indirect action of radiation   59, **247**
DNA damage   58–68, 72
    critical event in cell killing and mutation   61–2
DNA, structure   60
    FITC conjugation   17
    methods for detecting   63
    processing   65–7
    repair genes   71–5
Dose-modifying factor   155, **247**
Dose rate effect   47, 163–71, **247**
    in normal tissues   164–5
    mechanisms   163
    on cell survival   160, 163
Dose–recovery factor   164
Dose–response relationships   4–5, 78–85
    steepnesss   5, 81–3
Double-strand DNA breaks   61–2
'Double trouble'   114, **247**
Doubling time, volume   9, **247**
Duration of remission   59

Early reactions   3, 88, 123, **247**
    α/β ratios   110–11
    clinical characteristics   87–9
    pathogenesis   30–7
    repopulation   119
ED_{50}, definition   42, **247**
Electron-affinic radiosensitizers   141–6
Elkind recovery   46
Ellis formula   98–9
    limitations   100–3, 106
Enhancement of tumour response, in combined modality therapy   187–8
Epigenetic modulation of radiosensitivity   75
Excision-repair process   65
Exponential growth of tumours   9–10, **247**
Extrapolated tolerance dose (ETD)   112, **247**
Extrapolation number   53, **247**

'Flexible' tissues   24, **247**
Fibrosis   32
Field-size effect, *see* Volume effect

5 Rs of radiobiology   49, 163
Flow cytometry   17–22, **247**
Fractionation, accelerated radiotherapy and hyperfractionation   123–30
    sparing effect   47
    the LQ approach   106–20
    time–dose relationships   98–104
Free radicals   3, 58–9, **247**
    and the oxygen effect   133

Gastrointestinal tract, *see* Intestine
Genes up-regulated in hypoxia   139
Genetics of radiosensitivity   70–9
Glutathione   59
Gompertz equation   10
Growth delay, tumour   155, **247**
Growth fraction   14, **247**
Growth rate of tumours   2–13

Haemoglobin levels, and local tumour control   137, 144
Half-times for recovery   116
Hayflick limit   153
Head and neck cancer, combined radiotherapy/chemotherapy   198
Heart, tolerance and α/β ratio   32, 36
'Heavy' particles   173
'Hierarchical' tissues   24, **247**
High LET radiation, biological basis   177–80
    BNCT   230–1
    cell survival   175
    fractionation effect   177
    physical basis   180–2
Hyperbaric oxygen therapy (HBO)   141, **247**
Hyperfractionation   123–5, **247**
Hyperthermia   212–22, **247**
    biological effects   213–18
    clinical effects   219–21
    environmental and vascular effects   214
    interaction with radiation   215
    interaction with chemotherapy   221
Hypofractionation   101, 124, **247**
Hypoxia   132–51, **247**
    acute and chronic   134
    clinical meta-analysis   145, 147
    in tumours   133–9
    methods of exploiting   149–50
    methods of overcoming   141–9
    predictive tests   239
Hypoxic cell radiosensitizers   141–6
    clinical studies   144–7
Hypoxic fraction   134–8
    effect of chemotherapy   189
    measurement of   134–5
    relation to oxygen tension   136

Incomplete repair, and fractionation response   114–17, **247**
Incomplete repair model   115, 122

continuous irradiation 165–8
  H_m and g factors 116, 118
  isoeffect calculations 117
Independent cell kill, in combined modality therapy 186
Indirect action of radiation 59, **247**
Individualized radiotherapy, *see* Prediction of response
Inducible responses to irradiation 68, **247**
Initial slope, of cell survival curve 158–9, **247**
Interaction between drugs and radiation, the concept 185
  guidelines for minimizing toxicity 187
Interphase cell death **248**
  see also Apoptosis
Intestine, compensatory proliferation 26
  dose rate effect 165
  drug–radiation combinations, effect of timing 190–1
  early and late reactions 33
  neutron irradiation 178
  re-treatment tolerance 204
  tolerance and α/β ratio 32
Inverse dose-rate effect 169
Iododeoxyuridine (IUdR) 17–19, 240–1
Isobolograms 188
Isoeffect relationships 5, **248**
  calculations 112–14
  continuous irradiation 122, 165–9
  damage to mouse tissues 123
  history 98–105
  neutrons 178
  summary of formulae 122
  the LQ approach 106–22

Ki67 19–20
Kidney, compensatory proliferation 26–7
  high-LET effects 177–8
  radiation pathology 35
  retreatment tolerance 207
  tolerance and α/β ratio 32
  volume effect 38

Labelling index 15, **248**
Latent period or latency interval **248**
  in tumour growth 9–10
  to normal-tissue damage 25–6, 91–3
Late reactions 3, 125, **248**
  α/β ratios 110–11
  clinical characteristics 88–9
  lack of time factor 102
  pathogenesis 30–7
  proliferation 119
Lesion-interaction models 55–6
LET, *see* linear energy transfer
Lethal potentially-lethal (LPL) damage model 55, 164
'Light' particles 173
Limiting-dilution assay 43
Linear component, of cell survival curve 54, 159–61, 164

Linear energy transfer (LET) 173–6, **248**
Linear-quadratic (LQ) model **248**
  for cell survival 54–5, 106–21
  for fractionated radiotherapy 106–20
  isoeffect curves 123–5
Local multiply-damaged sites (LMDS) 59
Local tumour control 4
  effect of radiosensitizers 156
  relation to cell survival 45
Logistic dose–response relationship 81
Low dose rate radiotherapy, *see* Brachytherapy; Dose rate effect
Lung, compensatory proliferation 26
  dose rate effect 165
  drug–radiation combinations, effect of timing 192
  radiation pathology 35
  retreatment tolerance 205
  tolerance and α/β ratio 32
  volume effect 38
Lung-colony assay 43

Measurement of tumour size 8
Membranes, radiation damage 67–8
Meta-analysis of clinical hypoxia trials 142, 145, 147
Microdosimetry of particle tracks 59, 173–83
  boron neutron capture 230
  targeted radiotherapy 226, 229
MIBG 230
Micronucleus assay 44, 64
Misonidazole 141–6
Mitotic cell death 41, **248**
Mitotic delay 48, **248**
Models of radiation cell killing 52–7
Molecular targets for cell killing 52, 60–1
Multiple fractions per day 127–8
  spacing 115, 117, 127–8
Multi-target model 53, **248**
Mutations, radiation-induced 62

Neutral filter elution 63
Neutron therapy, clinical indications 180
  isoeffect relations 178
  RBE 182
Nominal standard dose (NSD) 99–100, **248**
Non-stochastic effects 89, **248**
Normal tissues, proliferative organization 24
Normal-tissue damage, assays for 4
  classification of clinical endpoints 89–90
  clinical manifestations 87–96
  dose rate effect 165
  effects of high-LET radiation 177–8
  prediction 241–3
  radiobiology 30–9
  reporting systems 89–91
  retreatment tolerance 203–11
Nucleoid sedimentation 63

'Overkill', by high LET radiation 174
Oxygen effect 132–9

Oxygen enhancement ratio (OER) 132–9, 174, 176, **248**
   and LET 176
   and oxygen tension 133
Oxygen fixation hypothesis 133
Oxygen supply to tumours, methods of improving 147–8

p53 74
Partial or complete tumour regression, significance 9, 49
Partial tolerance 113
Particle beams in radiotherapy 173–83
   therapeutic conclusions 182
PCNA 19–20
Percent labelled mitoses method 15–16
Perfluorocarbons 148
Plating efficiency 41, **248**
Pneumonitis, see Lung
Poisson statistics, for tumour control 79–80, 152
Potential doubling time (T_{pot}) 14–16, **248**
   and tumour response 241–2
Potentially-lethal damage (PLD) recovery 55, **248**
Prediction of normal-tissue tolerance 241–3
Prediction of tumour response 179, 234–43
   by hypoxia 239
   by proliferation 239–41
   by radiosensitivity 235–9
Proliferation, by flow cytometry 17–22
   during low dose rate irradiation 167–8
   during radiotherapy 125–8
   in normal tissues 24–9
   in tumours 14–22
   radiation dose to counteract 119
Proliferative organization of tissues 24–8
Protection of normal tissues, in combined modality therapy 187
Proton therapy 181–2
Pulsed brachytherapy 171
Pulsed-field gel electrophoresis (PFGE) 63

Quadratic component, of cell survival curve 54, 160–1
Quasi-threshold dose (D_q) 53, **248**
Quiescent tumour cells 14

Radiosensitivity, categories 70–2
   genetic control 72–4, 236
   human tumour cells 158–60
   sequence of events 66
Radiosensitizers 135, **248**
   clinical results 145
   effect of overall time 156
Radiotherapy, role in cancer management 1–2
RBE, see Relative biological effectiveness
Reassortment or redistribution 49, **248**
   and dose rate effect 163
Reciprocal-dose plot 109, 117
Recovery, split-dose 46–7, **248**
   dose rate effect 163, 166
   from potentially lethal damage 46–7

Recruitment into the cell cycle 14, 189
Regression of tumours, speed of 4, 12, **248**
Relative biological effectiveness (RBE) 174–6, **248**
Relative movement method 17–18
Reoxygenation 49, 136–8, **248**
   time-scale 163
Repair 46–7, 49, 65, 67, 71, **248**
   and dose rate effect 163
Repair-saturation models 55–7, **249**
Repopulation of treated tumours 12, 49
   accelerated 158
   and dose rate effect 163
   deduced from clinical studies 102, 125–6, 240
   enhanced by drugs 189
Retreatment tolerance of normal tissues 203–10
   clinical experience 207–9

Sensitizer enhancement ratio (SER) 142
SF_2, for human tumour cells 159, **249**
   for neutron irradiation 179
Silent interval in tumour growth 9–10
Skin, damage by neutrons 178
   damage by radiation plus heat 216, 218
   damage dependence on fraction size 101
   damage depending on penetration depth 31
   early and late effects 31–2, 91, 101
   isoeffect curves 123
   proliferative structure 24
   retreatment tolerance 203
   time-course of radiation effects 25, 31
   volume effects 37
Spatial cooperation 186, **249**
Specific growth delay 155
Spheroids, multicellular 153, **249**
Spinal cord, damage by neutrons 178
   pathogenesis 33–5
   proliferation after irradiation 26–8
   retreatment tolerance 205–6
   tolerance and α/β ratio 32
   volume effects 37
Spleen colony assay 43
Split-dose recovery 46, **249**
Stathmokinetic method 16, **249**
Stem cells, definition 24, 40, **249**
Stochastic effect 89, **249**
Strandqvist plots 98–9
Sublethal damage (SLD) 46, **249**
   see also Split-dose recovery
Sulphydryl (-SH) compounds 59
Supra-additivity or synergism 185–8, **249**
Survival curves 41–3
Synchronization, cellular 48, 189–91
Syndromes, radiation-sensitive 70–2
Synergism, see Supra-additivity

Target cells, for normal-tissue damage 30, **249**
Target theory 52, 57, **249**
Targeted radiotherapy 224–32, **249**
   agents 224

Targeted radiotherapy (*contd*)
   microdosimetry 229
   radionuclides 225
TDF 100
Telangiectasia 26, 79, 92, 102, 104
Therapeutic index or window 6–7, 83, 184, 196–7, **249**
Thermotolerance 213, **249**
Thiols 59–60
Thymidine analogues 16–22
Thymidine-^{3}H as a cell kinetic tool 15–17, 28
Time factor in fractionation 117
Time–dependence of drug-radiation combinations 190–2
Time–dose relationships in radiotherapy **249**
   formulae 122
   history 98–105
Time-line method 190–1
Time-scale of radiation effects 3
Tolerance of normal tissues 6, 30–9, 85, **249**
   calculations 112–14
Total effect (TE) of fractionation 112
Transient hypoxia 134, **249**
Tumour bed effect 158, **249**

Tumour control, as an endpoint of effect 155–6
   by targeted radioisotopes 227
   clinical picture 152
   dependence on treatment duration 156
Tumour control curves, steepness 78–86
Tumour growth delay 154–5
   relation to cell survival 45
Tumour growth rate 8–13
Tumour regression, speed of 12
   therapeutic significance 49, 158
Tumour response, endpoints 154–6
Tumour size, measurement of 8–9
Tumour size effect 157
Tumours, experimental systems 153–4
Two-component model of cell killing 53–4

Volume doubling time 9, **249**
   values for human tumours 11–12, 16
Volume effects 36–9, 94, **249**
   in brachytherapy 171

Xenografts, human tumour 154, **249**